ENVIRONMENTAL ENGINEERING AND SAFETY

Sangeeta Raut
Sudip Kumar Sen

SCIENTIFIC
PUBLISHERS

Published by
SCIENTIFIC PUBLISHERS (INDIA)

Jodhpur –

5 A, New Pali Road
P.O. Box 91
Jodhpur - 342 001 INDIA

Delhi –

4806/24, Ansari Road
Daryaganj
New Delhi - 110 002 INDIA

ISBN: 978-93-86102-40-9

Visit the Scientific Publishers (India) website at
http://www.scientificpub.com

Printed in India

Dedicated to

Our Parents, who encouraged us.

Our Teachers, who enabled us.

Our Students, who believed us.

Contents

Syllabus

HSSM3303 ENVIRONMENTAL ENGINEERING & SAFETY (3-0-0)

Objective: This course introduces the students to the environmental consequences of industries, developmental actions, etc., and the methods of minimizing their impact through technology and legal systems.

Module – I

Ecological Concepts: Biotic components, Ecosystem processes: Energy, Food chain, Water cycle, Oxygen cycle, Nitrogen cycle etc., Environmental gradients, Tolerance levels of environment factor, EU, US and Indian environmental law. Chemistry in environmental engineering: Atmospheric chemistry, Soil chemistry. Noise pollution: Noise standards, measurement and control. Water treatment: Water quality standards and parameters, Ground water, Water treatment processes, Pre-treatment of water, Conventional process, Advanced water treatment process

Module – II

Waste Water Treatment: DO and BOD of waste water treatment process, pretreatment, Primary and Secondary treatment of waste water, Activated sludge treatment: Anaerobic digestion, Reactor configurations and methane production

Air Pollution: Air pollution and pollutants, criteria pollutants, Acid deposition, Global climate change: Greenhouse gases, Non-criteria pollutants, Air pollution meteorology, Atmospheric dispersion. Industrial air emission control, Flue gas desulphurization, NO_x removal, Fugitive emissions.

Solid waste, Hazardous waste management, Solid waste management, Source classification and composition of MSW: Separation, storage and transportation, Reuse and recycling, Waste minimization techniques. Hazardous waste management, Hazardous waste and their generation, Transportation and treatment: Incinerators, Inorganic waste treatment, E.I.A., Environmental auditing

Module – III

Occupational safety and health acts, Safety procedures, Type of accidents, Chemical and heat burns, Prevention of accidents involving hazardous substances, Human error and hazard analysis. Hazard control measures in integrated steel industry, Petroleum refinery, L.P.G. bottling, Pharmaceutical industry. Fire prevention – detection, extinguishing fire, electrical safety, product safety. Safety management- safety handling and storage of hazardous materials, Corrosive substances, Gas cylinders, Hydro carbons and wastes. Personal protective equipments.

REFERENCE

Text Books

1. Environmental Engineering, Irwin/ McGraw Hill International Edition, 1997, G. Kiely.
2. Environmental Engineering by Prof B.K. Mohapatra, Seven Seas Publication, Cuttack.
3. Industrial Safety Management, L. M. Deshmukh, Tata McGraw Hill Publication.
4. Environmental Engineering by Arcadio P. Sincero & Gergoria A. Sincero PHI Publication.
5. Principles of Environmental Engineering and Science, M. L. Davis and S. J. Masen, McGraw Hill International Edition, 2004.
6. Environmental Science, Curringham & Saigo, TMH.
7. Man and Environment by Dash & Mishra.
8. An Introduction to Environmental Engineering and Science by Gilbert M. Masters & Wendell P. Ela - PHI Publication.
9. Industrial Safety Management and Technology, Colling. D A – Prentice Hall, New Delhi.

Preface

Human activities are affecting the environment, with numerous direct and indirect effects on ecosystems. The climate and atmospheric composition of earth are changing rapidly. Humans have directly modified half of the ice-free terrestrial surface and use 40% of terrestrial production. Our actions are causing the sixth major extinction event in the history of life on earth and radically modify the interactions among forests, fields, streams and oceans. This book is written to provide a conceptual basis for understanding ecosystem processes and their sensitivity to environmental and biotic changes. We believe that an understanding of ecosystem dynamics must underlie on our analysis of both the consequences and the mitigation of human – induced changes.

The present state-of-art is intended to introduce the science of ecological concept and ecosystem to advanced undergraduate students, beginning graduate students and practicing scientists from a wide array of disciplines. We define ecosystem to include aquatic and terrestrial ecosystem. We also included the description of water and waste-water treatment to provide a broader context for understanding water pollution. We provide access to some of the rapidly expanding literature in many disciplines that contribute to ecosystem understanding. This book incorporates materials that accounts for soil pollution and noise pollution as well as the chemistry of our atmosphere.

The first section of this book provides the understanding of ecosystem ecology. We introduce the science of ecosystem ecology and place it in the context of other components of the earth system – the atmosphere affects ecosystem processes and contributes to the global variation in terrestrial ecosystem structure and processes. We also consider the mechanisms by which ecosystems function and focus on the flow of energy and the cycling of nutrients. Along with this we then consider the important role of organisms in ecosystem processes through tropic interactions (feeding relationships) and environmental gradients.

In the second section of the book we consider air pollution, solid and hazardous waste management. Shortage of appropriate textbooks that could meet the need for training professionals on the nature and the magnitude of ambient and indoor air pollutions and their effects have been one of the outstanding problems in the existing higher learning institutions in India. Therefore, a well-developed teaching material to produce the required qualified professionals, who are considered to shoulder the responsibility of preventing and controlling of air pollutions by creating awareness and entertaining some interventional measures among the communities, is obvious. The present topic on air pollution is therefore, prepared to be used as a teaching material to train mainly environmental health and other students of health category in

India. It also focuses on sources, types and composition of solid wastes; functional elements of solid waste management program; community-based solid waste management. It is written for an integrated solid waste management system to minimize or if possible to eliminate public and environmental health problems so as to prevent diseases and promote health and well being of the community. Its scope is such that it will find application for diploma and degree students.

The third section of the book addresses the safety measures. It contains twelve chapters in which the major current environmental problems with their suggested solutions are discussed. Each chapter is presented in simple language and is provided with introduction, exercises, and suggested reading as appropriate. Text books, journals, internet sources and other lecture manuscript are used to develop this book. We have also incorporated the useful ideas of different instructors of the course to standardize it to its present status, which the authors hope to further improve the draft through the consultations, pretest and revisions. It is also hoped that this book will be of particular use not only for students of health category in colleges and universities, but to those graduates working in health care service institutions and environmental protection agencies.

We finish by considering the integrated effects of these processes at the global scale and their consequences for sustainable use by human societies.

Authors

Acknowledgements

We will be failing in our duty if we do not pay our thanks to those who have helped and encouraged us in writing this book. The authors are highly thankful to Dr. R.R Dash and Dr. R.I. Ganguly for their continuous motivation and encouragement from inception to completion of this manuscript. We express our sincere thanks to our student Er. Bhaswatimayee Sahu whose contribution in drawing the figures is reflected throughout this present state-of-art. We gratefully acknowledge Er. Nipon Sarmah for giving valuable exercises in all the chapters. Our heartfelt thanks to Mr. Tanay Sharma, Scientific Publishers (India), Jodhpur for facilitating the manuscript preparation process in all dimensions goes.

We would like to express our thanks to the Management of Gandhi Institute of Engineering and Technology for providing useful guidelines, technical and moral support during the development of this manuscript. Our parents deserve special thanks for their immense love, encouragement and multiple supports and the burden they shared during this manuscript preparation. Finally, we thank all the individuals who have in some ways contributed to this manuscript, either in conversations with us or through reviewing the draft.

Authors

Chapter - 1

Introduction

AN INTRODUCTION TO ENVIRONMENTAL ENGINEERING

Environmental engineering applies engineering principles to natural systems to solve and prevent environmental degradation. The discipline uses knowledge of soil, water, and air systems to engineer solutions from field to watershed scales.

Environmental engineering is the branch of engineering that is concerned with protecting people from the effects of adverse environmental effects such as pollution, as well as improving environmental quality. Environmental engineers work to improve recycling, waste disposal, public health, water and air pollution control, according to the U.S. Bureau of Labour Statistics.

The practice of environmental engineering dates back to the dawn of civilization. Ever since groups of people began living in semi-permanent settlements, they have had to deal with the challenges of providing clean water and disposing of solid waste and sewage. With the growth of cities and the advent of large-scale farming and manufacturing, people have also had to worry about air quality and soil contamination.

The first environmental engineer is said to have been Joseph Bazalgette. According to an article in the Postgraduate Medical Journal, Bazalgette oversaw the construction of the first large-scale municipal sanitary sewer system in London in the mid-19th century. This was prompted by a series of cholera epidemics, as well as a persistent unbearable stench, that were attributed to the discharge of raw sewage into the Thames river, which was also the main source of drinking water for the city. This great stink which was so noxious that it caused Parliament to evacuate Westminster, gave then Prime Minister Benjamin Disraeli grounds to ask for 3.5 million pounds to improve the city's sewage disposal system.

A drastic increase in environmental pollution across the globe has led to various research and development works in an effort to contain this menace and a relatively new discipline called **Environmental Engineering** has emerged which is providing excellent career opportunities. Environmental engineers apply combined methods of science and engineering to improve the environment so that people have access to healthy water, air and land suitable for human existence. Moreover, opportunities in environmental engineering have increased in last couple of decades and the professionals are recruited by companies working in chemical, geological, petroleum and mining sectors.

Environmental engineers specialising in basic sciences, ecology, economics, social sciences, and radiological science as well as civil, mechanical and electrical engineers have attractive career options. A qualified professional working for an environmental engineering firm is expected to design, construct and maintain waste management systems so that urban and rural areas are kept healthy. Job of environmental engineers in India is expected to widen up as owing to tremendous economic development the country is facing high degree of environmental pollution and environmental engineers will have to check it.

Environmental Attitudes

In recent decades, in the planetary time line, humans were just another animal trying to survive in an inimical, competitive environment. What allowed us to become the top of the animal kingdom? It could be a combination of brain development and the willingness to cooperate with each other. Whatever the reasons, small family groups eventually grew to larger, organized communities that successfully faced the daily challenges of obtaining food, surviving attacks by predatory animals, and countering some of the effects of natural forces like weather and disease. Although many people think that ancient people lived in harmony with their environment, science has uncovered evidence which contradicts that assumption. What we have discovered is that, while some societies thrived by using appropriate practices, other societies collapsed due to poor use of their environment. By conducting an internet search, you can find information about the collapse of an ancient society. The challenge for the modern world is to use our knowledge, both ancient and modern, to support the intelligent, sustainable use of our environment.

Environmental Ethics

There are many different attitudes about how people appreciate and interact with the environment. Most of these attitudes fall under one of three headings are described below:

The **development ethic** is based on the individual (egocentrism). It assumes that humans should be the dominant of nature and that earth and its resources exist for our benefits. This approach assumes that nature has no inherent value; that is, the environment has assessment only in so far as humans economically place value on it. The **preservation ethic** considers nature special in itself. Nature has inherent value or worth apart from human reliance on it. Conservationists have varied reasons for wanting to preserve nature. Some people have a strong respect for all life and respect the right of all creatures to live, no matter what the social or economic costs. Other conservationists' interest in nature is primarily aesthetic or recreational. They believe that nature is beautiful and should be available for picnics, camping, fishing, or just for peace or enjoying. Some preservationists value the scientific importance as well as aesthetics of nature. They argue that the human species depends on and has much to learn from nature. Rare and endangered species and ecosystems, as well as the more common ones, must be preserved because of their known or assumed long-range practical utility. The third attitude is referred to as the management or conservation ethic. It is related to the preservation point of view, but extends the consideration to the entire earth for all time. The **conservation ethic** stresses a balance between total development and absolute preservation. The goal of the conservation ethic is humans living together with a good quality of life, but in a way that sustains all life and protects earth.

Job Prospects

Environmental engineering has immense scope in India for professionals who are qualified in chemical, biological, thermal, radioactive or even mechanical engineering as environmental engineering is a diverse field and requires people from diverse background. Moreover, professionals from process engineering, environmental chemistry, water and sewage treatment, waste reduction management, pollution prevention, etc are required in large numbers.

Job prospects for professionals with qualification in environmental engineering are bright in research centres, NGOs and various governmental departments working towards green development. Candidates with M.Tech in environmental engineering have opportunities to work for government built assessment committees which study and analyse the environmental risks involved in certain projects. The report of such a committee is then submitted to the concerned authority for further observation. Scholars with PhD in environmental engineering can opt for teaching jobs in colleges and universities.

International Focus

International firms and supra-national bodies like UNO require candidates with qualification and exposure in environment technology for various projects. Jobs in such projects are not only high paying in terms of salaries but attractive in terms of facilities and amenities offered. The US on the other hand offers attractive salary packages for environmental engineers where median salary for an environmental engineer is 80,000$.

Demand and Supply

Environmental engineering is facing shortage of trained professionals. As it is a new branch focusing on various implications of environment pollution, lack of information among candidates prevents them from entering the profession. There is a shortage of environment engineers as more organisations embrace environment friendly processes. Most of the engineering students go for traditional engineering disciplines and do not wish to explore the yet not explored environment engineering. However, a rapid growth in environment pollution will require environmental engineers in large numbers so that the same can be mitigated effectively.

Market Watch

Various governmental and non-governmental bodies require environmental engineers who can work for designing of environmental policy and regulation development, contaminated land management and site remediation and ensure environmental health and safety of people. Candidates with specialisation in hazardous waste management and natural resource management have higher employability as the segment is in dire need of trained professionals. Similarly, professionals exposed to risk assessment of various mega projects can get attractive jobs in various panels aimed to study environmental impact of the concerned project.

Most of the demand in environmental engineering is coming owing to population growth, diminishing resources and risky exploitation of natural resources. Various state governments, mineral extracting companies and NGOs are looking to hire more environmental engineers so that public health problems can be solved. Moreover, more number of environmental engineers will be required to have sustainable development, environmental regulations, etc.

Positives/Negatives

Environmental engineering is a promising career; there are also certain things that one has to keep in mind. Jobs in government sector are filled with the due process established by the concerned authority. Similarly, the job requires lots of analysis and study of various facets and factors before reaching onto any conclusion.

There are also many advantages and one of them is that it is a promising career and still at a nascent stage. A lot of growth is expected in coming decades and early starters will have a certain advantage. Moreover, the job provides an opportunity to candidates to experience materialisation of conceived ideas towards green development. Environmental engineers help fellow humans by making the earth a better place.

■■■

Chapter 2

Ecological Concepts

2.1 ECOLOGY

Every aspects of life are influenced by the environment and the activities of surrounding organisms affect their environment. The term ecology is derived from the Greek term *Oekologie* which is a combination of two Greek words *oikos* meaning house or place of living and *logos* meaning study. It was Hans Reiter, who coined these two Greek words expressing *Oekologie* in 1885. The word ecology was coined by Ernst Haeckel. **Ecology** is the scientific study of how organisms interact with each other and with their environment. This includes relationships between individuals of the same species, between different species, and between organisms and their physical and chemical environments. Ecology can also be called **environmental biology.**

It is like an enormous puzzle in which each organism has requirements for life which interlock with many other individuals in the area (Fig. 2.1). Some of these individuals belong to the same species while some to different species most of the organisms have different ways of living or interacting.

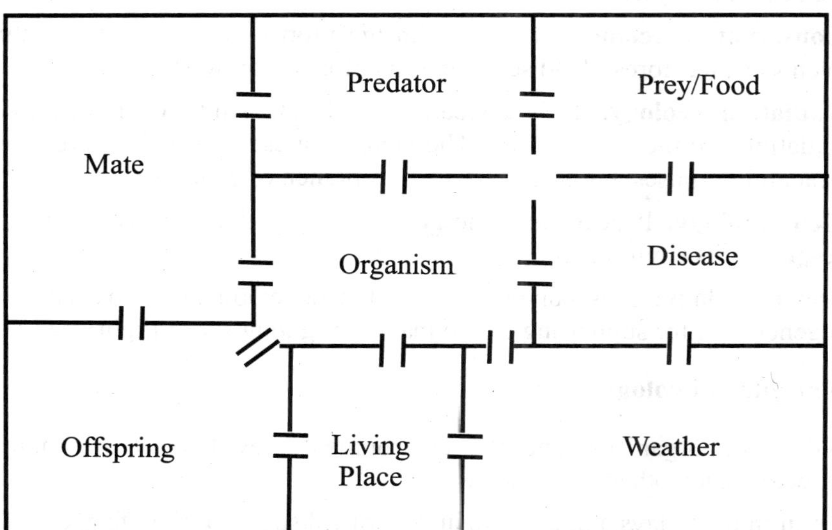

Fig. 2.1 Interlocking Nature of Ecology

2.1.1 Scope of Ecology

The scope of biology lies in the wide spectrum of its applications in regards to human interest. It has several practical applications in the field of agriculture, forestry, horticulture, conservation of biodiversity, biological indicators, meteorology, environmental physiology, pest control and population genetics. The ecology is broadly divided into two categories: autoecology and synecology. **Autecology** is the study of single species or individual organism or a factor in a population or community at all stages of its life span. It is concerned with the study of individual animal or plant species or its population in relation to the habitat in which it grows. **Synecology** is the study of group of organisms. It is concerned with the structure, nature, development and causes of distribution of communities. The various field of ecology has been discussed below:

(i) **Paleoecology:** It is concerned with the study of past geological environment and helps to find out the gaps in the evolutionary lines and origin of various fauna and flora.

(ii) **Cytoecology:** It is concerned with the study of cytological details of the species of populations in relations to different environmental conditions.

(iii) **Habitat ecology:** It deals with the nature of habitat and includes forest ecology, grassland ecology, fresh water ecology, marine ecology, desert ecology and cropland ecology.

(iv) **Population ecology:** It is concerned with the study of inter-relationships of different group of organisms. It also includes the number of the organism and what determines their number and distribution.

(v) **Ecosystem ecology**: It is concerned with the structure and working of ecological systems in relations to space and time and also analyse the structural and functional aspects of ecosystem.

(vi) **Conservation ecology:** It deals with the proper management of natural resources such as water, forest, land sea, mines etc. for human welfare.

(vii) **Radiation ecology:** It is concerned with the study of radioactive substances, radiation and the environment. The increasing use of atomic energy for military and peaceful purposes has given birth of this branch of ecology.

(viii) **Gene ecology:** It deals with the genetic constitution of species or population with respect to its environment.

(ix) **Space ecology:** It is concerned with the development of partially or completely regenerating for supporting life of man during long space flight.

2.1.2 Basic Concepts of Ecology

(i) All living organisms and the environment they live in are mutually reactive, affecting each other in various ways.

(ii) Environment plays a major role in the critical stages of the life cycle of the species.

(iii) The species reacts to the environmental changes and adjusts accordingly itself structurally and physiologically.

(iv) The environment also changes according to certain species- specific activities like growth, dispersal, reproduction, death, decay, etc.

(v) All plants and animals are related to each other by their coactions and reaction on the environment.

(vi) Under similar climatic conditions, there may simultaneously develop more than one community, some reaching the climax stage, and others under different stages of succession.

2.1.3 Six Major Levels of Ecological Organization

The six major levels of ecological organizations are discussed below:

(i) **Individual:** Organisms which belong to a species that includes all the individuals potentially able to interbreed with one another and produce fertile offspring.

(ii) **Population:** It consists of a group of individuals of the same species living in a particular area at a given time.

(iii) **Community:** Populations of different species living together and interacting with each other form a community like natural community of plants, animals and microbes forming a distinctive living system in a pond.

Communities occur in habitats, which refer to the kind of physical environments or places determined by the topography, vegetation structure, geology, and surrounding medium (air or water) such as forest habitat, seashore habitat etc. **Niche-** within each habitat, one can describe for a species, its place within the community – a combination of what it does and where it lives – this is termed as the **species niche**.

(iv) **Ecosystem:** It consists of a series of interacting biotic and abiotic sub-components and two important ecological processes acting as the major linking pathways between these components. These two processes, energy flow and nutrient cycling, are essential for the survival and maintenance of the biotic environment.

(v) **Biomes:** When environmental conditions (e.g. climate) are similar in different parts of the country or on a larger scale in different parts of the world, the habitats (e.g. in terms of vegetation type) and communities are also often similar. We can thus discern a higher level of organization, the biome such as tropical rainforest (high temperature and rainfall), coniferous forest (low winter temperature).

(vi) **Biosphere:** It is that part of the earth and atmosphere in which life exists.

2.1.4 Classification of Ecology

The ecologists has divided ecology into two major branches

(i) Aquatic ecology

(ii) Terrestrial ecology

2.1.4.1 Aquatic ecology

It includes the study of relationships of all aquatic environments, including oceans, estuaries, lakes, ponds, wetlands, rivers, and streams. The boundaries of an aquatic ecosystem are somewhat arbitrary, but generally enclose a system in which inflows and outflows can be estimated. Ecosystem ecologists study how nutrients, energy, and water flow through an ecosystem.

The physical characteristics of aquatic habitats affect the types of organisms living there. Living organisms in a particular environment are directly affected by environmental characteristics such as nutrient concentrations, temperature, water flow and shelter. Organisms those who are able to survive in the particular conditions and habitat and use the available resources will thrive only. Interactions between living organisms also affect the type of organisms found in an aquatic ecosystem, as competition for resources (e.g., food, habitat) and predation affects species abundance and diversity. In turn, the living organisms in an environment can influence some aspects of their environment. Understanding the basic components of aquatic ecosystems and its interaction among living organisms and their environment can lead to better management of human impacts on these systems.

2.1.4.2 Terrestrial ecology

Terrestrial ecology is the communities of species that interact in a land based habitat like a desert, grassland, a rainforest or tundra.

2.1.5 Concepts of Ecology: Structural and Functional Concept of Ecology

Based on structural components and their relationships, ecology and ecosystem can be explained in two ways: **structural concepts** and **functional concepts**.

2.1.5.1 Structural concepts

The structure is related to species diversity. Different types of organisms living in a particular environment are not only independent and mutually reactive but also react with the environment. Though organisms of a species maintain uniformity in their structure and functions through sharing a common gene pool, they possess sufficient plasticity to modify themselves according to changing environment by alternations in somatic characters (ecads) or genetic characters (ecotypes).

Due to their activities, sometimes organisms modify the environment to make it more congenial for their growth, development, reproduction and dispersal. Gradually, this modified environment may become less suitable for the community already living in it. This invites another community that also changes the environment and may again become less suitable for the community already living in it. The community also changes the environment further beyond its most favourable limit.

The development of different communities over a period of time at the same site is called **succession**. The process of succession and change in environment would continue till equilibrium is established between the changed environment and a community called **climax community**.

Under similar climatic conditions, different types of communities grow. Some of them have reached their climax stage while others occur in different stages of succession. The complex of many communities growing in a particular area and sharing a common climate is called **biome**.

Basic types of succession is as follows:

- **Primary succession:** It starts from the primitive substratum, where there was no previously any sort of living matter. The first group is known as pioneers, primary community or primary colonisers.
- **Secondary succession:** It starts from previously built substratum with already existing matter. The action of any external force such as climatic factors, biotic interventions, fire causes the existing community to disappear.
- **Autogenic succession:** After the succession has begun, the community itself as a result of its reaction with the environment modifies its own environment and thus causing its own replacement by new communities.
- **Allogenic succession:** When the replacement of the existing community is caused largely by any other external conditions and not by existing organisms.
- **Autotrophic succession**: It is characterised by early and continued dominance of auotrophic organisms like green plant. It begins in a predominantly inorganic environment and the energy flow is maintained indefinitely.
- **Heterotrophic succession:** It is characterised by early dominance of heterotrophs such as bacteria, actinomycetes and fungi.

The hypothetical representation of the stages of succession:

Pioneer ⇒ Seral stage I ⇒ Seral stage II ⇒ Seral stage III ⇒ Climax stage

2.1.5.2. Functional concepts

The biological community consists of a number of organisms and/or populations. Each population occupies a specific volume of the habitat circumscribed by the interaction of various environmental factors and trophic level of the organisms. It is called **ecological niche**. The degree of success of a particular population in an area is determined by the parameters of both abiotic factors as well as its interaction with other types of populations. The interactions amongst the populations can be positive, negative or neutral.

The flow of energy in the ecosystem is unidirectional or non-cyclic. Radiant energy is trapped by autotrophic plants or primarily producers. The energy is transferred to consumers and decomposers. Some energy is lost during its transfer from one trophic level to the next. Organisms use the energy in respiration.

A number of inorganic substances are taken by the living beings for their metabolism and body building. They are called **biogenetic nutrients**. The biogenetic nutrients keep on circulating between the biotic and abiotic components of the ecosystem.

The phenomenon is called **biogeochemical cycling**. Human beings exploit the ecosphere for their own benefits. As a result, only the economically important plants are allowed to grow in an ecosystem. Species diversity and natural interactions amongst the various components are reduced. When neglected, such an ecosystem deteriorates.

A disturbed or deteriorated ecosystem shows changes due to interactions inside, the assemblage of living being and their abiotic environment, modifying and changing both abiotic and biotic components. The change continues till a stable climax community develops. Where a disturbance continues, the deteriorated ecosystem changes the environment completely.

2.2 ECOSYSTEM

An ecosystem or an ecological system is the whole biotic community in a given area and its abiotic environment. It therefore includes a community of living organisms and their physical and chemical environment, linked by flows of energy and nutrients (Fig. 2.2).

An ecosystem can be of any size, from an area as small as a pinhead to the whole biosphere. The term was first used in the 1930s to describe the interdependence of organisms among themselves and with the living (biotic) and non-living (abiotic) environment. At ecosystem level, the units of study are comparatively very large and there are no practical units, if the nature is conceived as a single, giant ecosystem.

The overall view of this type of approach is that living organisms and their non-living environment are inseparably interrelated and interact with each other. Ecosystem ecology emphasises the movements of energy and nutrients (chemical elements) among the biotic and abiotic components of ecosystems.

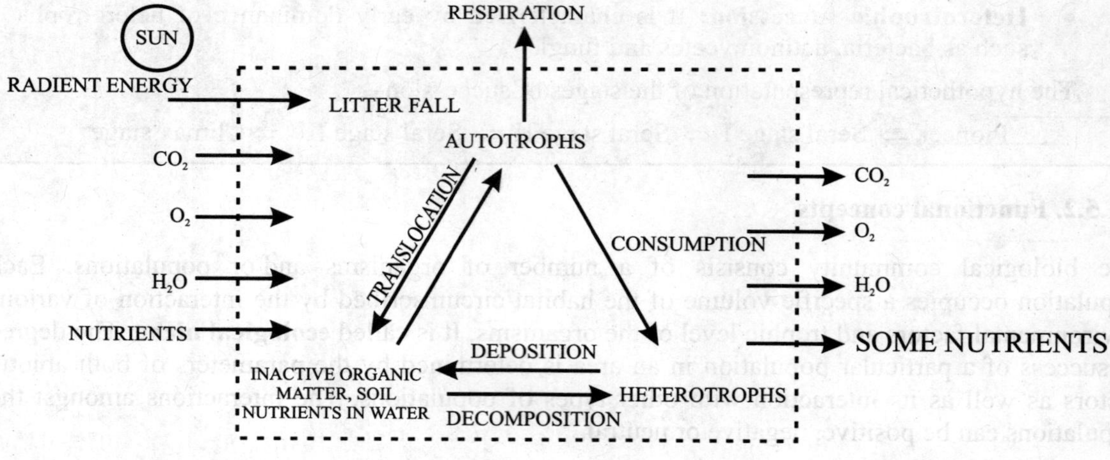

Fig. 2.2 Schematic Diagram of an Ecosystem, The dashed line represents the boundary of the system [Adapted from O'Neill 1976]

For ecosystems like streams, lakes, ponds and forests, stand surrounded by open land, the boundaries appear to be distinct; whereas for most ecosystems they do not. Apparent boundaries, however, are misleading, because no ecosystem stands alone, i.e., energy and materials flow from one ecosystem to another.

2.2.1 Concepts of an Ecosystem

The concept of ecosystem was first put forth by A.G. Tansley (1935) who states the system resulting from the integration of all the living and non-living factors of the ecosystem. He regarded the ecosystem as not only organism's complex but also the whole complex of physical factors forming the environment. According to E.P. Odum (1971) the ecosystem is the basic functional unit of organisms and their environment interacting with each other and with their own components. The ecosystem represents the highest level of ecological integration which is energy-based and this functional unit is capable of energy transformation, accumulation and circulation. The further concepts of ecosystem have been discussed:

(i) Ecosystem is the major structural and functional unit of ecology.

(ii) The structure of an ecosystem is related to the diversity of its species. Complex ecosystems have high species diversity.

(iii) The function of an ecosystem is related to the flow of energy among its biotic components and cycling of nutrients within and outside the system. The flow of energy is unidirectional but flow of nutrients is cyclic.

(iv) There exist varying degrees of positive, negative and even neutral interactions among organisms at both inter- and intra-specific levels.

(v) Energy is the driving force of an ecosystem which is unidirectional or non-cyclic.

(vi) The chemical components of the ecosystem move in a defined path called **biogeochemical cycles**.

(vii) Successful growth of the organism is governed by limiting factors. The minimal and maximum levels of tolerance for all species vary seasonally, geographically and according to the population.

(viii) Each ecosystem has its own energy budget. The relative amount of energy needed to maintain an ecosystem depends on its structure. Complex ecosystems, with species diversity, need less energy to maintain them.

(ix) Ecosystems maintain a homeostatic state. Certain negative feedback mechanisms help in this process.

(x) Under natural conditions, different kinds of population undergo succession.

2.2.2 Types of Ecosystem

The ecosystem is an overall integration of whole mosaics of interacting organisms and their environment. It is considered an open system with a continuous, but variable, influx and loss of materials and energy. These are categorised as under:

(i) Natural ecosystems

These operate under natural conditions without any major interference by man. Based on the kind of habitat, it is further divided into:

a. **Terrestrial ecosystem:** It is sub-divided into grassland, forest, mountain, desert and cave ecosystems.

- **Forest ecosystem:** It is considered one best example of terrestrial ecosystem. soil, moisture, air and sunlight form the abiotic or physical component. All the green plants of a forest are producers. There are number of consumers in an old dense forest. Consumers of first order in the forest are grasshoppers, rabbit, deer, monkey, birds etc which utilize plants directly as their food whereas secondary consumers are wolves, pythons, jackals etc which consume the flesh of herbivores.

- **Desert ecosystem**: They occur in the regions with less than 10% of annual rainfall or sometimes in hot region, where there is more rainfall, but evenly distributed in the annual cycle. The deserts cover about 17% of the total land.
- **Grassland ecosystem**: Natural grasslands occur where the rainfall is intermediate between that of desert and forest lands and occupies about 19% of area of land. The principal grassland ecosystems are Great Plains of Canada, United States of America.

b. **Aquatic ecosystem** : It is again sub-divided into the following types:

- **Freshwater ecosystem:** These may be lotic (running water such as spring, stream or river) or lentic (standing water such as lake, pond, pool etc.) ecosystems. There are three zones in the lakes or ponds: littoral, limnetic and profundal. The littoral zone is the shallow water region which is usually occupied by rotted plants. The limnetic zone ranges from the shallow to the depth of effective light penetration and associated organisms are small crustaceans, rotifers, insects and algae. The profundal zone is the deep water parts where there is no effective light penetration. The associated organisms are snails, mussels and crabs.
- **Marine ecosystem:** These include salt water bodies which may be deep bodies such as oceans or shallow ones such as a sea or estuary. The oceans have unusual stability in temperature, salinity and gaseous content. The upper portion of the oceanic region through which light penetrates is called **euphotic zone**. The region of the ocean below the euphotic zone is called **benthal** or **aphotic zone**. This zone is completely free of photosynthetic organisms.

(ii) Artificial ecosystems

These are maintained artificially by humans where, by addition of energy and planned manipulations, natural balance is disturbed regularly. For example, croplands like wheat, rice fields, etc., where humans try to control the biotic community as well as the physio-chemical environments, are artificial ecosystems.

2.3 STRUCTURE OF ECOSYSTEMS

The ecosystem has both structure and function. Functionally each ecosystem has two major components: abiotic or non-living components and biotic or living components.

(i) Abiotic or non-living components

The Abiotic part embraces the physical environment with which the organisms of the community interact (Fig. 2.3). The abiotic inputs are energy, inorganic substances (CO_2, C, N, O_2) and organic compounds (proteins, carbohydrates, humic acids). Radiant energy both heat and light imposes constraints on the system by influencing temperature, moisture regimes, seasonality and photosynthetic activity. Temperature and moisture determine in part which organisms can live in a system and affect its productive capability. The abiotic components also includes the amount and distribution of inorganic chemicals such as chlorophylls, etc and of organic materials, as proteins, carbohydrates, lipid, etc present either in the biomass or in the environment i.e. biochemical structure that link the biotic and abiotic components of the ecosystem,

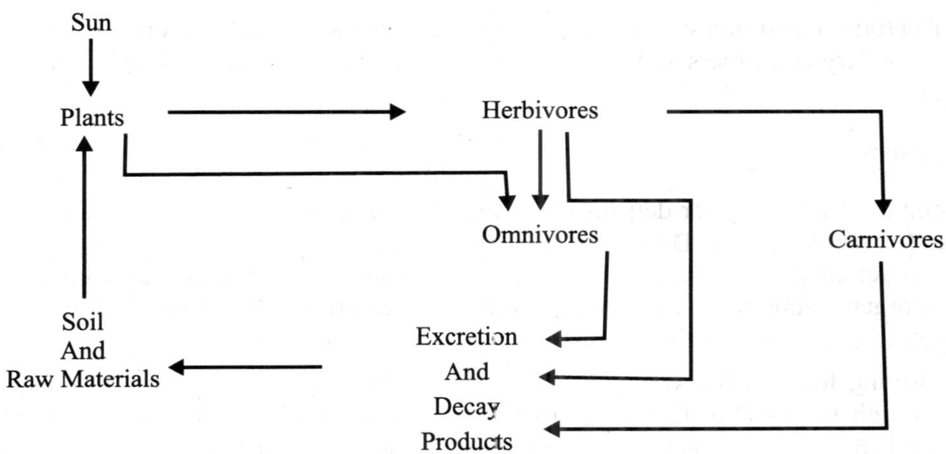

Fig. 2.3 Nutritional Relationships among Different Biotic Components of an Ecosystem

(ii) Biotic or living components

The living organisms of an ecosystem are the biotic components which include all forms of life including flora and fauna. On the basis of their nutritional relationships (i.e., trophic structure) they can be classified as follows:

a. Autotrophic components

These are self-nourishing organisms, the energy-capturing base of the systems are largely green plants and algae. They are also known as **producers** since they fix the energy of the sun and manufacture food from simple inorganic and organic substances. Autotrophic metabolism is greatest in the upper layers of the ecosystem, i.e., the canopy of the forest and the surface water of lakes and oceans.

b Heterotrophic components

They utilize the food stored by the autotrophs, rearrange it and finally decompose the complex materials into simple inorganic substances. In this role they influence the rate of energy flow and nutrient cycling. The heterotrophic component is often subdivided into two subsystems:

- **Consumers:** They feed largely on living tissue. According to their nature of food, they can be classified as:
- **Primary consumers**: These feed on producers (green plants) and are called *Herbivores* (plant eaters). Insects, rodents, rabbit, deer, cow goat are some of the common

herbivores in the terrestrial ecosystem. The herbivores of ecosystem are known as kee industry animals. They serve as the chief food source for carnivores.

- **Secondary consumers**: They are both carnivores and omnivores. They are flesh eating animals and the omnivores are the animals that are adapted to consume herbivores as well as plants as their food. They depend on primary consumers for food and are called carnivores (flesh eaters) e.g. cat, dog, crow, fox, snake, lizard etc
- **Tertiary consumers:** They are the top carnivores and survive on primary as well as secondary consumers and are not eaten by others. e.g. lion, tiger, hawk, vulture, leopard etc.

c. Decomposers

They depend on dead matter or detritus dissolved in organic and inorganic substances in aquatic systems and the soil matrix. Dead organic matter comes from plants and consumer remains, decomposers act on it. They depend on organic compounds of dead plants and animals, convert them into inorganic substances and release them to the environment. Thus they form the basis of internal cycling of nutrients in the ecosystem.

The driving force of the system is the energy of the sun, which causes all other inputs to circulate through the system. Outflows from one subsystem become inflow to another. While energy is used and dissipated as heat of respiration, the used and dissipated heat of respiration, the chemical elements from the environment are being recycled. Consumers regulate the speed at which nutrients turn over in the system.

2.4 ECOSYSTEM PROCESSES

The functional aspects of an ecosystem include the following:

(i) The rate of biological energy flow, i.e., the production and respiration rates of the community.

(ii) The rate of materials or nutrient cycles.

(iii) Biological or ecological regulation including both regulation of organisms by environment and regulation of environment by the organism.

An ecosystem in general performs the following functions as discussed below:

2.4.1. Food Chain

The transfer of food energy from the producers, through a series of organisms with repeated eating and being eaten is known as **food chain**. Net production is the base upon which the rest of life on earth depends. This energy stored by plants is passed through the ecosystem in a series of steps of eating and being eaten. Earth consumer population uses the food energy consumed to live and respire and the remaining energy can then be used to help produce new biomass by growth and reproduction. Ecosystems consist of a myriad series of such food chains. Species populations at each link in the various chains can in turn be grouped into what are known as **trophic levels**.

2.4.1.1 Types of food chains

Food chain in any ecosystem runs directly in which green plants are eaten by herbivores, herbivores are eaten by carnivores, and carnivores are eaten by top carnivores. Based on the type of food consumed by the consumers, food chains are of three types as discussed below:

2.4.1.2. Grazing food chain

This food chain starts from living plants, goes through herbivores and ends in carnivores.

Plants → Herbivores → Primary Carnivores → Secondary Carnivores

Size of food items and consumers has considerable influence on the direction a food chain takes, because there are upper and lower limits to the size of food an animal can capture. Some animals are large enough to defend themselves successfully. Some foods are too small to be collected economically; it takes too long to secure enough to meet the animal's metabolic needs. Thus the upper limit to the size of an animal's food is determined by its ability to handle and process, and the lower limit by its ability to secure enough.

Generally, energy transformed through the ecosystem by way of the grazing chain is reduced by a magnitude of 10 from one level to another. Thus if an average of 1000 kcal of plant energy is consumed by herbivores, about 100 kcal is converted to herbivore tissue, 10 Kcal to first – level carnivore production and 1 kcal to second – level carnivores. The amount of energy available to second and third level carnivores is so small that few organisms could be supported if they depended on that source alone. For all practical purposes, each food chain has from three to four links, rarely five (Prim 1982).

e.g. Grass → Grasshopper → Lizard → Hawk (Grassland)

Phytoplankton → Zooplankton → Fish → Snake (Pond)

2.4.1.2.1 *Detritus food chain*

It starts from dead organic matter (detritus) and ends in inorganic compounds e.g. mangrove ecosystems (Fig. 2,4) There are certain groups of organisms which feed exclusively on the dead bodies of animals and plants. These organisms are called **detritivores**.

e.g. Dead Plants / fallen leaves → Soils mites → Insects → Lizards

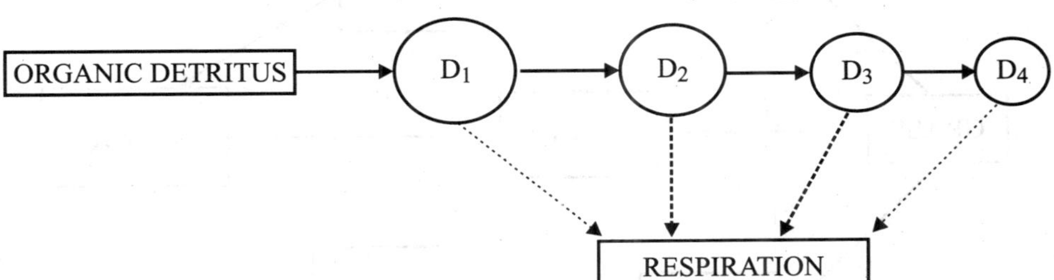

D_1 - D_4 are different levels in the chain. Size of circles indicate relative amount of energy.

Fig. 2.4 Schematic Diagram of Detritus Food Chain

Each link in the chain extracts nutrients and energy from the organic matter and loses energy through respiration before passing on the remaining organic matter to the next link in the food chain.

2.4.1.2.2 ***Parasitic food chain***

It is a food chain in which parasites live on or within the body of the host to derive energy. This food chain also starts from the green plants and animals to parasitic microbes.

e.g. Tree (bearing fruits) → birds → parasites

2.4.1.3 Significance of food chain

(i) The food chain studies help to understand the feeding relationship and the interaction between organisms in any ecosystem.

(ii) They also help us to appreciate the energy flow mechanism and matter circulation in ecosystem and understand the movement of toxic substances in the ecosystem and the problem of biological magnification.

(iii) The stability of an ecosystem is directly proportional to the number of links.

2.4.2 Food Web

In ecosystem, no organism lives wholly on another, the resources are shared, especially at the beginning of the chain. The marsh plants are eaten by a variety of invertebrates, birds, mammals and fish, and some of the animals are consumed by several predators. Thus food chains become interlinked to form a food web, the complexity of which varies within and between ecosystems (Figs 2.5 and 2.6). For example: grass → mouse → snakes → owls, sometimes mice are not eaten by snakes but directly by owls.

Food webs in constant environment are characterized by richness of greater species and more trophic interlinks. The food webs found in fluctuating environment contain only fewer trophic links. Thus the environmental conditions impose limitation and rigidity on the shape of the food web.

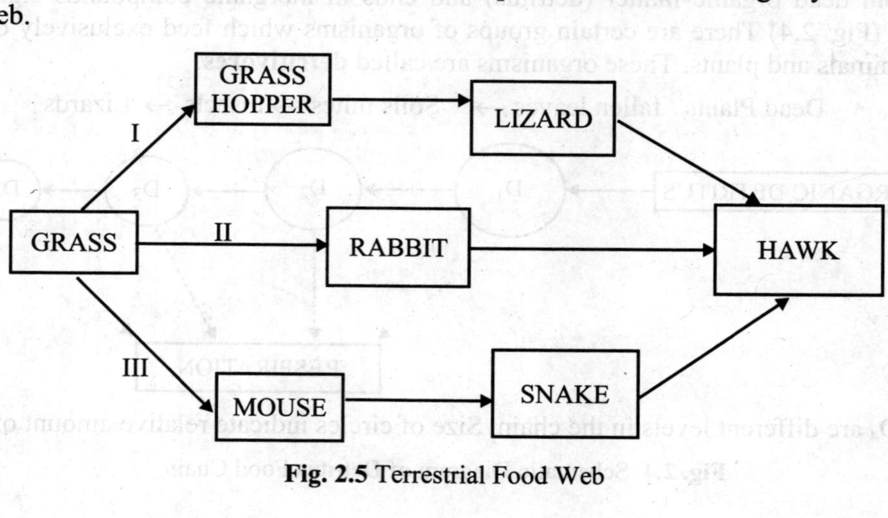

Fig. 2.5 Terrestrial Food Web

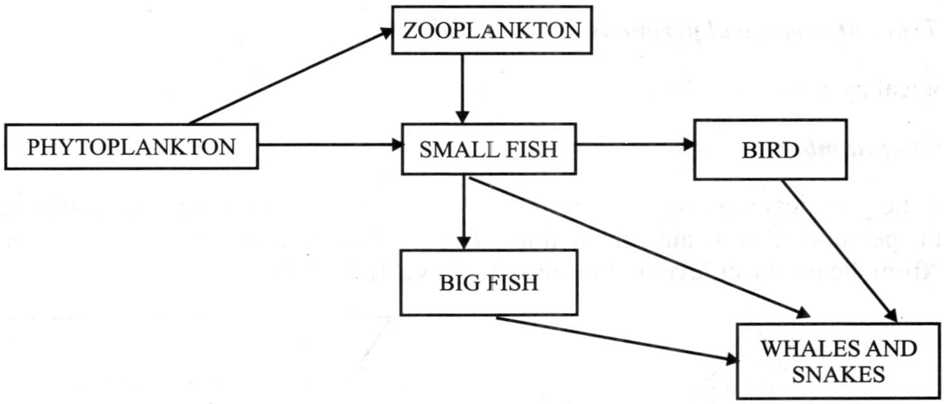

Fig. 2.6 Aquatic food web

2.4.2.1 Significance of food web

(i) Food webs are very important is maintaining the stability of an ecosystem.

(ii) Food web provides more than one alternative of food to most of the organisms in an ecosystem and therefore increases their chance of survival.

(iii) Complex food webs are more stable than simple ones because, if a species is removed, its position is taken by another species of similar tropic status so that the ecosystem is maintained.

(iv) The concept of food web is useful for analysis of a community.

2.4.2.2 Trophic levels & ecological pyramids

Feeding relationships in an ecosystem may be considered in terms of trophic or consumer levels. From a functional rather than a species point of view, all organisms that obtain their energy in the same number of steps from the producers belong to the same trophic level in the ecosystem.

2.4.2.2.1 *Different Trophic Levels*

(i) **First trophic level:** Producers belong to the first trophic level, e.g. plants, phytoplankton etc.

(ii) **Second trophic level:** Herbivores and higher levels of the carnivores belong to the second tropic level. Some consumers occupy a single trophic level, but many others occupy more than one trophic level.

2.4.2.3 Ecological pyramids

It is the graphical representation of trophic structures showing successive tropic levels of an ecosystem. This concept was given by Charles Elton (1927). Ecological pyramids are of three general types; pyramid of number, showing number of individuals, pyramid of biomass, showing total dry weight and other suitable measure of the total amount of living matter, and pyramid of energy, showing the rate of energy flow and productivity at successive trophic levels.

2.4.2.3.1 *Types of ecological pyramids*

The ecological pyramids may be of following three kinds.

(i) *Pyramid of numbers*

It refers to the graphical representation of the number of individuals in various trophic levels of a food chain, per unit area at any given time. The number of individuals at the trophic levels decreases from the producer level to the consumer level (Fig. 2.7a).

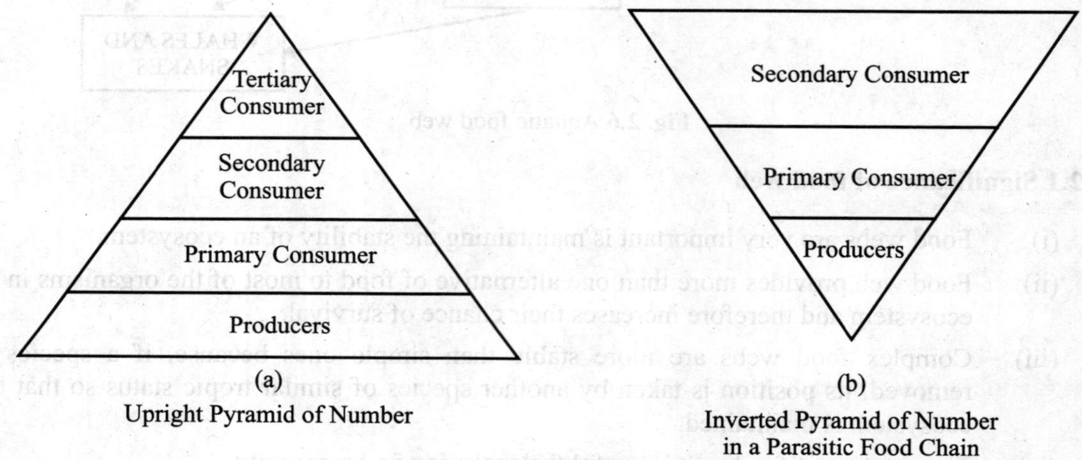

Fig. 2.7 Pyramid of Number

When the ecosystem contains lesser number of producers than those of consumers, the apex of the pyramid is directed downwards (Fig. 2.7b). Pyramids of this type are called **inverted pyramids** e.g. tree ecosystem. A single tree (producer) harbours many fruit eating birds (primary consumers) and these birds in turn host numerous parasites (secondary consumers).

However, consumer level based on numbers is misleading. Although consumers at the base of the pyramid may be most abundant, there is great deal of variation in their size and numbers. Plants are eaten by herbivores ranging in size from very small arthropods to elephants. The pyramid of number ignores the biomass of organisms and it also does not indicate the energy transferred or the use of energy by the groups involved.

(ii) *Pyramid of biomass*

It refers to the total weight or other measurement of living material the total bulk of organisms of fixed energy present at any one time – the standing crop in general, the biomass of producers is greater than the biomass of the herbivores they support, and the biomass of the herbivores is greater than that of the carnivores (Fig. 2.8a). This relationship results in a gradually sloping pyramid, particularly for terrestrial and shallow-water communities where the producers are large, accumulation of organic matter is great, life cycles are long and the rate of harvesting is low.

However, for some ecosystems, the pyramid of biomass is inverted. Primary production is aquatic ecosystems, such as lakes and open seas is concentrated in phytoplankton, planktonic algae have a short life cycle, multiply rapidly, accumulate little organic matter, and are heavily exploited by herbivorous for plankton. At any result the base of the pyramid is much smaller than the structure it supports (Fig. 2.8b).

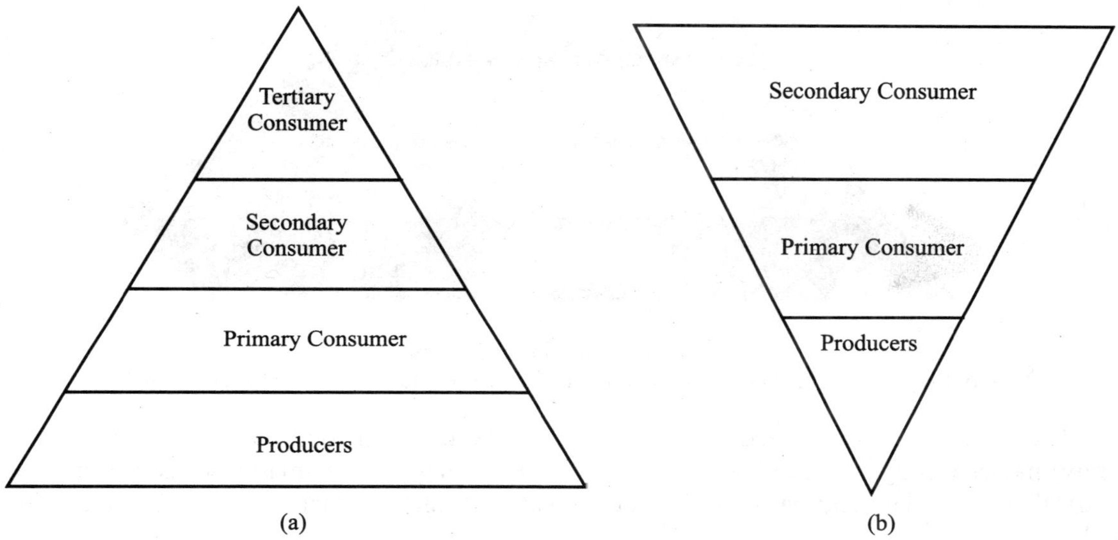

Fig. 2.8 Pyramid of Biomass

(iii) *Pyramid of energy*

This pyramid suggests the quantity of energy fixed, stored and passed on to the next trophic level and also emphasizes the role organisms play in the transfer of energy. At each trophic level substantial portion of energy is lost (Figs. 2.9 and 2.10). Hence the amount of energy decreases from the producer level to the consumer level. Since the flow of energy is always unidirectional, the pyramid of energy is upright. An energy pyramid illustrates how much energy is needed as it flows upwards to support the next trophic level.

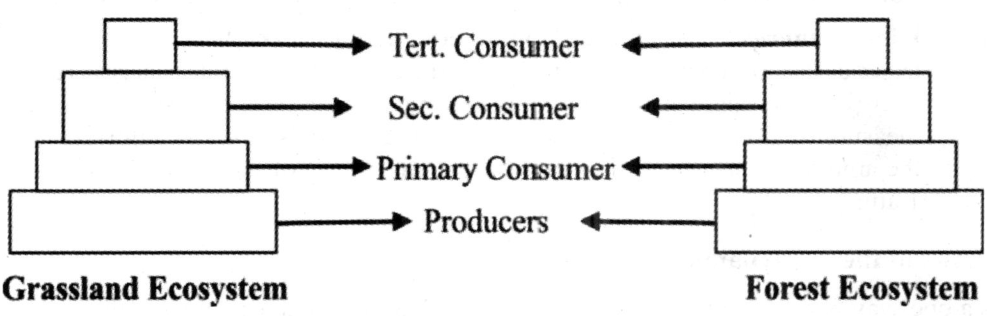

Fig. 2.9 Pyramid of Energy

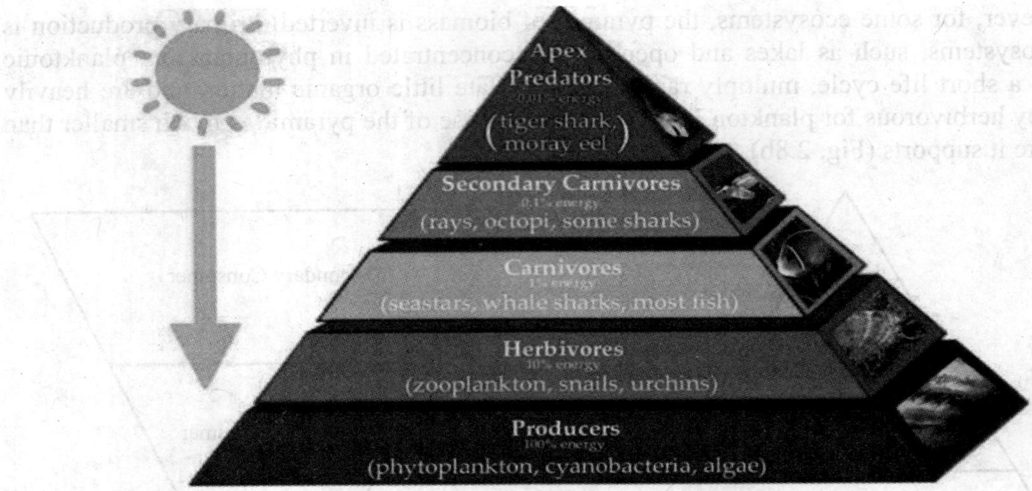

Fig. 2.10 Pyramid of Biomass
Courtesy: Great-barrier-reef-biome.wikispaces.com/file/view/energypyramid.jpg

The pyramid is constructed according to the rate at which food material (in the form of energy) passes through the food chain. Some organisms may have a small biomass, but the total energy they assimilate and pass on, may be considerably greater than that of organisms with a much larger biomass.

2.4.3 Energy Flow

Energy is the most necessary requirement as the very existence of the living world, including human life depends upon the flow of energy from one trophic to other trophic level. Energy is defined as the ability to do work.

Work = Force × Distance

2.4.3.1 Types of energy

Living organisms can use energy in several forms and various types have been discussed below:

(i) **Potential energy:** It is the energy at rest. It is capable of and available for work.

(ii) **Kinetic energy:** It is the energy in motion, during work. Work that results from the expenditure of energy can either be stored or concentrate energy (as potential energy) or arrange matter without storing energy. The SI (System International) measure of a unit of energy is the Joule.1 Joule equals to 4.184 calories. A calorie is the amount of heat necessary to raise 1 gram (1 ml.) of water by 1°C at a pressure of 1 atm.

2.4.3.2 Laws of thermodynamics

There is a one-way path along which energy moves. It has also been established that as the energy moves progressively through the various trophic levels it is no longer available to the previous level. Thermodynamics aims at studying phenomenologically the properties of material bodies characterized by macroscopic parameters without resorting to the microscopic phenomena under investigation. It deals with energy processes of a reacting system. Two laws describe the expenditure and storage of energy.

2.4.3.2.1 *1st law of thermodynamics*

It states that energy is neither created nor destroyed. It may change forms, pass from one place to another, or act upon matter, transforming it to energy in various ways. It is a measure of the change in the internal energy of a body in heat exchange.

2.4.3.2.2 *2nd law of thermodynamics*

Second law states that when energy is transferred or transformed, part of the energy in act is wasted; it assumes a form that cannot be passed on any further. The same thing happens to energy in an ecosystem. As energy is transferred from one organism to another in the form of food, a large part of the energy is degraded as heat through metabolic activity; with a net increase in entropy (the measure of relative disorder is called entropy). However, biological systems do not seem to confirm the second law, for the tendency of life is to produce order out of disorder, to decrease rather than increase entropy.

(i) Primary production: The flow of energy through the ecosystem starts with its fixation by plants and algae through photosynthesis. During that process solar energy hits chloroplasts within the cells, setting off reactions that produce carbohydrates and other organic molecules from carbon dioxide and water. The bonds of these organic molecules are energy-rich, containing energy from the sun. Thus all organic compounds of plants – cellulose, starch, proteins, nucleic acids and secondary compounds contain high amounts of chemical bond energy. The energy accumulated by autotrophs is called **primary product** because it is the first and most basic form of energy stored in an ecosystem. The rate at which energy is stored by photosynthetic activity is known as **primary productivity**. Primary production is divided into two parts:

(a) **Gross primary production (GPP):** It is the entire sun's energy that is assimilated by the photosynthetic activity of plants.

(b) **Net primary production (NPP):** Plants require energy for reproduction and maintenance. Energy remaining after respiration (R) and stored as organic matter is net primary production (NPP). It can be expressed as:

Net primary production (NPP) = Gross primary production (GPP) – respiration (R)

Production is usually measured as the rate at which energy or biomass is produced per unit area per unit time called **productivity**. This rate is expressed in terms such as kilocalories per square meter per year (Kcal/m^2/yr.), a measure of energy, or grams per square meter per year (g dry wt/m^2/yr.), a measure of biomass.

The accumulated organic matter found on a given area at a given time is the **standing crop biomass**. Biomass differs from productivity, which is the rate at which organic matter is created by photosynthesis.

(ii) Secondary production: Energy left over from maintenance and respiration goes into the production of new tissue, fat, growth and new individuals. This net energy of production is

secondary production or **consumer production**. Secondary production is greatest when the birth rate of the population and the growth rates of the individuals are the highest.

Herbivores are the energy source for carnivores, and when they are eaten, not all of the energy in their bodies is utilized. Part of it unconsumed, and again metabolic losses must be accounted for.

The energy budget of a consumer population can be summarized by:

$$C = A + F + U$$

Where, C is the energy ingested or consumed.

A is the energy assimilated.

F and U are the energy lost through faeces and nitrogenous wastes.

The term A can be refined further.

$$A = P + R$$

Where, P is secondary production

R is energy lost through respiration.

U, representing nitrogenous wastes, should be included as part of A (A=P+R+U) because they are involved in the homeostasis of organisms. Because of the difficulty of separating nitrogenous wastes from fecal wastes, U typically is not included in A.

Thus, $C = P + R + F + U$

or, secondary production is:

$$P = C - R - F - U$$

(iii) Net Production (NP): It is the rate of storage of organic matter not used by the consumers.

NP = NPP – consumption by consumers during the unit period as a season or year.

2.4.3.3 Energy flow in ecosystem

It is the transfer of net energy content of one trophic level to the next trophic level above and maintains the system in equilibrium. The study of energy transfer between different trophic levels in an ecosystem is known of **bioenergetics**.

Flow of energy is an ecosystem is based on the assumptions that the laws of thermodynamics hold for plants and animals, which can be arranged into at least three trophic levels namely producer, herbivore and carnivore. Whether at the producer or consumer level, energy flow through the ecosystem is mediated by the individual that consumes a quantity of energy in food; part of it is assimilated and part is lost as faeces, urine and gas. Part of the assimilated energy is used for respiration and part is stored as new tissue which can be used to some extent as energy reserve. Some is used for growth and the production of new individuals.

The population boxes can be filled into several trophic levels and linked to form a model of energy flow through the ecosystem (Fig. 2.11). In this model there is no attempt to separate out the decomposers and detritus feeders, because they must fall into one of the several tropic levels when the food web is collapsed.

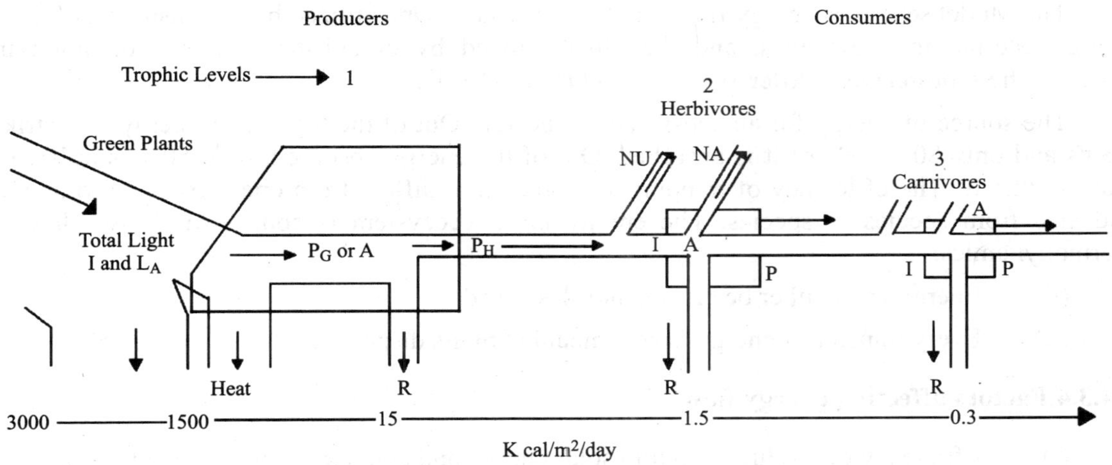

Fig. 2.11 Model of Energy Flow through the Ecosystem

To provide for the grazing and detritus paths of energy transfer, Wiegert and Owens in 1971 proposed a different model of energy flow, which transfers through two channels beyond the autotrophs (Fig. 2.12). They define all organisms utilizing living material as **biophages**, and all organisms utilizing nonliving matter as **saprophages**. First-order biophages, who eat living plants, are the traditional herbivores; whereas first-order saprophages feed on dead plant material as well as organic material egested by first-order biophages. First-order biophages in turn are utilized at death by second –order saprophages, which are functional carnivores. In turn, first-order saprophages may be preyed upon by second-order biophages. Although this model, like others, cannot separate out organisms that occupy several tropic levels, it does modify the Lindeman model so that the decomposers and detritivores are assigned to functional components.

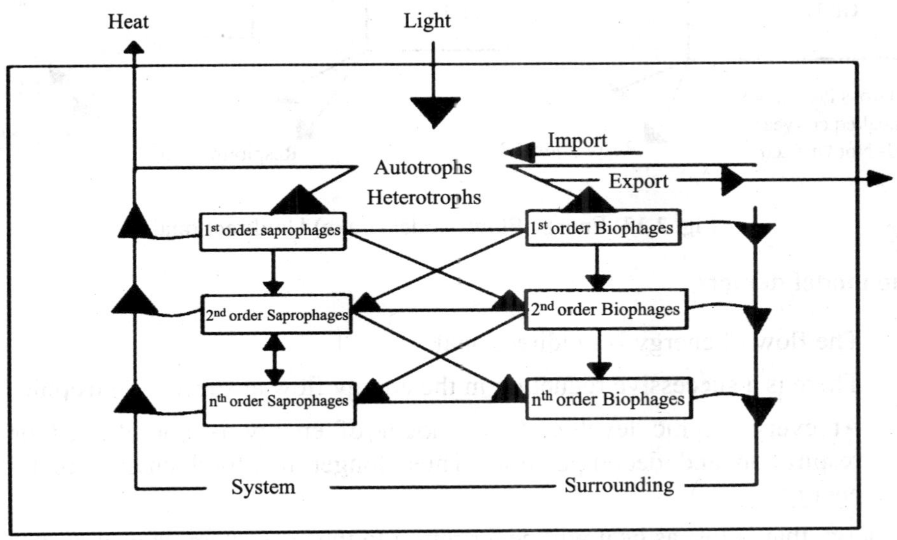

Fig. 2.12 Model of Two-Channel Energy Flow.

The Model separates energy flow into two pathways. One utilized by organisms that feed on living material, the biophages, and the other utilized by organisms that feed on nonliving material, the saprophages. (After Weigert and Owens, 1971).

The source of energy for an ecosystem is the sun. Out of the total solar energy, 8% strikes plants and only 80 – 85% of it is absorbed. Out of the energy absorbed, only 50% is utilized in photosynthesis. The efficiency of trapping solar radiation differs from one ecosystem to another and also from species to species. The energy in an ecosystem is controlled by two laws of thermodynamics:

(i) Energy can neither be created nor destroyed.

(ii) Every transfer of energy is accompanied by its dispersion.

2.4.3.4 Factors affecting energy flow

(i) Efficiency of producers to trap solar energy and convert it into chemical energy.

(ii) Use of chemical energy present in the producers by the consumers.

(iii) Amount of energy assimilated during metabolic activities.

(iv) Loss of energy in the form of unused energy, dead organisms and heat during respiration (Fig. 2.13).

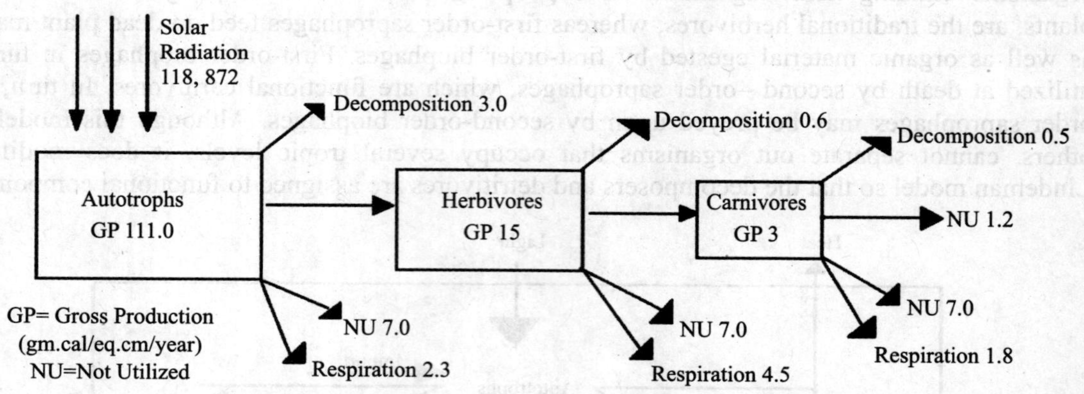

Fig. 2.13 Energy Flow Model (Lake) Unidirectional

2.4.3.5 The model depicts

(i) The flow of energy is unidirectional.

(ii) There is a successive reduction in the energy flow at successive trophic levels.

(iii) At every trophic level certain amount of energy is not utilized or lost due to respiration and decomposition. Thus, longer the food chain, greater the loss of energy.

The energy that is lost as heat will never return to the ecosystem. However, in such one way flow, the entire ecosystem would collapse if the primary source is cut off.

2.4.4 Nutrient Cycling

A **nutrient cycle** (or **ecological recycling**) is the association and exchange of organic and inorganic matter back into the production of living matter. Nutrient cycles occur within ecosystems. Ecosystems are interconnected systems where matter and energy flows are exchanged as organisms feed, digest, and migrate about. Minerals and nutrients gather in varied densities and uneven configurations across the planet. Ecosystems recycle locally, converting mineral nutrients into the production of biomass, and on a larger scale they participate in a global system of inputs and outputs where matter is exchanged and transported through a larger system of biogeochemical cycles. The pathways in which the chemical nutrients move through the biotic and abiotic components of the ecosystem are called ***biogeochemical cycles*** or ***nutrient cycles***.

2.4.4.1 Water cycle

The water cycle or hydrological cycle is a continuous cycle where water evaporates, rises up and becomes part of a cloud, falls down to earth as precipitation, and then evaporates again (Fig. 2.14). This repeats again and again to form a never-ending cycle. Water keeps moving and changing from solid to liquid to gas, over and over again.

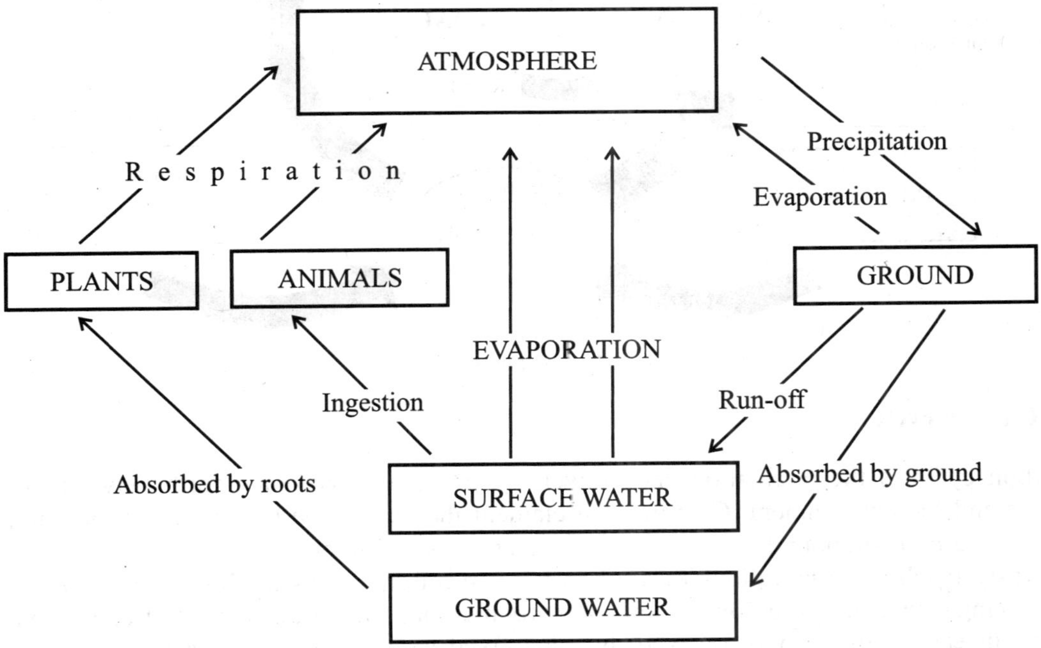

Fig. 2.14 Water Cycle in Nature

Precipitation creates runoff that moves over the ground surface and helps to fill lakes and rivers. It also percolates or moves downward through openings in the soil to replenish aquifers under the ground. These areas are usually close to oceans or large bodies of water that allow more water to evaporate and form clouds. As clouds move up and over mountains, the water vapour condenses to precipitate and freezes to fall as snow. Snow falls on the peaks and the melting of ice contributes to water flow.

2.4.4.2 Oxygen cycle

Circulation of various forms of O_2 occurs in nature. Free in the air and dissolved in water, oxygen is second only to nitrogen in abundance among uncombined elements in the atmosphere (Fig. 2.15). Plants and animals utilize oxygen to respire and return it to the air and water as carbon dioxide (CO_2). CO_2 is then taken by algae and terrestrial green plants and converted into carbohydrates during the process of photosynthesis, oxygen being a by-product. Water present in the world is the main oxygen generator of the biosphere; the algae are estimated to replace about 90% of all oxygen used. Oxygen is involved to some degree in all other biogeochemical cycles. For example, over time, detritus from living organisms transfers oxygen-containing compounds such as calcium carbonates into the lithosphere.

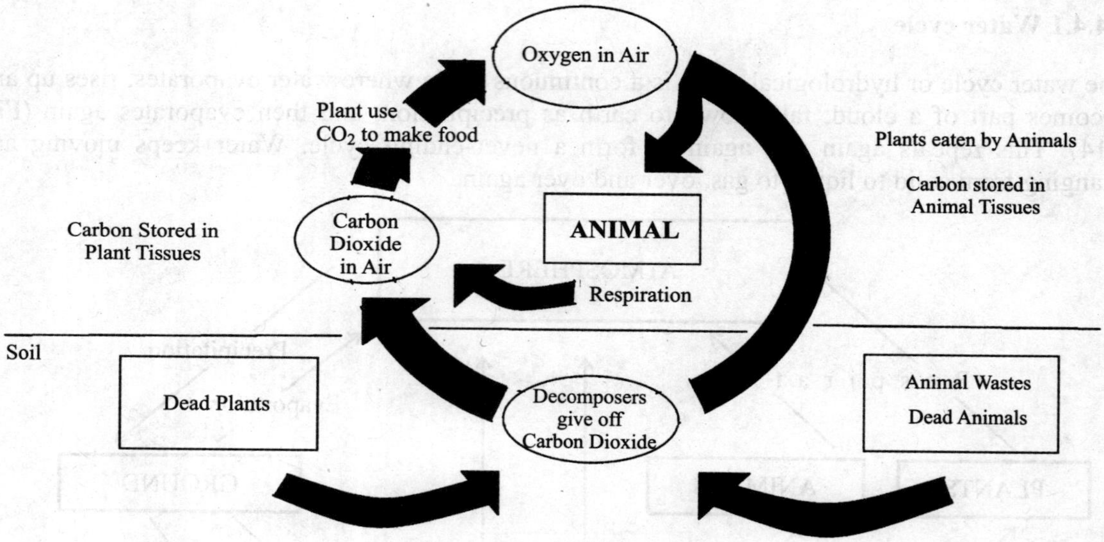

Fig. 2.15 Oxygen Cycle

2.4.4.3 Carbon cycle

The carbon cycle is the circulation and conversion of carbon back and forth between living organisms and the environment. Carbon is an element that cannot be broken down into simpler substances. Carbon compounds can be as solids (such as diamond or coal), liquids (such as crude oil), or gases (such as carbon dioxide). Carbon is often referred to as the **building block of life** because living things are based on carbon and carbon compounds. The amount of carbon on the earth and in earth's atmosphere is fixed, but that fixed amount of carbon is dynamic, always changing into different carbon compounds and moving between living and nonliving components. Carbon is released to the atmosphere from carbon sources and stored in plants, animals, rocks, and water known as **carbon sinks**. This process occurs in a number of steps. In the first step, through photosynthesis (the process by which plants traps the solar energy and synthesise it to grow), plants take carbon dioxide from the atmosphere and release oxygen. The carbon dioxide is converted into carbon compounds that makes the plant tissue, which are stored in both the above ground parts of the plants (shoots and leaves), and the below ground parts (roots). In the next step, animals eat the plants, breathe in oxygen, and exhale carbon dioxide. The carbon dioxide emitted by animals is then available for plants to use in photosynthesis. Carbon stored in plants that are not eaten by animals eventually decomposes after the plants die, and is either released into the atmosphere or stored in the soil. Large quantities of carbon can be released to the atmosphere through geologic processes like volcanic eruptions and other natural changes that destabilize carbon sinks. For example, increasing temperatures can cause carbon dioxide to be released from the ocean.

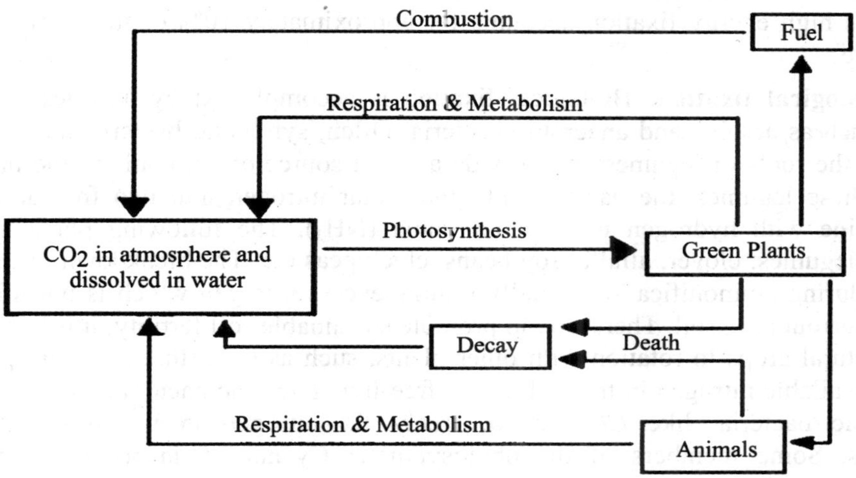

Fig. 2.16 Carbon Cycle

2.4.4.4 Nitrogen cycle

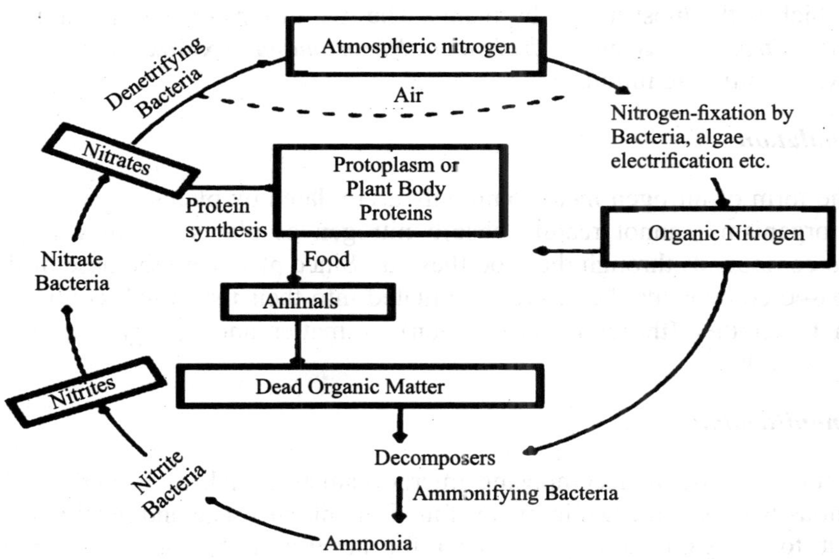

Fig. 2.17 Nitrogen Cycle

2.4.4.4.1 *Nitrogen fixation*

Nitrogen fixation is the conversion of atmospheric nitrogen (N_2) into reactive compounds such as ammonia (NH_3) and nitrate (NO_3). The breaking of the bonds between the nitrogen atoms requires a great deal of energy and occurs naturally in two primary ways:

(i) **Abiotic fixation**: Nitrate is the result of high energy fixation in the atmosphere from lightning and cosmic radiation. In this process, N_2 is combined with oxygen to form nitrogen oxides such as NO and NO_2, which come down to the earth's surface in rainfall as nitric acid (HNO_3). This high energy fixation accounts for approximately 10% of the nitrate entering the nitrogen cycle.

(ii) **Biological fixation:** Biological fixation is accomplished by a series of soil micro-organisms such as aerobic and anaerobic bacteria. Often, symbiotic bacteria such as *Rhizobium* are found in the roots of legumes and provide a direct source of ammonia to the plants. In root nodules of these legumes, the bacteria split molecular nitrogen into two free nitrogen atoms, which combine with hydrogen to form ammonia (NH_3). The following plants are common examples of legumes: clover, alfalfa, soy beans, chick peas etc. The breakdown of these legumes by bacteria during ammonification actually returns excess nitrogen which is not utilized by the plant to the surrounding soil. Therefore, to promote sustainable soil fertility, it is beneficial to use these agricultural crops in rotation with other plants, such as corn, that are more profitable but deplete the available nitrogen in the soil. Some free-living aerobic bacteria, such as *Azotobacter*, and anaerobic bacteria, like *Clostridium*, freely fix nitrogen in the soil and in aquatic environments. Some members of the photosynthetic Cyanobacteria fix nitrogen in aquatic environments as well.

2.4.4.4.2 *Nitrification*

Nitrification is the process by which ammonia is oxidized to nitrite ions (NO^{2-}) and then to nitrate ions (NO^{3-}), which is the most usable by plants. The two groups of microorganisms involved in the process are *Nitrosomonas* and *Nitrobacter*. *Nitrosomonas* oxidize ammonia to nitrite and *Nitrobacter* oxidize nitrite to nitrate.

2.4.4.4.3 *Assimilation*

Nitrates are the form of nitrogen most commonly assimilated by plants through root hairs. Since heterotrophic organisms cannot readily absorb nitrogen as plants do, they rely on acquiring nitrogen-based compounds through the food they eat. Since plants are the base of the food chain, the nitrogen-based compounds they have assimilated into their tissue will continue to pass from one organism to another (through consumption) as matter and energy transfers through the ecosystem's food web.

2.4.4.4.4 *Ammonification*

In ammonification, a host of decomposing microorganisms, such as bacteria and fungi, break down nitrogenous wastes and organic matter found in animal waste and dead plants and animals and convert it to inorganic ammonia (NH_3) for absorption by plants as ammonium ions. Therefore, decomposition rates affect the level of nutrients available to primary producers.

2.4.4.4.5 *Denitrification*

Denitrification is the process by which nitrates are reduced to gaseous nitrogen (N_2) and lost to the atmosphere. This process occurs by facultative anaerobes in anaerobic environments. Farmers with waterlogged fields and soils that have high clay content are especially vulnerable to nitrogen losses due to denitrification

2.4.4.5 Phosphorus cycle

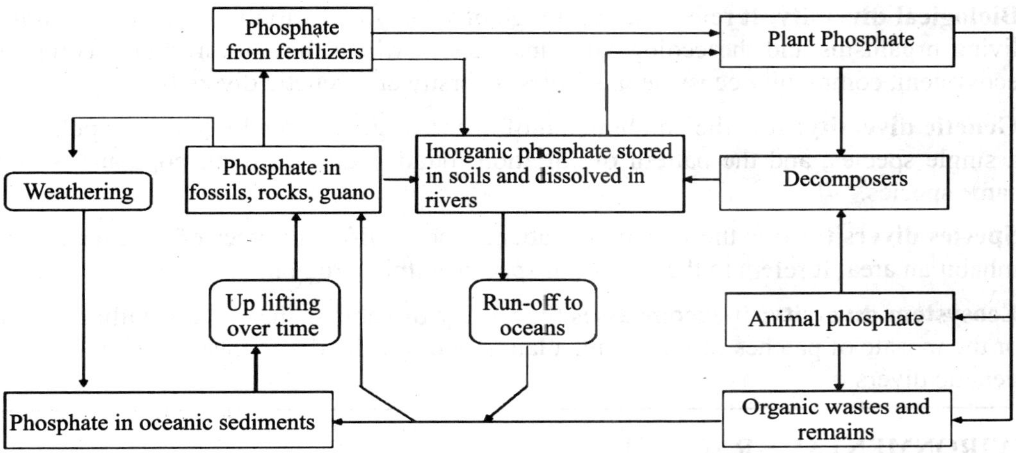

Fig. 2.18 Phosphorus Cycle

Major environmental reservoir: rocks

(i) Leaching: water dissolves phosphates in rocks and carries to lake, stream, etc.

(ii) Dissolved phosphate: used by plants and passed through food chain.

(iii) Animals return phosphorus to environment by: excretion, death and decay.

Phosphorus Cycle and Humans (Fig. 2.18):

(i) Phosphates mined for fertilizers → returns P to soil.

(ii) Erosion: P in soil and rocks washed away into water systems.

2.4.5 Control

It regulates the modification of the environment by the organisms and vice versa. The nitrogen fixing by bacteria is supposed to be environment regulation by organisms and photoperiodism is the organism's regulation by environment.

2.4.6 Biodiversity

Biological diversity or biodiversity is the variety and variability of life on earth and it is generally expressed at three levels like genetic diversity, species diversity and ecosystem diversity. Biodiversity ensures food, fuel, shelter, medicines and other vital resources for our survival.

Biodiversity forms the web of life of which we are an integral part, also it forms the soul of our survival. It depends on the wide variety of plants, animals and microorganisms.

The biological diversity encompasses different ecosystem, species, genes, and their relative abundance as discussed below:

- **Natural diversity:** It refers that the number of different native species and individuals in a habitat or geographical area. It is also the variety of different habitats within an area and the range of genetic variation among individuals within a species.
- **Biological diversity**: It refers the full range of variety and variability within and among living organisms and the ecological complexes in which they occur, and encompasses ecosystem, community ecosystem, species diversity and genetic diversity.
- **Genetic diversity:** It is the combination of different genes found within a population of a single species, and the pattern of variation found within different populations of the same species.
- **Species diversity:** It is the variety and abundance of different types of organisms which inhabit an area. It refers to the variety of species within a region.
- **Ecosystem diversity:** It encompasses the variety of habitats that occurs within a region, or the mosaic of patches found within a landscape. It is harder to measure than species or genetic diversity.

2.5 ENVIRONMENTAL GRADIENTS

The environmental factors that affect the survival and reproduction of the organisms are called **environmental gradients**. These include:

(i) Physical and chemical factors like light, temperature and pH known as **conditions**.

(ii) Factors the organism uses like food, water, shelter etc. known as **resources.**

(iii) The presence of other organisms like predators and competitors with which organisms interact.

(iv) The level of most environmental factors varies over some limited range, with a gradient between the extremes: high to low, large to small.

(v) To live and thrive in a given situation, each species must have the essential resources and conditions necessary for growth and reproduction.

(vi) A species population tends to inhabit those areas offering a particular combination of suitable conditions and resources. This is because each species only functions efficiently over a limited part of each environmental gradient – its **tolerance range.**

2.5.1 Tolerance Levels of Environment Factor

The bell-shaped curve illustrates a universal tolerance curve (Fig. 2.19). Every environmental factor has two zones discussed as below:

(i) **Zone of tolerance**: This zone is favorable for the growth and development of the organism which may be further subdivided into two sub-zones:

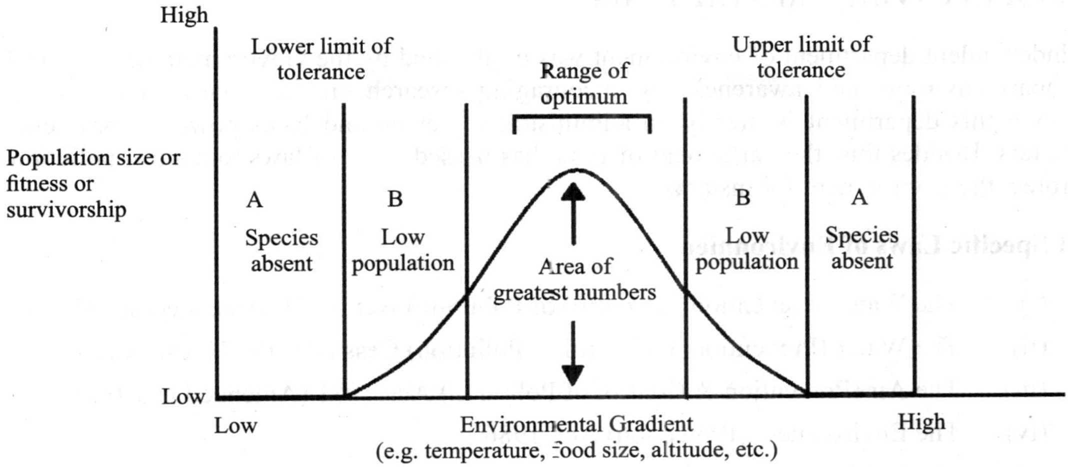

Fig. 2.19 Schematic Tolerance Curve for a Single Species Population Existing on a Single Environmental Gradient

a. ***Range of optimum***: Within this zone, majority of individuals of the species population can survive and reproduce.

b. ***Zone of physiological stress***: Towards the high and low ends of the environmental gradient most individuals suffer increasing physiological stress where conditions are less than optimal such that most individuals can stay alive but cannot function efficiently nor reproduce. Not all individuals are genetically identical and show exactly the same levels of tolerance to same environmental parameter. A few individuals with a slightly different genotype and tolerance level may in fact survive fairly well under conditions where the majority of individuals do badly.

(ii) ***Zone of intolerance***: The areas where no individuals of the species can survive because physical conditions are too extreme for that species to permit survival or lack of food resources.

a. There is a variation of tolerance from species to species with respect to a particular factor. An organism may have narrow range of tolerance for one factor and wide range of tolerance for another factor.

b. Organisms having wide range of tolerance for all factors, have better chance of survival and are widely distributed in the world.

c. Thus in the environment the organisms are controlled by:

- The quantity variability of materials for which there is a minimum requirement.
- The physical factors which are critical.
- The limit of tolerance of the organisms to different components of environment.

2.6 INDIAN ENVIRONMENTAL LAWS

An independent department of environment was established by the Government of India in 1980 to impart environmental awareness by encouraging research on the environmental problems. However this department is merely an administrative set up and lacks power to prosecute the defaulters. Besides this, the parliament of India has passed a no. of laws to control pollution and to protect the environment for instance.

2.6.1 Specific Laws of Environment

(i) The Water (Prevention & Control of Pollution) Act, 1974 (Amended in 1978, 1988)

(ii) The Water (Prevention & Control of Pollution) Cess Act, 1977 (Amended in 1991)

(iii) The Air (Prevention & Control of Pollution) Act, 1981 (Amended in 1987)

(iv) The Environment (Protection) Act, 1986

(v) The Wildlife (Protection) Act, 1972

(vi) The Forest (Conservation) Act, 1980.

2.6.2 The Regulatory Structure

The Regulatory Structure

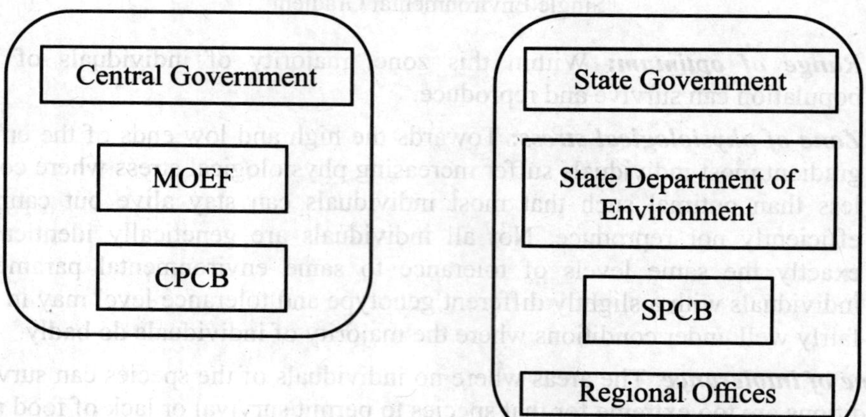

Note:

MOEF – Ministry of Environment and Forest, Govt. of India.

CPCB – Central Pollution Control Board.

SPCB – State Pollution Control Board.

2.6.2.1 Functions of MOEF

(i) Environmental policy planning and research.

(ii) Implementing legislation and monitor / control pollution.

(iii) Environmental clearance for projects.

(iv) Promotion of environmental education, training and awareness.

(v) Forest conservation, development and wildlife protection.

(vi) Biosphere reserve programme.

(vii) Co – ordination with concerned National and International Agencies.

(i) The Water Act, 1974

Objectives:

a. Prevention and control of water pollution.

b. Maintaining or restoring the wholesomeness of water.

c. Establishment of boards for prevention and control of water pollution.

(ii) The Air Act, 1981

Objectives:

a. Prevention, control and abatement of air pollution.

b. Maintaining the quality of air.

c. Establishment of boards for the prevention and control of air pollution.

(iii) The Wild-life Protection Act, 1972

Objectives:

a. To maintain essential ecological processes and life supporting systems.

b. To preserve the biodiversity.

c. To ensure a continuous use of species i.e. protection and conservation of wildlife.

4. The Indian Forest Act, 1980

Objectives:

a. Protection and conservation of forests.

b. To ensure judicious use of forest products.

5. The Environmental (Protection) Act, 1986

Objectives:

a. Protection and improvement of environment (Water, Air and Land).

b. Prevention of hazards to all living creatures and property.

c. Maintenance of harmonious relationship between human beings and their environment.

REVIEW QUESTIONS

1. What is a community? What is the difference between the concepts of community and population?

   A community is a set of populations of living beings that live in the same region and interact with each other.

   In ecology, a population is a set whose members (living in a given place in a given time) are part of the same species. A community is a set of populations of different species (living in a given place in a given time.

2. **What is the difference between an ecological niche and a habitat?**

   An ecological niche is a set of peculiar activities, resources and strategies that a species explores to survive and reproduce.

   An habitat is the place where the species lives to explore its ecological niche. In other words it can be said that the habitat is the "address" of the species and the ecological niche is the "profession" of the species.

3. What is a biosphere?

   A biosphere is a set of all of the ecosystems of the planet.

4. **What are the processes that autotrophic beings use to produce organic material from inorganic substances?**

   Autotrophic beings make organic material by photosynthesis or by chemosynthesis. There are photosynthetic autotrophs, like plants, and chemosynthetic autotrophs, like some bacteria.

5. **What is a biome?**

   A biome is a prevailing ecosystem constituted by similar biotic and abiotic factors present in one or more regions of the planet.

6. **What are the major terrestrial biomes?**

   The major terrestrial biomes are: tundra, taigas (or boreal forest), temperate forests, tropical forests, grasslands and deserts.

7. **How are the heterotrophic beings divided in the ecological study of food interactions?**

   Heterotrophs are divided into consumers and decomposers. An ecosystem can exist without consumers but it cannot be sustained without decomposers. Without the decomposers the organic material would accumulate causing environmental degradation and later death of the living beings.

8. **What are trophic levels? How many trophic levels can a food chain have?**

   Trophic levels correspond to positions on a food chain. Therefore producers always belong to the first trophic level and decomposers to the last trophic level; consumers that directly eat the producers belong to the second trophic level and so on.

There is no limit regarding the number of trophic levels on a chain, since many orders of consumers can exist.

9. What are primary consumers? Can a food chain present quaternary consumers without having secondary or tertiary consumers? Can a tertiary consumer of one chain be a primary or secondary consumer of another chain?

Primary consumers are living beings that eat autotrophic beings, i.e., they eat the producers. Primary consumers always belong to the second trophic level of a chain.

A food chain cannot have consumers of superior orders without having the consumer of the inferior orders. A consumer however can participate in several different chains not always belonging to the same consumer order in each of them.

10. What is the difference between the concepts of food chain and food web?

The chain concept is a theoretical model to study the energy flux in ecosystems. Actually in an ecosystem the organisms are part of several interconnected food chains, forming a food web. Therefore the chain is a theoretical linear sequence and the web is a more realistic representation of nature in which the food chains interconnect forming a web.

11. Can the amount of available energy in a given trophic level be larger than the available energy in inferior trophic levels? What does that condition means to the conformation of the energy pyramids?

A superior trophic level always has less available energy than inferior trophic levels. This is because in each trophic level only a fraction of the organic material of the level below is incorporated into the consumers (into their bodies). The other part is eliminated as waste or is used in the metabolism as energy source. Therefore, it is never possible to have energy pyramids with inverted conformation, i.e., with the tip to the bottom and the base to the top. It is also not possible to have superior trophic levels with a variable dimension larger than inferior ones. In every energy pyramid, from the base to the top, the size of the variable dimension decreases.

12. What is the gross primary production of an ecosystem? How does GPP relate to photosynthesis?

Gross primary production of an ecosystem, or GPP, is the quantity of organic material found in a given area in a given period.

Since only autotrophs produce organic material and photosynthesis is the main production process, GPP is a result of the photosynthesis.

13. What is the respective importance of water, carbon and nitrogen for living beings?

Water is the main solvent of living beings and it is necessary practically for all biochemical reactions, including as reagent of photosynthesis. Many properties of water are very important for life.

Carbon is the main chemical element of organic molecules; carbon dioxide is also reagent of photosynthesis and product of the energetic metabolism of living beings.

Nitrogen is a fundamental chemical element of amino acids, the building blocks of proteins that in their turn are the main functional molecules of living beings; nitrogen is also part of the nucleic acid molecules, the basis of reproduction, heredity and protein synthesis.

14. What is biodiversity?

Biological diversity is the variety of species of living beings of an ecosystem. In ecosystems which are more biodiverse, like tropical forests, a great variety of plants, microorganisms and animals live; in ecosystems less biodiverse, like deserts, there are less variety of living beings.

15. How does biological diversity relate to the characteristics of the abiotic factors of an ecosystem?

The availability of abiotic factors like light, moisture, mineral salts, heat and carbon dioxide, more or less conditions the biodiversity of an ecosystem. Photosynthesis depends on water and light, and plants also need mineral salts, carbon dioxide and adequate temperature for their cells to work. In environments where these factors are not restrictive the synthesis of organic material (by photosynthesis) is at a maximum, plants and algae can reproduce easier, the population of these beings increase, potential ecological niches multiply and new species emerge. The large mass of producers makes viable the appearing of a diversity of consumers of several orders. In environments with restrictive abiotic factors, like deserts, the producers exist in small numbers and less diversity, a feature that thus extends to consumers and conditions fewer ecological niches to be explored.

16. Write two objectives of Air Act, 1981 and what is its jurisdiction?

(i) To maintain the quality of air and prevent air pollution

(ii) Establishment of boards for control of air pollution

(iii) Its jurisdiction is central pollution control board and state pollution control board.

■■■

Chapter 3

Chemistry in Environmental Engineering

3.1 INTRODUCTION

The environment consist of water, the essential component of hydrosphere air, essential component of troposphere, soil, the essential component of lithosphere living organisms etc. In this chapter we will focus on the chemistry of water (aquatic chemistry), air (atmospheric chemistry) and soil (soil chemistry), altogether called **environmental chemistry** and of living organisms, especially micro-organisms (microbiology).

3.2 AQUATIC CHEMISTRY

Water covers about 70% of earth's surface and is essential for life. Water is the only substance that exists naturally on earth and presents in all three physical states of matter e.g. gas, liquid, and solid. Most of the water on earth, 97%, is salt water stored in the oceans; only 3% is fresh water. Total freshwater on earth, 68% is protected in the icecaps of Antarctica and Greenland, 30% is in the ground, and only 0.3% is contained in surface waters such as ponds, lakes and rivers. Pure water is colourless, odourless, and tasteless. Water molecules have a simple structure- two hydrogen atoms covalently bonded to one oxygen atom and empirical formula is H_2O. Many substances (organic and inorganic) dissolve in water and it is commonly referred to as the universal solvent.

3.2.1 Physical Properties of Water

Water has a number of unique physical properties that makes it essential for life (Table 3.1). Most plants and animals contain more than 60% water by volume. Without water, life would not have been created on earth.

Table 3.1 Physical Properties of Water

Properties	Concentration
Molar mass	18.015
Molar volume	55.5 moles/liter
Boiling point (BP)	100°C at 1 atm.
Freezing point (FP)	0°C at 1 atm.
Triple point	273.16 K at 4.6 torr

Surface tension	73 dynes/cm at 20°C
Vapor pressure	0.0212 atm. at 20°C
Heat of vaporization	40.63 kJ/mol
Heat of fusion	6.013 kJ/mol
Heat capacity (cp)	4.22 kJ/kg.K
Dielectric constant	78.54 at 25°C
Viscosity	1.002 centipoise at 20°C
Density	1 g/cc
Density maxima	4°C
Specific heat	4180 J kg^{-1} K^{-1} (T=293…373 K)
Heat conductivity	0.60 W m^{-1} K^{-1} (T=293 K)
Melting heat	3.34 x 105 J/kg
Evaporation heat	22.6 x 105 J/kg
Critical temperature	647 K
Critical pressure	22.1 x 106 Pa
Speed of sound	1480 m/s (T=293 K)
Relative permittivity	80 (T=298 K)

3.2.2 Chemical Properties of Water

Water has many unique chemical characteristics that make it ideal for nurturing life and has been discussed below:

(i) Evaporation and condensation provide energy to run the hydrological cycle.

(ii) The hydrogen atoms create a positive electrical charge while the oxygen atom creates a negative charge. The attraction to one another is what causes water to form droplets.

(iii) Reactions with acids and bases.

(iv) It has the ability to act as a oxidizing and reducing agent.

(v) It has a corrosion activity.

(vi) Decomposition into simpler substances.

(vii) Water ionizes into H^+ and OH^-.

(viii) Acids dissociate in water to increase the concentration of H^+.

(ix) Water has the second highest surface tension of all common liquids; only mercury is higher.

3.2.2.1 pH

pH is a measure of the amount of acid or base in water. The pH scale ranges from 0 (acidic) to 14 (alkaline or basic) with neutrality at 7. The pH of lakes generally ranges between 6 and 9. The concentration of gases such as oxygen and carbon dioxide, directly influence pH. Most organisms tolerate only very narrow ranges in pH; therefore, large amounts of alkalinity are needed as natural buffers to changes in pH.

3.2.2.2 Alkalinity

Alkalinity is the measure of the pH-buffering capacity of water. Lakes that have high alkalinity (over 100 mg/l as calcium carbonate) are able to sustain large inputs of acid with little change in pH. Addition of acid can occur naturally (e.g., during bacterial decomposition of organic material in the sediments; during natural diffusion of carbon dioxide into the surface waters), or because of pollution (acid deposition, both wet and dry fall). The ability of the lake to maintain a stable pH is crucial to the survival of its aquatic inhabitants. In lakes where soils are rich in calcium carbonate, groundwater seepage can contribute significantly to lake alkalinity.

3.2.2.3 Turbidity

Turbidity is a measure of the light-transmitting properties of water and is comprised of suspended and colloidal material. It is important for health and aesthetic reasons.

3.2.2.4 Hardness

Hardness is correlated with total dissolved solids (TDS). It represents total concentration of Ca^{2+} and Mg^{2+} ions, and is reported in equivalent $CaCO_3$ whereas other ions (Fe^{2+}) may also contribute (Table 3.2). Hardness expressed as mg/l $CaCO_3$ is used to classify waters from soft to very hard.

Table 3.2 Classification of water based on hardness

Hardness mg/l $CaCO_3$	Classification
0 – 60	Soft
61 – 120	Moderately hard
121 – 180	Hard
>180	Very hard

Example 3.1 Calculate hardness of the water sample containing Ca^{2+}, Mg^{2+} and HCO_3^- as 60 mg/l, 29.3 mg/l and 122 mg/l, respectively.

Solution.

Hardness due to Ca^{2+} in mg/l as $CaCO_3$

$$= \text{concentration (mg/l)} \times \frac{50}{\text{Eq. wt. of metal ion}} = 60 \times \frac{50}{20} = 150 \text{ mg/l as } CaCO_3$$

$$\text{Hardness due to } Mg^{2+} \text{ in mg/l } CaCO_3 = 29.3 \times \frac{50}{12.2} = 120 \text{ mg/l as } CaCO_3$$

Hardness due to HCO_3^- as $CaCO_3 = 122 \times \frac{50}{61/2} = 200$ mg/l as $CaCO_3$.

HCO_3^- ions are only considered for bivalent cations of Ca and Mg

Hence equivalent mass of HCO_3^- is taken as molecular mass/2

Total hardness (TH) = 150+120=270 mg/l as $CaCO_3$

Carbonate hardness (CH) = 200 mg/l as $CaCO_3$

Non-carbonate hardness = 270-200 = 70 mg/l as $CaCO_3$

Example 3.2 Calculate different types of hardness of a given water sample with the following analysis.

Ca^{2+} = 60 mg/l

Mg^{2+}=12 mg/l

Na^{2+}= 10 mg/l

HCO_3^- = 122 mg/l

Solution

Calcium hardness in meq. $= \frac{60}{20} = 3$ meq.

= 3x50 = 150 mg/l as $CaCO_3$

Magnesium hardness in meq. $= \frac{12}{12} = 1$ meq

= 1x50 = 50 mg/l as $CaCO_3$

Sodium does not contribute towards hardness

Total hardness = Ca hardness + Mg Hardness = 200 mg/l as $CaCO_3$

Carbonate hardness is due to carbonate and bicarbonate only

HCO_3^- = 122 mg/l $= \frac{122}{61/2} = 4$ meq.

= 4x50 = 200 mg/l as $CaCO_3$

Non-carbonate hardness = 200-200 = 0 mg/l as $CaCO_3$ due to $[OH^-]$

There shall be no non-carbonate hardness.

3.2.2.5 Electrical conductivity

Electrical conductivity is the measure of the ability of an aqueous solution to carry an electric current. From the concentration of the ions (cations and anions), the conductivity (i.e., the electrical conductivity, etc) can be calculated as:

$$\sum_{i=0}^{n} C_i f_i$$

Where, C_i is the concentration of the ionic species, i, in the solution in mg/l or meq/l.

f_i is the conductivity factor of the ionic species in $\frac{\mu S}{cm}$.

The parameters such as salinity, total solids (TS) and TDS can be calculated from the conductivity as,

Salinity in mg NaCl/L = eC × f_s, where f_s is a conversion factor ≈ 0.52 to 0.55

Total solids : in mg TS/L = eC × f_{ts}

Where f_{ts} ≈ 0.52 to 0.7 determined experimentally for the sample of water

TDS in mg TDS/L = eC × f_{tds}

Where f_{ts} ≈ 0.52 to 0.7 determined experimentally for the sample of water

The ionic strength is a measure of the intensity of the electric field that can be computed from eC and TDS as:

$$I = 1.6 \times 10^{-5} \times eC$$

$$I = 2. \times 10^{-5} \times TDS$$

$$I = \frac{1}{2}\sum C_i Z_i^2$$

Where Z_1 is the charge of the species and C_i is the molar concentration of the species

3.2.2.6 Solubility and solubility product

In a saturated solution of a sparingly soluble salt there exists a state of equilibrium between the excess of the solid salt and the ions produced from the salt which goes into solution (Table 3.3). The solubility product of the sparingly soluble salt in the given solvent and at a definite temperature can be obtained by applying the law of chemical equilibrium.

A saturated solution of a sparingly soluble salt such as AgCl contains very small amount of dissolved silver chloride. Since AgCl is a strong electrolyte, like all other salts, all the dissolved AgCl is present as ions. This can be represented as,

$$\text{AgCl (Solid)} \Rightarrow \text{AgCl (Solution)}\ Ag^+\ (aq) + Cl^-\ (aq)$$

At equilibrium, the salt continues to dissolve at a rate that exactly equals to the reverse process, the return of the salt from the solution. So the solution is said to be saturated. Thus a saturated solution is a solution in which equilibrium exists between the solid AgCl and its ions in solution. A saturated solution represents the limit of the ability of a salt to dissolve in a solvent. This is a measure of the solubility of the sparingly soluble salt.

As the solution is dilute and contains no other ions, the activity of the ions is more or less equal to their concentrations. The solubility (s) of a substance in a solvent is the concentration in the saturated solution. Solubility of a substance may be represented in grams per 100ml of the solution. It can also be expressed in moles per litre. Hence, molar solubility is defined as the number of moles of the substance per on litre of the solution. The value of solubility of substance depends on the solvent and temperature.

Applying the law of mass action to the above equilibrium for AgCl, we have

$$K_{eq} = \frac{[Ag^+]\,[Cl^-]}{[AgCl]}$$

The solution being saturated, the amount of AgCl in contact with saturated solution does not change with time hence *[AgCl]* remains constant. Thus, the equilibrium expression becomes

$$K_{eq} \times [AgCl] = [Ag^+]\,[Cl^-]$$

$$K_{sp} = [Ag^+][Cl^-]$$

Where K_{sp} is called the **solubility product constant** (simply the solubility product) of the salt (Table 3.3). The solubility product may be defined as the product of the ionic concentration of a sparingly soluble salt in its saturated solution at a certain temperature. In general, for any sparingly soluble salt of the type, M_mA_n dissociating as:

$$M_mA_n = nM^{n+} + nA^{m-}$$

And the solubility product,

$K_{sp} = [M^{n+}]^m[A^{m-}]^n$, where $[M^{n+}]^m[A^{m-}]^n$ is called as the **ionic product** or **ion product**.

Table 3.3 The solubility products of some sparingly soluble salts

Sparingly Soluble Salt	Solubility Equilibria	Solubility Product Expression
Ag_2CrO_4	Ag_2CrO_4 (solid) ⇒ 2 Ag^+ (aq) + CrO_4^2 (aq)	$K_{sp} = [Ag^+]^2[CrO_4^{2-}]$
As_2S_3	As_2S_3 (solid) ⇒ 2 As^{3+} (aq) +3S^{2-} (aq)	$K_{sp} = [As^{3+}]^2[S^{2-}]^3$
Ca F_2	CaF_2 (solid) ⇒ Ca^{2+} (aq) + 2 F^- (aq)	$K_{sp} = [Ca^{2+}][F^-]^2$
$Zn_3(PO_4)_2$	$Zn_3(PO_4)_2$ (solid) ⇒ 3 Zn^{2+} (aq) + 2 PO_4^{3-} (aq)	$K_{sp} = [Zn^{2+}]^3[PO_4^{3-}]^2$

It is worthwhile mentioning that the solubility product is not the ionic product in the general sense but only in the particular case when the salt solution is saturated. The solubility product is also called the **concentration solubility product**.

How are the product of concentration of ions and the solubility product related?

From the definition of the solubility product, it is clear that the product of concentrations of ions in a saturated solution of a sparingly soluble salt is equal to its solubility product, i.e.,

$K_{sp} = [Ag^+][Cl^-]$ for silver chloride. If the product of concentrations of ions is greater than the solubility product constant, then the salt will be precipitated out i.e., when $[Ag^+][Cl^-] > K_{sp}$ (AgCl), silver chloride will be precipitated out.

Relation between solubility and solubility product of a sparingly soluble salt

For a saturated solution of a sparingly soluble salt, MB, we have the equilibrium

$$\text{MB (s, standard solution)} = M^+ \text{ (aq)} + B^- \text{ (aq)}$$

If s is the solubility of the salt in mol dm^{-3}, then the concentration of $[M^+]$ and $[B^-]$ will each equals to s gm ions dm^{-3}.

Hence, $K_{sp} = [M^+][B^-]$

$$= (\text{s gm ions dm}^{-3}) \times (\text{s gm ions dm}^{-3}) = s^2\,(gm\ ions)^2\ dm^{-6}$$

$$(\text{s gm ions dm}^{-3}) = (K_{sp})^{1/2}$$

In other words, the solubility of a sparingly soluble binary salt is equal to the square root of its solubility product. The solubility product is expressed in gm moles dm^{-3} for the sparingly soluble salt. The solubility and solubility product differ in the following facts as given below:

(i) The solubility is applicable to all substances whereas the solubility product is used in the case of sparingly soluble electrolytes.

(ii) The solubility of a sparingly soluble electrolyte decreases in presence of a common ion while its solubility product remains constant.

- The principle of solubility product is also applied in the precipitation of soluble salts from their saturated solutions in pure state. This phenomenon is known as salting out and is used in the purification of sodium chloride.
- The knowledge of solubility product of a sparingly soluble substance can help to predict whether in a given condition a substance would get precipitated or not.

3.2.2.7 Solution of gases in a liquid

The solubility of a gas in a solvent depends on the pressure and the temperature. When a gas enclosed over its saturated solution the equilibrium, gas $\rightleftharpoons$ gas in solution, exists. According to the Le Chatelier principle, if pressure is increased on the system, the equilibrium will shift in the direction which will reduce the pressure. The pressure can be reduced when more gas will be dissolved in the solvent. Thus, the solubility or concentration of a gas in a given solvent is increased with increase in pressure. William Henry (1803) put forth the relationship between pressure and solubility of a gas in particular solvent and this generalization is known as **Henry's law**. It may be stated as for a gas in contact with a solvent at constant temperature, the concentration of a gas in particular solvent at constant temperature, the concentration of the gas that dissolves in the solvent is directly proportional to the pressure of the gas.

Mathematically,

$$c \propto p \quad \text{or} \quad c = kp$$

where k is the proportionality constant called **Henry's law constant**. The value of k depends upon the nature of the gas and solvent, and the units of p and c used. Limitations of Henry's Law:-

Henry's Law holds good to the gases behaving nearly ideally under the following conditions:

(i) At moderate temperature and pressure.

(ii) If the solubility of the gas in the solvent is low.

(iii) The gas does not react with the solvent. Thus ammonia or HCl which reacts with water does not obey Henry's Law.

(iv) The gas does not associate or dissociate on dissolution in the solvent.

- The solubility of a gas in water varies considerably with the nature of the gas. The gases which are easily liquefied are more soluble in water. Therefore, carbon dioxide is more soluble than hydrogen or oxygen in water.
- Under constant pressure, the solubility of a gas diminishes with rise in temperature.
- The solubility is generally expressed in terms of absorption coefficient (a), introduced by Bunsen.

Example 3.3

(a) The solubility of a sparingly soluble salt MX_2 is 1×10^{-4} m water. Find out its solubility product.

(b) The solubility of calcium phosphate is 5 mol dm^{-3}. What is its solubility product?

Solution

(a) MX_2 dissociates as:

$$1MX_2 = M^{2+} + 2X^{-}1$$

$$K_{SP} = [M^{2+}][X^{-}]^2$$

Since solubility of MX_2 is $1x10^{-4}$ M,

$$[M^{2+}] = 1 \times 10^{-4}\ M, [X^-] = 2 \times 1 \times 10^{-4} M$$

$$K_{sp}(MX_2) = (1 \times 10^{-4} M)(2 \times 10^{-4} M)^2 = 4 \times 10^{-12} M^3$$

(b) $Ca_3(PO_4)_2 = 3Ca^{2+} + 2PO_4^{3-}$

Solubility of calcium phosphate =5 mol dm^{-3}

Hence, $[Ca^{2+}] = 3 \times 5$ mol dm^{-3} = 15 mol dm^{-3}

$$[PO_4^{3-}] = 2 \times 5 \text{ mol dm}^{-3} = 10 \text{ mol dm}^{-3}$$

$$K_{sp}(Ca_3(PO_4)_2) = [Ca^{2+}]^3 [PO_4^{3-}]^2 = (15 \text{ mol dm}^{-3})^3 (10 \text{ mol dm}^{-3})^2$$

$$= (376 \text{ mol}^3 \text{ dm}^{-9}) \times (10 \text{ mol}^2 \text{ dm}^{-6}) = 337.5 \times 10^3 \text{ mol}^5 \text{ dm}^{-15}$$

Example 3.4 Find the pH of a saturated solution of $Ca(OH)_2$ at 298 k and its solubility product is 4 x 10^{-6}.

Solution. $Ca(OH)_2$ dissociates as

$$Ca(OH)_2 = Ca^{2+} + 2OH^-$$

$$K_{sp} = [Ca^{2+}] \times [OH^-]^2$$

If s is the solubility of $Ca(OH)_2$, them $[Ca^{2+}]$=s, and $[OH^+ -]$ =2s

Hence, $K_{sp} = s(2s)^2 = 4s^3 = 4 \times 10^6$

or $s^3 = 4x\ 10^6/4 = 10^{-6}$

Hence, $s = (10^{-6})^{1/3} = 10^{-2}$

Thus $[OH^-] = 2s = 2 \times 10^{-2}$

$$\log[OH^-] = \log(2 \times 10^{-2}) = \log 2 \times \log 10^2 = 0.3010 - 2 = -1.7$$

$$pOH = -\log[OH^-] = -(-1.7) = 1.7$$

$$pH = pK_w - pOH = 14 - 1.7 = 12.3 \text{ (since } pk_w \text{ at 298K} = 14)$$

Example 3.5 Calculate the solubility of $Mg(OH)_2$ in 0.01 M NaOH ($K_{sp} = 1.0 \times 10^{-11}$).

Solution.

$$Mg(OH)_2 = Mg^{2+} + 2OH^-$$

$$K_{sp} = [Mg^{2+}]\,[OH^-]^2$$

If s is the solubility, then $[Mg^{2+}]$ = s, and $[OH]^-$ = 2s

Hence, $K_{sp} = (s)\,(2s)^2 = 4s$

$$4s^3 = 1.0 \times 10^{-11}$$

or $$s^3 = \frac{1.0\times10^{-11}}{4} = 0.25\times10^{-11} = 2.5\times10^{-11} = 2.5\times10^{-12}$$

Hence, $$s = (2.5\times10^{-12})^{1/3} = 1.44\times10^{-4}\text{ M}$$

In presence of 0.01 M NaOH $[OH^{-1}] = 0.01 + 1.44\times10^{-4} \sim 0.01$

$$K_{sp} = [Mg^{2+}][OH^-]^2 = [Mg^{2+}](2\times0.01)^2$$

$$[Mg^{2+}] = \frac{K_{sp}}{(0.02)^2} = \frac{1.0\times10^{-11}}{(0.02)^2} = \frac{1.0\times10^{-11}}{4\times10^{-4}} = 2.5\times10^{-7}\text{ M}$$

Example 3.6 Calculate the concentration of oxygen at 20ºC and partial pressure 0.2 atm from the solubility of pure oxygen water at 20ºC and 1 atm, which is 1.38 x 10^{-3} mol dm^{-3}

Solution.

According to Henry's law

C=kp or k= c/p

Hence, from the data, $k = \frac{1.38\times10^{-3}\text{ mol dm}^{-3}}{1.00\text{ atm}} = 1.38\times10^{-3}\text{ mol dm}^{-3}\text{ atm}^{-1}$

The concentration of oxygen at the given pressure,

$$c = kp = 1.38 \times 10^{-3}\text{ mol dm}^{-3}\text{ atm}^{-1} \times 0.2\text{ atm}$$
$$= 2.76 \times 10^{-4}\text{ mol dm}^{-3}$$

Example 3.7 At 25ºC the solubility of nitrogen gas in water is 0.015 gdm^{-3} when the partial pressure of nitrogen is 580 torr. Find its solubility in water at this temperature when its partial pressure is 800 torr.

Solution

Henry's law at two different concentrations and pressure,

$$\frac{c_1}{p_1} = k \text{ and } \frac{c_2}{p_2} = k$$

Since k has the same value of the two cases,

We have $\frac{c_1}{p_1} = \frac{c_2}{p_2}$

$c_1 = 0.015 \text{ gdm}^{-3}$ $c_2 = ?$

$p_1 = 580$ torr $p_2 = 800$ torr

Hence, $$\frac{0.015 \text{ g dm}^{-3}}{580 \text{ torr}} = \frac{c_2}{800 \text{ torr}}$$

or $$c_2 = \frac{800 \text{ torr} \times 0.015 \text{ g dm}^{-3}}{580 \text{ torr}} = 0.0207 \text{ g dm}^{-3}$$

3.2.3 Organic Chemical Properties of Water

Organic substances may be natural or man-made. Most natural organic compounds are produced by plants and animals. The presence of organic compounds (whether natural or man-made) in water acts as contaminant for which physical, chemical or biological treatment processes has been adopted to minimize these compounds.

3.2.3.1 Classification

(i) **Hydrocarbons** containing C and H only. These are methane (CH_4), ethane ($CH_2 - CH_3$), ethylene ($CH_2 = CH_2$), benzene (C_6H_6) and toluene ($C_6H_6 - CH_2$).

(ii) **Halogenated compounds** containing C, H and halogen, e.g., $CHCl_2, CH_2Cl_2, CCl_4$ etc.

(iii) **Carboxylic acids-** contain esters, ketones, acetone (CH_3COCH_3), aldehyde, (RCHO), ether, HCOOH, etc.

(iv) **Organometallic compounds**, some of these are very toxic.

3.2.3.2 Naturally occurring organic matter

The naturally occurring organic substances present in water, wastewater and bottom sediments are as follows:

(i) **Proteins**: These contain C, H, O, S and N with basic building blocks of amino acids.

(ii) **Lipids**: These are fats, oils, waxes and hydrocarbons insoluble in water but soluble in some organic solvents. These are slowly biodegradable.

(iii) **Carbohydrates**: They contain C, H, O, cellulose, hemicelluloses, starch and lignin are the examples of carbohydrates. These expecting lignin are readily biodegradable.

(iv) **Plant pigments**: These are made up of chlorophyll, and carotenes and include alcohols, ketones and carotenoids.

3.2.3.3 Synthetic organic chemicals

The synthetic organic chemicals, which are the contaminants, are the products of the chemical petroleum and agrochemical industries. These chemicals include:

(i) **Pesticides and agrochemicals**: These include pesticides such as dichlorodiphenyltrichloroethane (DDT), aldrin, dieldrin and endrin, organic phosphates such as parathion, melathion and dimethoate and herbicides and fungicides such as simazine, propazine, 2,4,5 – T and diquat.

(ii) **Surface active agents**: These substances when put in water can lower the surface tension of water that is why these are used in washing, emulsifying, wetting, and foaming etc. The molecule of a surfactant (i.e., detergent) has two parts – a hydrophobic part, hydrocarbon residue which is insoluble in water but soluble in non-polar liquids (oils) and a hydrophilic part, which is soluble in water but in soluble in non-polar liquids. Lauryl sulphate ($C_{12}H_{25} - OSO_3^- \, Na^+$), etc are used as detergents. These are harmful to aquatic environment as they cause foaming and reduce the diffusion of atmospheric oxygen into water. In sewage, concentrations of surfactants may be upto 20 mg/l and in industrial waste waters, concentrations may be upto 1000 mg/l in rivers, the values range from 0 to 1 mg/l.

3.2.3.4 Halogenated hydrocarbons

The significant hydrocarbons are trihalometanes (THMs) or haloforms. These are represented by CHX_3 where X is F, Cl, Br or I. $CHCl_3$, $CHBr_3$, $CHCl_2Br$, $CHClBr_2$ are the examples of these compounds. Chlorinated aromatic compounds, dioxins and polynuclear aromatic hydrocarbons (PAH) are also included.

- Chloride (Cl_2) is a pungent reactive gas with high electronegativity and electron affinity. It is one of the major inorganic anions in water and wastewater (NaCl). It is used to destroy pathogens and has many industrial applications.
- ClO_2 has been used as a bleaching agent in the paper mills and pulp industry.
- Iodine is used in medicine as antiseptic and in number of pharmaceutical preparations and in photochemicals.

3.2.3.5 Measuring parameter of organic matter in water

The organic matters present in water and waste water can be measured because of their oxygen demanding characteristics. Being rich in energy, the organic matter undergoes oxidation to form stabilized low energy mineral and end products. For this degradation/ oxidation work, it takes up dissolved oxygen (DO) in water causing depletion in DO. That is why, organic wastes are called **oxygen demanding waste**.

3.2.3.5.1 Biochemical oxygen demand (BOD)

Degradation of organic matter is usually carried out by host of microscopic living beings (microbes) called **decomposer**. Thus, biodegradation is essentially the biochemical metabolic activities of the microbes, for which organic matter acts as food/ substrate. In municipal wastewater, around 85% organic matter is biodegradable which corresponds to maximum BOD value called **ultimate BOD**. Biodegradation being a first order rate process, ultimate BOD (L_0) cannot be determined in infinite time. That is why, as agreed upon internationally, a BOD value in 5 days at 20°C is taken as the standard measurement value. In India BOD_3 at 27°C is taken as standard measurement in mg/l or ppm.

The BOD is the most important measurement for evaluating the characters of wastewater, predicting the impacts on receiving water bodies. It defines the respiratory needs of a biological community over a specified period (mg of O_2/l).

BOD is taken up in two distinct phases. In first phase, organic compounds are decomposed and in the second phase, known as nitrification phase, ammonium ion is oxidized to nitrite and then to nitrate by the action of bacterial species.

Dissolved oxygen is measured at the beginning and at the end of the incubation period by Winkler titration (iodometric) method or by oxygen- membrane probes (Clark cell). Reduction in dissolved oxygen is taken as the measure of BOD of the sample.

$$\text{BOD (mg/l)} = (D_2 - D_1)/P$$

Where, D_1 and D_2 are dissolved oxygen of the sample at the beginning and at the end of the test and p is the sample dilution.

3.2.3.5.2 Chemical oxygen demand (COD)

COD is another oxygen measuring parameter which is based on exclusive chemical oxidation principle. In COD, a strong oxidizing agent such as potassium dichloromate in acid medium is used for chemical oxidation of organic matter. It is measured in mg/l or ppm.

$$\text{Organic matter} \quad + Cr_2O_7^{2-} + H^+ + \text{(acid)} \rightarrow Cr^{3+} + CO_2 + H_2O$$

$$\text{Oxidant} \qquad Cr^{6+} \xrightarrow{reduction} Cr^{3+}$$

$$\text{Organic matter} \qquad \xrightarrow{Oxidation} CO_2 + H_2O$$

COD theoretically is not related to BOD

Empirically for municipal wastewater, $BOD_5 \approx$ 0.4 to 0.6 COD. Some refractory organics like benzene do not even respond to chemical oxidation. Even though COD does not represent exactly the biodegradable organics, it is sometimes taken as measuring parameter of organic waste as a COD value can be determined in 3-4 hours as against 3-5 days for a BOD value. In many industrial samples, COD testing may be the only feasible course because of the presence of bacterial inhibitors or chemicals which would interfere with BOD determination.

3.2.3.5.3 Theoretical value of oxygen demanding parameter

$$C_xH_yO_z + \left(x + \frac{y}{z} - \frac{z}{2}\right)O_2 \rightarrow xCO_2 + \frac{y}{2}H_2O$$

The total oxygen that can be determined from the balanced stochiometric equation as above. For municipal wastewater

$$COD \approx 0.92\ \text{ThOD. BOD}_u = 0.85\ \ \text{ThOD and BOD}_5 \sim 0.4\ \text{to}\ 0.6\ \text{COD}$$

3.2.3.5.4 Total organic carbon (TOC)

TOC measures all organic carbon as CO_2 for which all inorganic carbons (HCO_3^-, CO_3^{2-}, CO_2, etc.) must be removed before the test by acidifying and aeration of the sample. TOC test for glucose:

$$C_6H_{12}O_6 + 6O_2 \xrightarrow{\text{Catalyst at } 950°C} 6CO_2 + 6H_2O$$

TOC test measures the mass of carbon per liter of sample.

$$\text{For glucose,}\ \ COD \approx \text{ThoD} = \frac{6\times 32}{180} = 1.066$$

$$TOC = \frac{6\times 12}{180} = 0.4$$

$$\frac{COD}{TOC} = 1.\frac{066}{0}.4 = 2.66$$

i.e., COD=2.66 x TOC

Example 3.8 **A wastewater contains 376 mg/l of phenolic waste (*C₆H₅OH*). Calculate its ThOD, TOC, COD, BODᵤ and BOD₅. Assume 85% biodegradability and 92% chemical oxidation.**

Solution.

$$C_6H_5OH + O_2 \rightarrow CO_2 + H_2O$$

$$x = 6, y = 6, z = 1. x + \frac{y}{4} - \frac{z}{2} = 6 + \frac{6}{4} - \frac{1}{2} = 7$$

$$C_6H_5\,OH + \ 7O_2 \quad \rightarrow \quad 6CO_2 + 3H_2O$$

1millimole 7 milimole

Theoretical/stoichiometric oxygen requirement is

Number of millimoles of phenol x 7 x molecular mass of oxygen

$$= \frac{376}{94} \times 7 \times 32 = 896\ \text{mg/l} = \text{Th OD}$$

Number of gram atoms in phenol in 1 millimole=6, number of millimole of phenol=4.

Number of mg atoms of carbon in wastewater = 24.

TOC in mg/l = 24 x 12 = 288 mg/l TOC

COD =0.92 x ThOD =896 x 0.92=824.32 mg/l

Ultimate BOD=BOD_U= 0.85 x ThOD = 0.85 x 896 = 761.6 mg/l

BOD_5=0.4 to 0.6 forms COD for municipal wastewater

= 0.5 x 824.32= 412 mg/l

Example 3.9

(a) Determine COD and TOC in phenol and ethylene glycol of concentration 120 mg/l and 100 mg/l respectively.

(b) Determine pH of rainwater if the concentration of CO_2 in the atmosphere is 330 ppm at 25°C and 1 atm.

(c) Determine pH and alkalinity of the water sample at 15^0C containing HCO_3^- and CO_3^{2-} 260 mg/l and 30 gm/l, respectively $\left(K_2 = 3.72 \times \frac{10^{-11}\ \text{mol}}{1}\right)$

(d) What is a buffer? Explain how does the carbonic acid behave as a buffer?

Solution:

(a) Phenol, C_6H_5OH, molar mass =94 g/mole

$$C_6H_6O + 7O_2 \longleftrightarrow 6CO_2 + 3H_2O$$

$$COD = \frac{7 \times 32}{94} \times 120 = 286 \text{ mg/l}$$

$$TOC = \frac{6 \times 2}{94} \times 120 = 92.4 \text{ mg/l}$$ (Where concentration of phenol is 120 mg/l)

Ethylene glycol, $\begin{matrix} CH_2\ OH \\ | \\ CH_2OH \end{matrix}$ or $C_2H_6O_2$ molar mass = 62 g/mole

$$C_2H_6O_2 + 2.5\,O_2 \Rightarrow 2CO_2 + 3H_2O$$

$$COD = \frac{2.5 \times 32}{62} \times 100 = 129.3 \text{ mg/l}$$

$$TOC = \frac{2 \times 12}{62} \times 100 = 38.7 \text{ mg/l}$$ (Where concentration of ethylene glycol is 100 mg/l)

(b) For electroneutrality, in the rain water,

$$[H^+]=[HCO_3^-]+2[CO_3^{2-}]+[OH^-]$$

pH of rain water is <7, hence the concentrations of the bases, CO_3^{2-} and OH^- are negligible.

Thus, $[H^+]=[HCO_3^-]$

$$CO_2' + H_2O \rightarrow H_2CO_3 \text{ (rain water)}$$

$$H_2CO_3 \leftrightarrow H^+ + HCO_3^-$$

$$K_1 = \frac{[H^+][HCO_3^-]}{[H_2CO_3]} \sim 4.47\times10^{-7} \text{ mol/l}$$

Since $[H^+]=[HCO_3^-]$, $K_1 = \dfrac{[H^+]^2}{[H_2CO_3]} = 4.47\times10^{-7}$

$$[H_2CO_3] \sim 1.12\times10^{-5} \text{ mol/l}$$

Hence, $[H^+]=(4.47\times10^{-7}\times1.12\times10^{-5}\text{ mol}^2/\text{L}^2)^{1/2} = 2.24\times10^{-6}$

$pH = -\log[H^+] = 5.66$ where

$$[CO_2(aq)] = [H_2CO_3] = 1.12\times10^{-5} \text{ mol/l}$$

$[(CO)_2\,(g)] = 330$ ppm, then $Pco_{2(g)} = 0.00033$ atm.

0.00033 atm = (0.164 x 10^4 atm)

$x_{CO_2(g)}$ (mole fraction) $= 2.01\times10^{-7}$

$$X_{CO_2(g)} = \frac{n_{CO_2}(g)}{n_{CO_2(g)} + n_{H_2O'}} \quad \text{where, } n_{H_2O} = 55.6 \text{ mol/l}$$

Hence, $CO_2(aq) = n\,CO_{2(g)} \approx n_{H_2O}\cdot x_{CO_2(g)} = 55.6\times2.01\times10^{-7} = 1.12\times\dfrac{10^{-5}\text{ mol}}{1}$

(c) $HCO_3^- = CO_3^{2-} + H^+$

$$K_2 = \frac{[H^2][CO_3^{2-}]}{[HCO_3^-]} = 3.72\times\frac{10^{-11}\text{ mol}}{1}$$

Molar mass of $HCO_3^- = 61$ g/mol

$$[HCO_3^-] = \frac{260}{1000\times61} = 4.26\times\frac{10^{-3}\text{ mol}}{1}$$

Molar mass of $CO_3^{2-} = 60 \dfrac{g}{mol}$

$$[CO_3^{2-}] = \frac{30}{1000 \times 60} = 50 \times \frac{10^{-4}\ mol}{1}$$

$$[H^+] = K_2 \frac{[HCO_3^-]}{[CO_3^{2-}]} = 3.72 \times 10^{-11} \times \frac{4.26 \times 10^{-3}}{5.0 \times 10^{-4}} = 3.17 \times 10^{-10}$$

$$pH = -\log[H^+] = 9.5$$

Alkalinity due to $HCO_3^- = 260 \dfrac{mg}{l}$

Alkalinity due to $CO_3^{2-} = 30 \times 2\ (valency) = 60 \dfrac{mg}{l}$

Alkalinity $= 260 + 60 = 320 \dfrac{mg}{l}$

(d) Buffers are the solutions which resist changes in pH even on addition of a small amount of acid or base. Buffers possess buffering capacity. A riverine system is said to have buffering capacity even when subjected to acid rain.

Weak bases, such as HCO_3^-, CO_3^2 and OH^- all will resist pH changes if a strong acid is added. Similarly, weak acids, such as $H_2CO_3^-$ and H_3O resist pH changes if a strong base is added. That is why, in riverine system, there is buffering capacity. Carbonic acid, $H_2CO_3^-$ behaves as a buffer in the pH range 6 to 9 due to the following reactions.

$$CO_2(aq) + H_2O \rightarrow H_2CO_3$$

$$H_2CO_3 + H_2O \rightarrow H_3O^+ + HCO_3^-$$

$$HCO_3^- + H_2O \rightarrow H_3O^+ + CO_3^{2-}$$

In moving to right, more HCO_3^- is produced with reduction of CO_2, thus reducing pH only slightly.

3.2.3.5.5 *Total dissolved solid (TDS)*

TDS are a measure of salt dissolved in a water sample after removal of suspended solids. TDS is residue remaining after evaporation of the water. The TDS load carried in streams throughout the world has been estimated by Livingston (1963) to 120 mg/l.

3.2.3.5.6 *Conductivity*

The concentration of TDS is related to electrical conductivity (EC; mmho/cm) or specific conductance. The conductivity measures the capacity of water to transmit electrical current. The conductivity is a relative term and the relationship between the TDS concentration and conductivity is unique to a given water sample and in a specific TDS concentration range. The conductivity increases as the concentration of TDS increases.

TDS and conductivity affect the water sample and the solubility of slightly soluble compounds and gases in water (e.g. $CaCO_3$, and O_2). In general, the corrosiveness of the water increases as TDS and EC increase, assuming other variables being kept constant.

3.2.4 Biological Characteristics of Water

Water quality in developed and developing countries continues to deteriorate due to increased industrialization in developing countries and natural disasters like flooding and droughts. Nowadays waterborne diseases are bigger than wars and AIDs.

Most important biological organisms in water and wastewater are pathogens, as they transmit diseases not native to aquatic systems and usually require an animal host for growth and reproduction, which unfortunately, can be transported by water, becoming a temporary member of the aquatic community. Many species of pathogens are able to survive in water and maintain their infectious capabilities for significant periods of time. It includes species of bacteria, viruses, protozoa and helminthes.

3.2.5 Modes of Expressing Concentration

Two modes of expressing concentration of a constituent in water solutions are used:

(i) **Mass/volume** – The mass of solute per unit volume of water solution. This is the weight per unit volume mg/l- parts per million (ppm)

(ii) **Mass/ mass or weight/weight**- The mass or weight of a solute in a given mass or weight of a solution mg/kg or ppm

(iii) **Parts per billion** (ppb) = 1000 ppm

(iv) **Parts per trillion** (ppt) = 1000 ppb

3.2.5.1 Concentrations in terms of mole concept

A mole of a constituent is the mass of the constituent which is numerically equal to the molecular weight of the constituent. A mole has the Avogadro number (= 6.023×10^{23}) of constituents (e.g., atoms, molecules or ions of the constituent). It follows that the mole notation refers to a fixed number of particles, but not to a fixed weight.

1 g. atom of oxygen = 6.023×10^{23} numbers of oxygen atoms.

1 mole of O_2 = 6.023×10^{23} numbers of oxygen molecules.

1 mole of SO^{2-}_4 = 6.023×10^{23} numbers of SO^{2-}_4 ions.

1 mole of CO_2 = 44 g of CO_2 (= molecular weight of CO_2).

$$\text{Number of moles} = \frac{\text{weight in gram}}{\text{gram molecular weight}}$$

3.2.5.2 Modes of concentration in mole concept

(i) Molarity (m) $\text{mole/l} = \frac{\text{moles of solute}}{\text{1L of solution}}$

(ii) Molality (m) $\text{mole/kg} = \frac{\text{Moles of solute}}{\text{1 kg of solvent}}$

(iii) Normality (N) $\text{eqv./l} = \frac{\text{equivalent of solute}}{\text{1 L of solution}}$

Where $\text{equivalent of solute} = \frac{\text{weight in g}}{\text{Equivalent weight}}$

$$\text{Equivalent weight} = \frac{\text{atomic or molecular weight}}{\text{valency of the atom or valency of the cation equivalent to protons}}$$

(iv) $\text{Mole fraction (X)} = \frac{\text{number of moles of solute}}{\text{Total number of moles of solute and solvent in solution}}$

Example 3.10 A solution contains 2% by weight of $CaSO_4$ in water. Express the concentration in (a) molarity, (b) molality and (c) mole fraction.

Solution. Molecular weight of $CaSO_4$ = (40+32+4 x 16) g/mole = 136 g/mole

$$2\% \text{ by weight} = \frac{2\text{g}}{100\text{ g}} = \frac{20\text{ g}}{1000\text{ g}} = 20 \text{ g/kg}$$

$$2\% \text{ by volume} = \frac{2\text{g}}{100\text{ cc}} = \frac{20\text{ g}}{1000\text{ cc}}\ 20 \text{ g/L}$$

(a) $\text{Molarity (M)} = \frac{mole}{lL} = \frac{20\text{ g/lL}}{126\text{ g/mol}} = 0.147\text{ M}$

(b) $\text{Molaity (m)} = \frac{mole}{kg} = \frac{20\text{ g/kg}}{136\text{ g/mole}} = 0.147\text{ m}$

(c) Mole fraction, $X_{caSO_4} = \frac{20/136}{\frac{20}{136}+\frac{980}{18}} = \frac{0.147}{54.587} = 0.0026$

Where weight of water is 980 *g*

$$X_{H_2O} = \frac{980/18}{\frac{20}{136}+\frac{980}{18}} = \frac{54.44}{54.587} = 0.9974$$

Example 3.11 The concentration of Ca^{2+} in a solution is 90 *mg/l.* Express the concentration in eqv/l and in *mg/l* as *$CaCO_3$*.

Solution. The equivalent weight of Ca^{2+}= 20/*g/geq* or 20 *mg/meq*.

$$\text{Concentration in } eqv/l\ (N) = \frac{\text{concentration in mg/L}}{\text{equivalent in mg/meq}} = \frac{90 \text{ mg/L}}{20 \text{ mg/meq}} = 4.5 \text{ meq/L}$$

$$\text{Equivalent weight of } Ca \text{ as } CaCO_3 = \frac{100}{2} = 50 g/meq$$

Concentration of Ca in mg/l as $CaCO_3$=50 mg/meq x 4.5 meq/l = 225 mg/l

3.2.6 Stoichiometry

Stoichiometry deals with the quantitative composition of the reactants and the products and their relative weights in the chemical reaction. The study of quantities of products and their relationship to quantities of reactants in a chemical reaction is called **stoichiometry**. A balanced chemical equation is not just a qualitative description of a reaction, it is also quantitative. It tells us not just what the reactants and products are, but also the relative amounts of reactants and products that are involved in the reaction.

$$MgCl_2 + 2AgNO_3 \rightarrow 2AgCl + Mg\ (NO_3)_2$$

One mole of magnesium chloride reacts with two moles of silver nitrate to give two moles of silver chloride and one mole of magnesium nitrate. For example, to calculate the amount of silver chloride produced when 7.35 mol of magnesium chloride reacts with silver nitrate, we use the ratio 2 mol AgCl per 1 mol $MgCl_2$

$$7.35 \text{ mol } MgCl_2 \text{ X } \frac{2\ mol\ Ag\ Cl}{1 \text{ mol Mg } Cl_2} = 14.7 \text{ mol AgCl}$$

Two reactants undergo a combination reaction to form a new compound, for example:

$$H_2 + Cl_2 = 2\ HCl$$

The formation of $AgCl^-$ when NaCl and $AgNO_3$ solutions are mixed gives rise to an exchange reaction:

$$NaCl + AgNO_3 = AgCl + NaNO_3$$

Reaction with oxygen or air is often called **combustion**

$$C_2H_5OH + O_2 = 2\ CO_2 + 3\ H_2O$$

A combustion reaction is often an oxidation and reduction reaction. Since the oxidation of one substance involves the reduction of another, this type of reaction is often called **redox reaction**. In the following reaction:

$$2\ Al + Fe_2O_3 = 2\ Fe + Al_2O_3$$

The element Al is oxidized, but Fe is reduced. This reaction is also called **displacement reaction** because Al displaces Fe in the oxide.

The units in which the mass of an atom are expressed as atomic mass units (amu). At one time, the lightest atom was assigned a mass of 1 amu and the mass of any other atom was expressed in terms of this standard. Isotopes are atoms of the same element with different numbers of neutrons such as the ^{20}Ne and ^{22}Ne isotopes of neon or the ^{35}Cl and ^{37}Cl isotopes of chlorine. The atomic weight of an element is the weighted average of the atomic masses of the different isotopes of an element. Naturally occurring carbon, for example, is a mixture of two isotopes, ^{12}C (98.89%) and ^{13}C (1.11 %). The molecular weight of a compound is the sum of the atomic weights of the atoms in the molecules that form these compounds. The term mole literally means a small mass. It is used as the bridge between chemistry on the atomic and macroscopic scale. If the mass of a single ^{12}C atom is 12.000 amu, then one mole of these atoms would have a mass of 12.000 grams. By definition, a mole of any substance contains the same number of elementary particles as there are atoms in exactly 12 grams of the ^{12}C isotope of carbon.

Stoichiometric calculations can be made on the basis of mass-mass relationship or mass balance in which the left-hand side of the chemical reaction calculated in grams, milligrams or kilograms must be equal to the right-hand side of the reaction. For calculation involving mass balance, the following steps are taken into consideration:

(i) First write down the balanced molecular equation involved in the chemical reaction

(ii) Write down the atomic mass or molecular mass of the reactants and products below each formula

(iii) Coefficients of reactants and products are on molar basis

(iv) Calculate the unknown factor or factors by applying the unitary method (i.e., the rule of the three)

For the reaction involving the oxidation of methane gas by oxygen producing carbon dioxide and water, the stoichiometric reaction is:

$$CH_4 + 2SO_2 \rightarrow CO_2 + 2H_2O$$

$$16g + 64g \rightarrow 44g + 36g$$

$$\text{Mass balance } (16+64)\text{ g} = (44+36)\text{ g}$$

$$\text{Or } \quad 80\text{ g} = 80\text{ g}$$

Water always contains impurities; the constituents may be soluble solids, liquids or gases. It is therefore, required to determine the amount of the impurities present in a given quantity of water, i.e., the concentration of the solution.

Example 3.12 The composition of air is given in % volume as 78.1% N_2, 20.95% O_2, and 0.95% Ar. Calculate the average molecular weight of air and its composition by % weight.

Solution. The % volume gives the number of relative moles.

Component	% Volume = % moles	Molecular weight, g/mole	Grams	% Mass
N_2	78.10	28	78.10 x 28 = 2186.8	75.53
O_2	20.95	32	20.95 x 32 = 670.4	23.15
Ar	0.95	40	0.95 x 40 = 38	1.32
Total	100	100	2895.2	100

Average molecular weight = 2895.2 g/100 mole = 28.952 g/mole.

Example 3.13 In the treatment of potable water, an aluminum sulphate solution is used as a coagulant to produce aluminum hydroxide sludge as per the reaction.

$$Al_2(SO_4)_3\ 14H_2O + 3Ca(HCO_3)_2 \rightarrow Al(OH)_3 + 3CaSO_4 + 14H_2O + CO_2.$$

Calculate the amount of sludge produced if 100kg of aluminum sulphate coagulant is used.

Solution. Stoichiometric equation is:

$$Al_2(SO_4)_3\ 14H_2O + 3Ca(HCO_3)_2 \rightarrow 2Al(OH)_3 + 3CaSO_4 + 14H_2O + 6CO_2$$

Molecular weights:

$$Al_2(SO_4)_3\ 14H_2O = (2x27)+(3x32)+(12x16)+14x18 = 594 \text{ g}$$

$$2\ Al(OH)_3 = (2x27) + (6x16) + (6x1) = 156 \text{ g}$$

594 g of Aluminum sulphate produce 156 g of aluminum hydroxide sludge.

Hence, 100 kg of Aluminum sulphate will produce $\frac{156g}{594g}$ x 100 x 100g = 26.26 kg of chemical sludge.

3.3 MICROBIAL CHEMISTRY

Antonie van Leeuwenhoek in 1674 first discovered microbes by using a microscope of his own design. Microbes are single-cell organisms and so tiny that millions can fit into the head of a needle. The study of microorganisms is called **microbiology**. Micro-organisms (or microbes for short) play a very important role in our lives. Some microbes cause disease but the majorities are completely harmless. In fact we couldn't live without them, but they could live without us. These microscopic organisms play a vital role in sustaining life on earth, fixing gases and breaking down dead plant and animal matter into simpler substances that are used at the beginning of the food chain. Microrganisms occur in an amazing variety of shapes, sizes and they are divided generally into one of 6 groups: Bacteria, Virus, Fungi, Arachea, Protozoa and Algae. In this chapter, we will discuss only bacteria which play key role in environmental engineering.

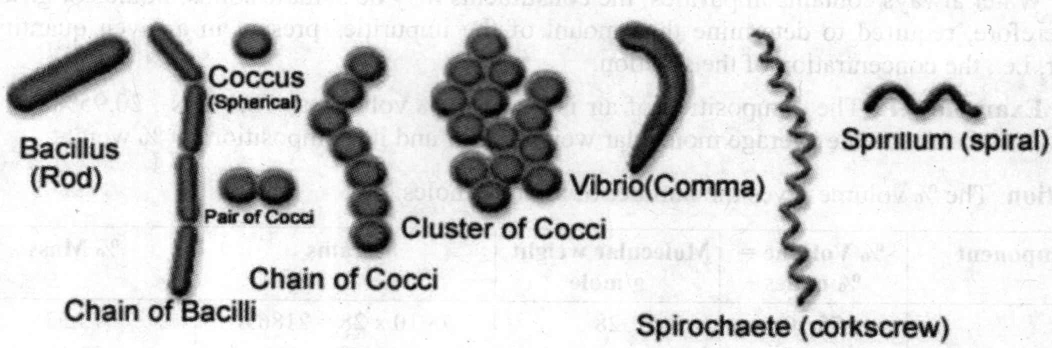

Fig. 3.1 Classification of Bacteria Based on Shapes

Bacteria are single celled microbes and constitute a large sphere of prokaryotic microorganisms. It is present in almost all the habitats like water, soil, hot spring, radioactive environment, space and bottom of the earth. Bacteria also live in parasitic and symbiotic relationships with animals and plants. There are approximately 5×10^{30} bacteria on earth. More than a million bacterial cells are present in a millilitre of fresh water and approximately 40 million bacterial cells in one gram of soil. Based on shapes, bacteria can be classified into 5 groups: spherical (cocci), rod (bacilli), spiral (spirilla), comma (vibrios) or corkscrew (spirochaetes). They can exist as single cells, in pairs, clusters or chains (Fig. 3.1).

Bacterial growth occurs through asexual reproduction or binary fission. Generally it follows four phases of growth (Fig. 3.2). The first phase of growth is called **lag phase**. In this phase bacterial cells need to adapt to their new environment. The second phase of growth is the **log phase** or **logarithmic phase**. During this phase the rate of population increase doubles with each consecutive time period. The third phase is caused by depletion of nutrient and called **stationary phase**. The final phase is the **death phase** where the bacteria run out of nutrients and die. The final phase is the death phase where the bacteria come to an end of nutrients and die.

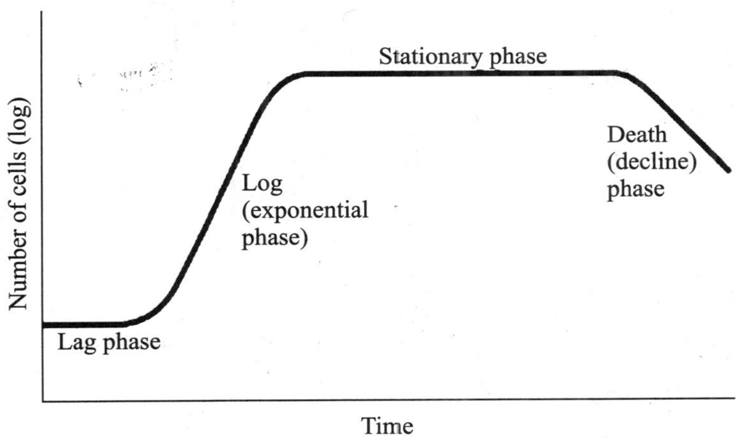

Fig. 3.2 Growth Curve of Bacteria

3.3.1 Bacterial Thermodynamics

The most fundamental property of bacterial cell system is ability to utilize and transfer energy. For this process thousands of enzymes catalyzed chemical reactions are involved. Since every chemical reaction involves gain or loss of electrons, the amount of energy used or released depends on oxidized compound (electron donor) and reduced compound (electron acceptor). The energy must be balanced through electron transport and energy transformation which follows the law of thermodynamics. Based on electron donor and acceptor availability, bacteria generate energy through photosynthesis, aerobic respiration and anaerobic respiration

3.3.2 Fermentation

Biochemical reactions in bacteria occur in super systematic manner, called **metabolic pathways.** Two basic types of metabolic pathways: pathways start from large to smaller molecules (high oxidative state to low oxidative state) which is called **catabolism** or **catabolic pathways**, it begins from small to large molecules (low oxidative state to high oxidative state) which is essential for the formation of cellular material, called **anabolism** or **biosynthetic pathway**. All catabolic pathways are oxidative and thus energy producing, bio-synthetic pathways are energy consuming and reductive. Entire metabolic process based on thermodynamics which control transformation and transfer mechanisms (Fig. 3.3).

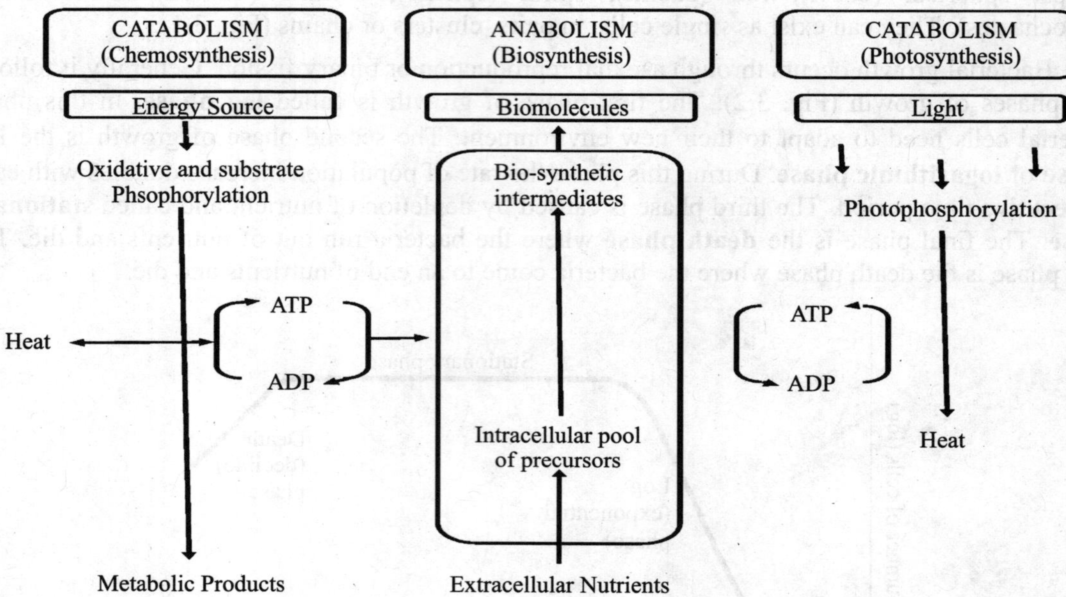

Fig.3.3 Schematic Diagram of Metabolic Energy Formation and Usage

3.3.3 Importance of Bacteria

3.3.3.1 Water treatment process

Almost all oxidative sewage treatment processes depends on microorganisms to oxidise organic constituents which are not generally willing for sedimentation or flotation. For reduced solids sludge, anaerobic microorganisms are also used to produce methane gas (amongst other gases) In potable water treatment (the slow sand filter method) employs a complex gelatinous layer composed of a wide range of microorganisms to remove both particulate material and dissolved solids from untreated water.

3.3.3.2 Energy production

Bacteria are used to convert various forms of agricultural and urban waste into usable fuel. Microorganisms are used in fermentation to produce ethanol and in biogas reactors to produce methane.

3.3.3.3 Industrial enzymes and chemicals

Streptococcus synthesized streptokinase which is used as a clot buster for removing clots from the blood vessels of patients who have undergone myocardial infarctions leading to heart attack.

Different bacteria produce so many industrially important chemicals like citric acid, acetic acid, butyric acid, lactic acid etc.

3.3.3.4 Science and technology

Microrganisms are crucial utensils in biotechnology, biochemistry, genetics, molecular biology, space research etc. Thus many microorganisms are used as model organisms for development of science and technology.

3.3.3.5 Bioleaching

Many microorganisms are used for recovery of precious metals where chemical process is not viable like copper, aluminum, gold etc.

3.3.3.6. Food technology

Wild or genetically modified microorganisms are enormously used for different food processing like brewing, baking, pickling, wine making etc. It also plays key role in drug production like antibiotic, insulin etc.

3.4 ATMOSPHERIC CHEMISTRY

3.4.1 Structure of the Earth's Atmosphere

The earth's atmosphere is divided into regions based on temperature gradients. The major regions of atmosphere are (Table 3.3):

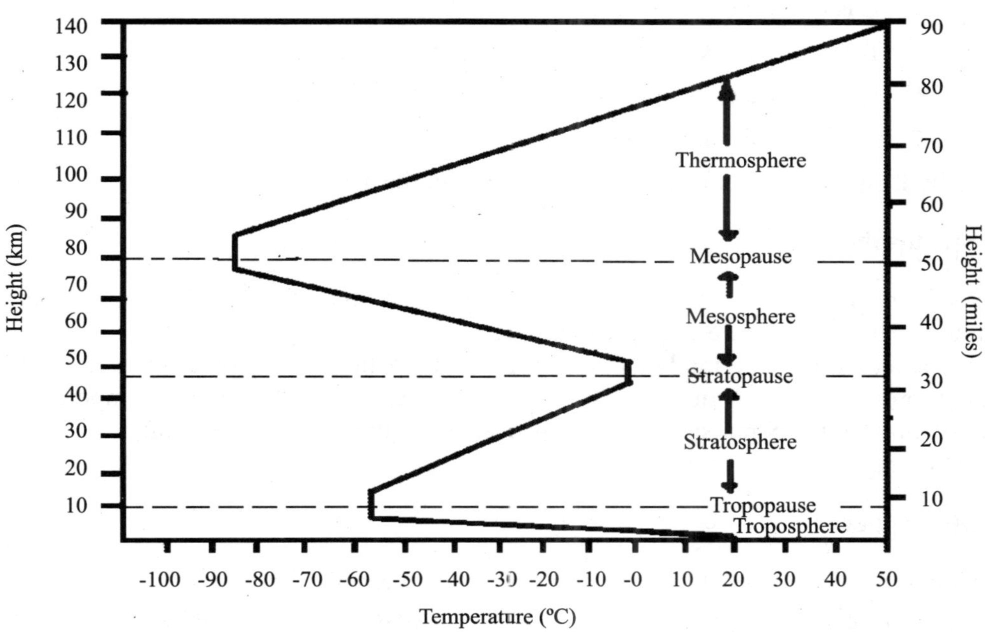

Fig. 3.4 Vertical Temperature Profile of Earth's Atmospheric Regions

3.4.1.1 Troposphere

It is the mid latitude region where the temperature falls with increasing attitude. This is known as negative lapse rate. The increase continues to an altitude known as the **tropopause**. The height of troposphere is about 10 km above the earth's surface (Fig. 3.4).

The lower 0 to 2 km of the troposphere is called **atmospheric boundary layer** (ABL). The ABL is that region where the wind velocity is affected by the shear resistance of the earth's surface. At the earth's surface, the wind velocity is lowest and increases gradually (non-linearly) to the top of the ABL. Above the ABL, the wind velocity is approximately constant, being unaffected by the shear resistance of the earth's surface. Atmospheric pollution occurs within ABL. In higher regions within troposphere, large-scale air circulation and global climate circulation modeling occurs.

The atmosphere is computed to weigh 4.5×10^{15} tones. The density of the atmosphere at ground level is 1.29 g/l (1.29 Kg/m^3) and decreases with altitude. It also decreases with increasing temperature.

Gas volume – 1 gm mole of an ideal gas, at STP (Standard Temperature and Pressure) occupies 22.414 liters. An ideal gas is one that obeys the ideal gas law:

$$PV = nRT$$

Where,

P =Pressure, Pa (N/m^2)

V = Volume, m^3

n = Number of moles

R = Gas constant = 8.314 J /K moles

T = Temperature, K

$R = R_d$ (in dry air) = 287.04 J/Kg. K

$R = R_w$ (for moist air) = 461.5 J/Kg. K

The hydrosphere is about 230 times heavier than the atmosphere.

3.4.1.2 Stratosphere

It extends from 11 – 50 km from the earth's surface. The upper boundary of this region is called **stratopause.** The temperature increases with increase in altitude with a maximum of -2°C. This tendency leads to strong thermodynamic stability, with little turbulence and vertical mixing. The ozone layer present in that region absorbs deadly UV radiation and raises the temperature causing a positive lapse rate. Stratosphere is significant for the flight of jet planes and for radio communications.

3.4.1.3 Mesosphere

It shows negative lapse rate i.e., temperature falls with increasing altitude. This is due to absence of UV absorbing species, particularly ozone. In mesosphere, temperature decreases with increasing altitude. The upper boundary of the mesosphere is the **mesopause,** which can be the coldest naturally occurring place on earth with temperatures below 130 K (−226°F; −143°C). The exact upper and lower boundaries of the mesosphere vary with latitude and with season, but the lower boundary of the mesosphere is usually located at heights of about 50 kilometres (160,000 ft) above the earth's surface and the mesopause is usually at heights near 100 kilometres, except at middle and high latitudes in summer where it descends to heights of about 85 kilometres.

3.4.1.4 Thermosphere

The thermosphere is the layer of the earth's atmosphere directly above the mesosphere and directly below the exosphere. Within this layer, ultraviolet radiation (UV) causes ionization. The temperature rises once again giving a positive lapse rate. Here the atmospheric gases particularly O, NO split into atoms and also undergo ionization after absorption of solar radiation in the far UV region. Due to the photoionization reactions this region is rich in electrically charged particles and electrons. Therefore, it is also called **ionosphere.** Ionosphere is the environment for earth-orbiting satellites and ballistic missiles, therefore, it is becoming important for atmospheric pollution.

Table 3.3 Major Regions of Atmosphere c

Region	Altitude Range, Km	Temperature Range °C	Important Chemical Species
Troposphere	0 – 10	15 to -56	N_2, O_2, CO, H_2O
Stratosphere	11 – 50	- 56 to -2	O_2
Mesosphere	51 – 85	- 2 to – 92	O_2^+, NO^+
Thermosphere	86 – 500	-92 to 1200	O_2^+, O^+, NO^+

3.4.2 Chemical Composition of Earth's Atmosphere

Earth's atmosphere close to ground level contains:

78% Nitrogen (N_2)

21% Oxygen (O_2)

1% Argon (Ar) with fractions of a percent for CO_2, CH_4, H_2, CO etc.

In this ABL most living matter exists. Microbial life has been found in the stratosphere, but only in minute quantities.

3.4.3. Contaminant Gases in the Earth's Atmosphere

(i) **Sulphur oxides:** Pollution from sulphur oxides is primarily from two colourless gaseous compounds, SO_2, SO_3. altogether they are called SO_x.

$$S + O_2 \rightarrow SO_2$$

$$2SO_2 + O_2 \rightarrow 2SO_3$$

$$SO_3 + H_2O \rightarrow H_2SO_4$$

In atmosphere, SO_2 is converted to SO_3 and then to H_2SO_4 by photolytic and catalytic processes.

- Sulphur is present in coal and sulphur-containing minerals (iron pyrites, FeS_2, lead sulfide, PbS).
- Sulphur in fuel oils can be removed in high-pressure catalytic reactors, in which hydrogen combines with sulphur to form H_2S,
- H_2S being toxic is adsorbed by activated carbon and is converted to SO_2 for further disposal

(ii) **Carbon Monoxide:** CO is a colourless, odourless and tasteless gaseous compound. It is only slightly lighter than air and is weakly soluble in water. Automobiles are the greatest contributors of CO to atmosphere and soils provide a source for removing the CO.

$$2C + O_2 \rightarrow 2CO$$

$$2CO + O_2 \rightarrow 2CO_2$$

$$CO_2 \rightarrow CO + O$$

$$CO_2 + CO \rightarrow 2CO$$

- CO is due to incomplete burning of carbonaceous fuel. Automobiles are major sources of pollution of CO.
- In automobiles, during expansion, the combustion process becomes kinetically linked and CO concentration is much higher than the equilibrium value.
- Stratification of the engine with two compartments, the first for pre-ignition of air/fuel mixture and the second for providing a broad flame for efficient burning, helps in better fuel burning and CO reduction.

(iii) **Oxides of Nitrogen:** Nitric oxide (NO) is a colourless, odourless gas, while NO_2 is a red – brown gas with a pungent, choking odour.

$$N_2 + O_2 \rightarrow 2NO$$

$$2NO + O_2 \rightarrow 2NO_2$$

Photolytic process is:

$$NO_2 + hv \rightarrow NO + O^*$$

Where, hv = Photon of solar radiation of energy

O^* = Very reactive atomic oxygen

$$O^* + O_2 \rightarrow O_3$$

$$O_3 + NO \rightarrow NO_2 + O_2$$

$$HC + NOx + Sunlight \rightarrow Photochemical\ Smog$$

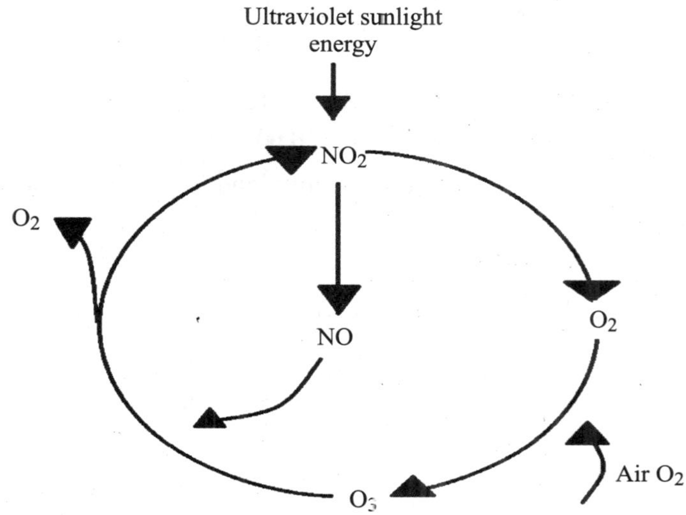

Fig. 3.5 Photolytic NO_2 Cycle

$$2NO_2 + H_2O \longrightarrow + HNO_3 + HNO_2$$

(Nitric Acid)

- Electric power generators and automobiles, which use fossil fuels, are the major sources for NO_X and hydrocarbon (HC) pollutants due to incomplete combustion.
- The thermal fixation of atmospheric N_2 and O_2 produces thermal NO_X, while the combustion of chemically bound N_2 in the fuel produces fuel NO_X.
- Both nitrogen monoxide and nitrogen dioxide are formed in higher temperature combustion of air. Oxidation of NO in the atmosphere enhances its NO_2 content.

(iv) Particulates: Particulate matter (PM), also known as particle pollution, is a complex mixture of extremely small particles and liquid droplets. Particle pollution is made up of a number of components, including acids (such as nitrates and sulfates), organic chemicals, metals, and soil or dust particles. There are also many particulate pollutants. Particulate matter ranges in size from small as 10^{-6} mm to as large as 1 mm. Because of their very small size, they remain in the atmosphere for long periods and can travel great distances. They will on their own or on agglomeration with water droplets, settle on the earth's surface such as fly ash, dust storms, volcanic ash, forest fires etc. They interfere with plant growth when deposited on their leaves, and it interferes with the balance of CO_2 between the plant and the atmosphere.

(v) Hydrocarbons: They are compounds of hydrogen and carbon and classified into 3 types based on molecular structure:

(a) *Aliphatic or Acyclic HCs:* All carbon atoms are arranged in chains without branching e.g:

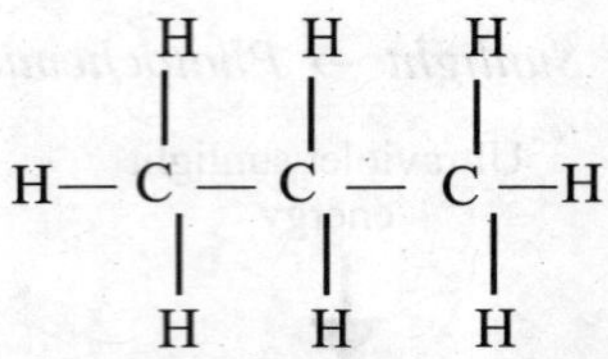

Propane (C_3H_8)

(b) *Aromatic:* HCs contain six carbon rings and each carbon atom has an H or C attached to it.

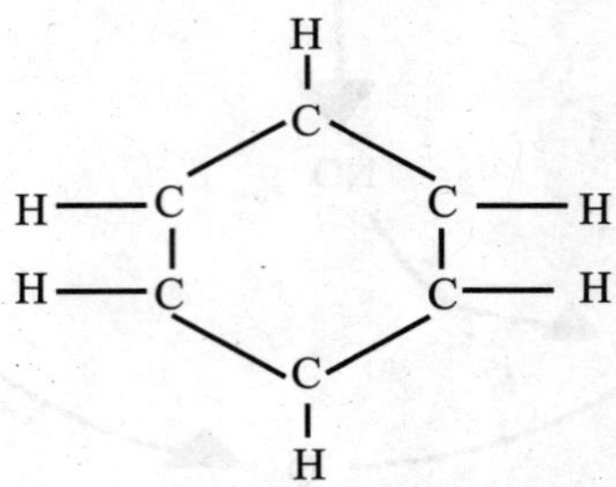

Benzene (C_6H_6)

(c) *Alicyclic*: HCs containing a ring structure (six Cs) other than benzene.

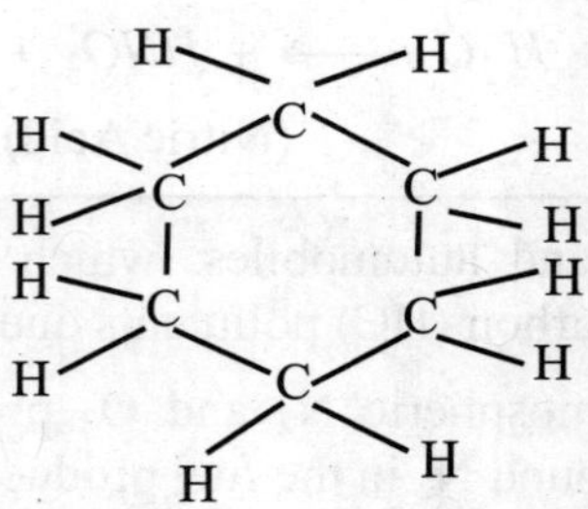

Cyclohexane (C_6H_{12})

Automobiles account for most of the HCs in the urban air which is undesirable because of their toxicity, particularly the more common benzene compounds such as C_6H_6 and C_7H_8.

3.5 SOIL CHEMISTRY

3.5.1 Soil

The word *'Soil'* is derived from a Latin word *'Solum,'* which means earthy material in which plants grow. Soil can be defined as the weathered layer of the earth's crust with living organisms and their products of decay intermingled. Soil study is commonly referred to as soil science or pedology (pedos=earth) or edaphology (edaphos = soil).

3.5.2 Soil Formation

It involves two stages: weathering and soil development or pedogenesis.

3.5.2.1 Weathering

It involves the breakdown of bigger rocks into fine, smaller mineral particles.

(i) Physical weathering: Agents such as temperature, water, ice, gravity, wind etc. exert mechanical effect on the rocks as a result of which fragments are broken down to regolith.

(ii) Chemical weathering: The presence of moisture and air are essential. It takes place through the following reactions such as: Water soluble minerals, like gypsum, limestone etc. of the rock get weathered by the solvent action which increases in the presence of CO_2 and organic acids formed by the decay of organic remains of plants and animals. The solution of these minerals gets absorbed on the surface of negatively charged colloidal particles or removed by leaching.

(a) **Hydrolysis:** It mostly takes place in combination with other reactions (such as oxidation, reduction or carbonation) and involves the chemical action of water with strong bases producing hydroxides of iron, magnesium, aluminum, calcium etc.

$$K_2Al_2Si_6O_{16} + 2H_2O + CO_2 \longrightarrow Al_2O_3.\ 2SiO_2.\ 2H_2O + 4\ SiO_2 + K_2CO_3$$

Orthoclase (Primary mineral) Kaolin (Secondary clay mineral)

(b) **Oxidation:** Oxygen reacts with minerals to produce oxides. The latter when dissolved in water weakens the rock and brings about weathering such as:

$$4FeO + O_2 \longrightarrow 2Fe_2O_3$$

Oxides and sulphides of iron, aluminium, manganese etc. which are easily oxidized and cause the chemical weathering of rocks.

(c) **Reduction:** It mainly occurs in deep zones of earth crust, which are poorly aerated. Eg.

$$2Fe_2O_3 \longrightarrow 4FeO + O_2$$

(d) **Carbonation:** It is the combination of CO_2 and H_2O to form carbonic acid. The latter combines with hydroxides of Ca, Mg and other minerals of the rock to form carbonates as well as bicarbonates.

$$CO_2 + H_2O + CO_2 \rightarrow H_2CO_3, Ca(OH)_2 + CO_2 \rightarrow CaCO_3 + H_2O, CaCO_3 + H_2O + CO_2 \rightarrow Ca(HCO_3)_2$$

(e) **Hydration**: Water molecules get attached to the rock. So the volume of the original material increases and hydrated mineral becomes soft. The soft rock can be readily weathered.

$$2Fe_2O_3 + 3H_2O \rightarrow 2Fe_2O_3.3H_2O$$

Hematite Limonite

$$CaSO_4 + 2H_2O \rightarrow CaSO_4.2H_2O$$

(iii) Biological weathering

A number of microorganisms such as lichens and mosses extract mineral nutrients such as Mg, Ca, Na, K, Fe, P, Si, Al etc. from the rock. These nutrients are combined with organic matter and thereby return to the developing soil when the vegetation decomposes.

The soil formed by weathering of rocks is known as **primary soil** or **embryonic soil**. The latter may mature into **residual soil** or **sedimentary soil**.

3.5.2.2 Pedogenesis

It is modification of mineral matter through interaction between biological, topographic and climatic effects which leads to the development of a number of layers, i.e., horizons of soil, known as **soil profile**.

3.5.3 Factors Affecting Soil Formation

(i) ***Active factors:*** Rainfall, temperature, wind, humidity and evaporation

(ii) ***Passive factors:*** Parent materials and topography which influence aeration, texture and chemical characteristics of the soil

(iii) ***Biospheric factors:*** Micro-organisms in soil modify the physico-chemical processes in the soil.

3.5.4 Classification of Soil

The soils are classified into three types and are discussed below:

(i) Zonal soils: The climate is the important factor over the process of soil formation.

(ii) Intrazonal soils: Here local geological conditions preclude the normal pattern of soil formation.

(iii) Azonal soils: It is rich in colloids and because it is not leached, it may also be rich in nutrient materials. It lacks humus, which is finely divided, amorphous, incompletely decomposed black coloured organic matter added to the mineral matter of soil. Depending on the organic matter content, soils are classified into:

- **(i) Mineral soils:** Rich in mineral particles.
- **(ii) Peat & muck:** Rich in organic matter accumulated in wet areas.
- **(iii) Mor:** Low in basic minerals.
- **(iv) Mull:** Rich in basic minerals.

3.5.5 Composition of Soil

The soil consists of four major components are discussed below:

(i)	Inorganic or mineral matter:	45%
(ii)	Organic matter:	5%
(iii)	Soil water:	25%
(iv)	Soil air:	25%

3.5.5.1 Mineral or inorganic matter

Minerals are naturally occurring inorganic solid homogenous substances e.g. quartz (SiO_2), calcite ($CaCO_3$), gypsum ($CaSO_4$. $2H_2O$), orthoclase ($KAlSi_3O_8$), etc.

3.5.5.2 Organic matter

It represents partially decayed and partially synthesized plant and animal residues. It contains living and dead microbial cells, microbial synthesized compounds and derivates of microbial decay.

Important characteristics of organic matter in soil are discussed as follows:

(i) It improves the physical condition of the soil.

(ii) It increases the water holding capacity of the soil.

(iii) It is a major source of nutrients, i.e., nitrogen, phosphorus and sulphur.

(iv) It is the main source of energy for soil organism.

(v) It is responsible for most desirable surface soil structure.

(vi) It promotes a greater proportion of large pore sizes.

(vii) It improves aeration status of the soil.

3.5.5.3 Soil water

It not only acts as solvent and transporting agent but also maintains soil texture, arrangement and compactness of soil particles. It is a main regulator of physical, chemical and biological activities in the soil. Soil water may be physically classified into:

(i) Gravitational water: It is the free water which moves down into the soil because of gravity till it reaches water table.

(ii) Capillary water: It is present in the interspaces of soil particles.

(iii) Hygroscopic water: It is the water held firmly on the surface of soil particles.

The total amount of water present in the soil is called **holard**. The amount of water that can be absorbed by plants out of holard is known as **chesard**. The remaining unabsorbed water is called **echard**.

3.5.5.4 Soil air

It occupies the pore space between soil particles which is not water filled. The exchange of CO_2 and O_2 between the soil pore space and the air is called **soil aeration**. The crumb and inter crumb pores are involved in soil aeration. A loam soil with humus containing 34% of air and 66% of water is considered best for majority of crops. Most of the gaseous inter change in soil occurs by diffusion. Importance of soil air are discussed below:

(i) Soil air is essential for the respiration of soil micro-organisms and plants.

(ii) Poor aeration results in the formation of toxic substances such as H_2S, oxalic acid and formic acid in soil.

(iii) Soil aeration increases the permeability of roots to water.

3.5.6. Types of Indian Soil

(i) Red soil

(a) The colour is due to high proportion of iron components.

(b) These are mainly sandy to loamy in texture with gravels on upper slopes, sandy soils and deeper loamy soils on lower slopes and clay in the valley bottoms.

(ii) Laterite soils or latosols

(a) These soils are porous clay rich in hydroxides of iron and aluminum.

(b) At low elevations, such soils are suitable for paddy cultivation.

(c) At higher elevations they favour the growth of coffee, rubber, tea and Cinchona.

(iii) Black soil

(a) These soils are predominantly clay with patches of loams and sandy loams.

(b) Fertile since they contain humus.

(iv) Skeletal (mountain) soils: They are stony, sandy hill foot fans.

(v) Desert soils: They are sandy.

(vi) Alluvial Soils (in river beds):

(a) The soil is generally alkaline or neutral.

(b) The soil is rich in loams and clay components.

3.5.7. Causes of Soil Degradation

Soil degradation is a complex issue. It is caused by the interaction of physical forces and human activities. Its impact is increasing and is having a negative effect on food production. Some areas are more badly affected than others but in a globalised world the impacts are felt worldwide. The method of dealing with soil degradation depends on the cause of the problem, but also the resources available to the host country.

Soil becomes prone to degradation whenever inappropriate land use and management practices are adopted.

(i) Deforestation: Deforestation or over cutting of vegetation, woodlands or scrublands to obtain timber, fuel wood or other products at a pace exceeding the rate of natural re-growth of biomass renders the soil to become erosion prone due to action of rainwater or wind that contributes to removal and transport of valuable surface soil.

(ii) Fuel wood: Fuel wood is usually converted into charcoal which burns clearly. In this process all 'H' and 'O' are removed leaving behind Carbon.

(iii) Overgrazing: Overgrazing of natural pastures resulting from increasing livestock population and poorly managed grazing exceeding the land's carrying capacity, causes decrease in vegetative cover. When insufficient amounts of grass litter are left for the soil, the soil organisms die and the soil loses fertility.

(iv) Agriculture: Agricultural activities and practices can cause soil degradation in a number of ways depending on land use, crops grown and management practices adopted. Soil degradation problems are serious when fragile (e.g. desert) or marginal (e.g. sloping) lands are cultivated in the absence of adequate conservation measures. Many of our rain fed areas are subject to extensive degradation caused by runoff and soil erosion.

In high productivity irrigated regions, excessive and imbalanced use of fertilizers often leads to decline in soil fertility. Use of heavy machinery is another factor that contributes to soil degradation resulting from physical degradation of soils.

(v) Industrialization: Due to rapid industrial development, the natural soil environment quality degrades day by day. Almost all industrial activities cause some pollution and produce xenobiotic (human-made) chemicals waste. The most common chemicals are agricultural chemicals, petroleum hydrocarbons, polynuclear aromatic hydrocarbons (such as naphthalene and benzo(a)pyrene), solvents, pesticides, lead, and other heavy metals or improper disposal of waste. Contamination is correlated with the degree of industrialization and intensity of chemical usage.

- Soil pollutants are minerals, nitrates, nitrites, sulfates, phosphates, humic substances, and anthropogenic pollutants.
- Measurements of contaminants flux from the soil (earth's surface) to the atmosphere are used for monitoring and assessing air emission and the environment impact of waste sites.
- Derivatization method is used to separate trace elements from thermatrices and concentrate analyte species by hydride generation, ethylation or phenylation.
- Solvent extraction method is a traditional solvent extraction process. In this method, the sample is placed in a cavity that is gradually filled with the extracting liquid phase by condensation from the distillation matrix.

3.6 NOISE POLLUTION

3.6.1 Noise

In scientific terms, noise (or sound) is a pressure oscillation in the air or water of any medium which conducts and travels (radiates) away from the source. Noise, is an unwanted sound energy and is also considered as a pollutant when it exceeds certain limits. Noise has a short residence and decay time and hence does not remain in the environment for long periods, like air or water pollutants. By the time the average individual is spurred to action to abate, control or at least, complain about sporadic environmental noise, the noise may no longer exist.

The physiological and psychological effects of noise on us are often subtle and insidious, appearing so gradually that it becomes difficult to associate cause with effect. There occurs a vast variation in individual sensitivity to noise and people are affected differently when they are at home or outside or at work.

3.6.2 Sources of Noise Pollution

Noise can be emitted from a point source (electric fan), an aerial source (discotheque) or a line source (moving train). Noise pollution comes from a wide variety of sources, including traffic, industrial equipments, construction activities, sporting and crowd activities, low flying aircrafts etc.

3.6.3 Physical Properties of Sound

(i) Sound waves

Sound is a mechanical, longitudinal wave characterized by pressure amplitude (A), wavelength (λ), Time Period (T), frequency (f) $= \frac{1}{T}$, $\lambda = \frac{C}{F}$, C = velocity of sound in air at sea level at 20ºC is approximately 340m/s under ambient condition.

Sound waves result from the vibration of solid objects or the separation of fluids as they pass over, around or through holes in a solid object. Audible vibrations occur in the hearing range which varies from person to person and is also dependent on age, extent of hearing loss and even physiological condition. The hearing range extends from a frequency of about 20 to 20,000 Hz. Sound is defined as any pressure variation that the human ear can detect. The simplest sound pressure variation (caused by a pure tone) produces the sinusoidal wave pattern (Fig. 3.6).

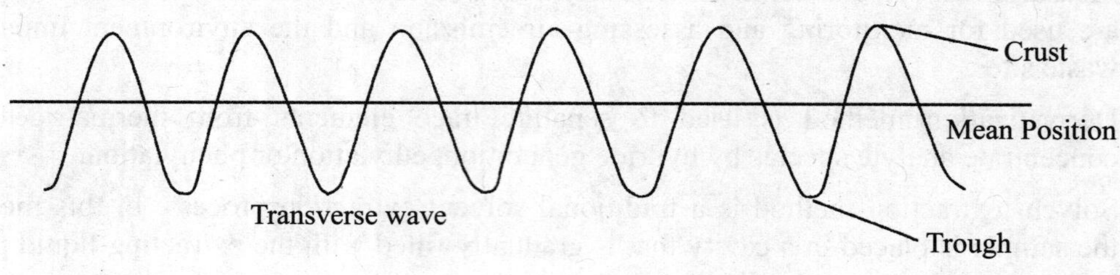

Fig. 3.6 Parameters of a Sinusoidal Wave Pattern

Some of the elemental noise terms are discussed below:

(i) Amplitude (A) - the maximum or minimum pressure

(ii) Wavelength (λ) - the distance between successive troughs or crests

(iii) Period (P) - the time between successive peaks or troughs

(iv) Frequency (f) - the number of complete pressure variations or cycles per second

(ii) Sound Power (w) and Intensity (I)

The rate at which energy is transmitted by sound waves is called **sound power** (w) measured in watts. The average sound power per unit area normal to the direction of propagation of a sound wave is termed **acoustic** or **sound intensity (I)**.

$$I = \frac{w}{4\pi r^2} watt/m^2$$

R = Locus radius

Intensity is proportional to the square of the sound pressure

$$I = \frac{P^2}{\rho C} watt/m^2$$

Where,

I = Acoustic intensity, w/m^2

P = Sound pressure (Pa)

ρ = Density of the medium, kg/m^3

(for air ρ = 1.185 kg/m^3 at 20°C at standard pressure)

C = Speed of sound in the medium, m/s

(Note: 1 watt = 1 joule/sec. = $1\frac{N.m}{Sec.}$)

(iii) Measurement of sound level in Decibel (dB)

Sound power level (L_w) in decibel is defined as ten times the logarithm of the ratios of the measured sound power to reference sound power for noise measurement purposes.

$$L_w = 10\log_{10}\frac{W}{W_0}$$

Since sound power is proportional to the square of sound pressure, the Sound pressure level (L_p) in decibel is defined as ten times the logarithm of the ratios of square of the measured sound pressure to square of reference sound pressure

$$L_p = 10\log_{10}\frac{P^2}{P_0^2} = 20\log_{10}\frac{P}{P_0}$$

Where, P_0 = Reference pressure, is the threshold of hearing (20 μ Pa)

P = Measured pressure (Pa)

***(iv)* Combination of sound pressure levels**

Decibel values cannot be added directly because they are logarithms.

$$L_{comb} = 10\log_{10}\left[\frac{w_1 + w_2 + \ldots\ldots + w_n}{w_o}\right]$$

$$= 10\log\frac{1}{w_o}\left[10^{Lw_1/10} + 10^{Lw_2/10} + \ldots\ldots\ldots 10^{Lw_n/10}\right]$$

$$Lw = 10\log\sum_{i=1}^{n} 10\frac{Lwi}{10}$$

(v) Equivalent continuous sound level (L Aeq)

Equivalent continuous sound level (L Aeq) is the constant noise level applied to a source with fluctuations in sound level over a period of time.

$$L_{Aeq} = 10\log_{10}\left[\frac{1}{T}\sum_{i=1}^{n} 10^{0.1Li\times ti}\right]$$

n = Number of samples.

L_i = Noise level of ith sample

t_i = Fraction of the total time T which is the time period.

(vi) Frequency

Audible sound frequencies range from 0.015 to 15 kHz. The wide range of audible sound frequency is divided into Octave Bands. An octave band is the frequency interval between a given frequency and twice that frequency. Eg. 0.1-0.2 kHz, 1-2 kHz.

3.6.3.1 Noise standards

Noise standards are prescribed by the Central Pollution Board (India) under the Noise Pollution Act 2000 (Table 3.4 and 3.5).

Table 3.4 Permissible Sound Levels

Area	Day (in dB)	Night (in dB)
Industrial	75	70
Commercial	65	55
Residential	55	45
Silence Zone	50	40

Table 3.5 Sound Levels of Common Noise Sources

Source of Noise	Sound Level in dB
Jet Engine	120 – 140
Rock Music	100 – 110
Heavy Truck	80 – 90
Business Office	60 – 70
Library	40 – 50
Bed Room	10 – 20

3.6.4 Classification of Sounds

Sounds are classified as follows:

(i) **Continuous sound:** It is an uninterrupted sound level that varies less than 5 dB during the period of observation. E.g. a household fan.

(ii) **Intermittent sound:** It is a continuous sound that lasts for more than one second but then is interrupted for more than one second. E.g. a dentist's drill.

(iii) **Impulsive sound:** It is of short duration, less than one second. e.g. Typewriter or hammering noises.

3.6.5 EU Noise Directive

The EU Directive (86/188 EEC) is on the protection of workers from the risks related to exposure to noise at work. The objective of the directive is to reduce the level of noise experienced at work by taking action at the noise source. Two exposure levels are used:

Daily personal noise exposure of a worker $L_{EP,\,d}$.

$$L_{EP,\,d} = LAeq_{\cdot Te} + 10 \log_{10} \frac{Te}{To}$$

Where,

$$LAeq_{\cdot Te} = 10 \log_{10} \left\{ \frac{1}{Te} \int_0^{Te} \left[\frac{P_A(t)}{P(O)} \right]^2 dt \right\}$$

T_e = daily duration of a worker's exposure to noise

T_o = 8h

P_A = A – weighted instantaneous sound pressure in Pa

Weekly average of the daily values, $L_{EP,\,w}$

$$L_{EP,w} = 10 \log_{10} \left\{ \frac{1}{5} \sum_{k=1}^{m} 10^{0.1(L_{EP,d})} K \right\}$$

Where, $(L_{EP,d})k$ = the values of $L_{EP,d}$ for each of the m working days in the week being considered.

The EU directive specifies that when the daily exposure level exceeds 85 dBA, the worker is to be advised of the risks and trained to use ear protectors. If the daily exposure level exceeds 90 dBA, a programme to reduce levels should be put in place.

Example 3.14

If a sound source has a pressure of 3000 μPa at 10 m distance, compute:

(a) the sound pressure level in dB

(b) the sound intensity in W/m^2

(c) the sound power in W

Solution

$P = 3000\ \mu Pa$

$P_0 = (reference\ level) = 20\ \mu Pa$

(a) $SPL = 20 \log_{10} \dfrac{P}{P_0} = 20 \log 150 = 43.5\ dB$

(b) Sound intensity $I = \dfrac{p^2}{pC} = \dfrac{(3000\times 10^{-6})^2}{1.185\times 340} = 2.2\times 10^{-8}\ W/m^2$

(c) Sound power $= 4\pi r^2 I = 4\times 3.14\times 10^2 \times 2.2\times 10^{-8} = 2.76\times 10^{-5}\ W$ *Ans.*

Example 3.15

If two sources have equal pressure of 3000 μPa, determine the sound pressure in dB.

Solution

$$L_p = 20 \log \frac{P}{P_0} = L_w = 10 \log \frac{I}{I_0}$$

as $p\alpha\sqrt{1}$ (*i.e.* sound pressure is proportional to square root of intensity). The new sound pressure equals to the original pressure multiplied by $\sqrt{2}$.

$$L_p = 20 \log \sqrt{2}\ P/P_0 = 20\log\sqrt{2}\ \frac{3000\ \mu Pa}{20\mu Pa} = 20 \log\sqrt{2}\times 150 = 46.5\ dB$$

Example 3.16

Determine the sound power level form combining the four sound levels of 56, 68, 71 and 48 dB.

Solution

$$L_{w1} = 10 \log {}^{I_1}\!/_{I_0}, I_1 = I_0^{\ LP_1/10}, I_2 = I_0 \times 10^{LP2/10} \text{ etc.}$$

$$I_1 = I_0 \times 10^{5.6}, I_2 = I_0 \times 10^{6.8}, I_0 \times 10^{7.1}, I_4 \times 10^{4.8}$$

$$I = I_1 + I_2 + I_3 + I_4 = I_0\ (10^{5.6} + 10^{6.8} + 10^{7.1} + 10^{4.8})$$

Resultant SPL $= 10[\log(10^{5.6} + 10^{6.8} + 10^{7.1} + 10^{4.8})] \approx 73\text{dB}$

Example 3.17

If an industrial fan generates a noise level of 65 dB for 10 minutes out of every hour compute the equivalent continuous level (L_{Aeq}) if the background level is 55*dB*.

Solution

L_1=65dB L_2 (background) = 55 dB. $\Delta T_1 = 10\,\text{minutes}$, $\Delta T_2 = 50\,\text{minutes}$. Total time 1hr

$$L_{Aeq} = \text{(Equivalent noise)} = 10\log\left[\frac{1}{T}(10^{L_1/10} \times \Delta T_1 + 10^{L_1/10} \times \Delta T_1)\right]$$

$$= 10\log\left[\frac{1}{60}(10^{6.5} \times 10 + 10^{5.5} \times 50)\right] = 76.8\,\text{dB}$$

Example 3.18

Calculate L_{dn} (SPL for day night 24 hrs) when L_d (day time SPL) average 67dB (6 AM to 10 PM) and Night time noise L_n 43 *dB* (10 PM to 6 AM)

Solution

$$L_{dn} = 10\log\left[\frac{1}{24}(10^{L_d/10} \times 16 + 10^{L_n/8})\right]$$

$$= 10\log\left[\frac{1}{24}(10^{6.7} \times 16 + 10^{4.3} \times 8)\right]$$

$$= 10\log\left[\frac{1}{24}(80189957 + 159621)\right] = 65.24\,\text{dB}$$

Example 3.19

Calculate the mean sound power level of four sound power level of 46 bD, 65 dB 72 dB and 58 dB in a place.

Solution

$L_{w1} = 46\,\text{dB}$ $\quad I_1 = 10^{4.6} \times I_0$

$L_{w1} = 65\,\text{dB}$ $\quad I_1 = 10^{6.5} \times I_0$

$L_{w2} = 72\,\text{dB}$ $\quad I_1 = 10^{7.2} \times I_0$

$L_{w4} = 58\,\text{dB}$ $\quad I_4 = 10^{5.8} \times I_0$

$$L_W \text{ (mean SPL)} = 10 \log \frac{I_0}{I_0 \times 4}(10^{4.6} + 10^{6.5} + 10^{7.2} + 10^{5.8}) = 66.92 \approx 67 \text{ dB}$$

3.6.6 Noise Measurement

(i) Sound level meters

Although sound power cannot be measured directly, sound intensity can be measured with modern instrumentation (ISO 9614).

Sound level meters are used to measure the sound pressure level. Noise meters are classified as follows:

(i) Type 0: For laboratory reference situations

(ii) Type 1: Precision grade, used for accurate field measurements – 0.7 dB

(iii) Type 2: Industrial grade, for non-critical survey work – 1.5 dB

(iv) Type 3: Survey grade, low cost sound level indicators – 2.5 dB

Microphone is one of the most important elements of the sound level meter system and it typically determines the type of instrument. Digital audio tape (DAT): DAT systems are available which can be used to record sounds directly in the field (instead of using sound level meters). All analysis and calculations of noise parameters are carried out by subsequent computer analysis.

(ii) Calibration

Sound level meters must be calibrated by a Pistonphone or a sound level calibrator both before and after use.

(iii) Measurement procedures

The recommended procedure varies with the purposes of the noise measurement are discussed as below:

(a) Measurement of workplace sound pressure should preferably be made in the undisturbed sound field (the person concerned being absent), with the microphone located at the position normally occupied by the ear exposed to the highest value of exposure (EC, 1986).

(b) Environmental noise measurements are normally taken at locations where the maximum noise nuisances are likely to occur.

(c) The microphone should be placed at a height of 1.2 m above the ground surface.

(d) Presence of adjacent sound reflecting surfaces must be taken into account.

(e) When noise intrusions are being measured, it is recommended that the background noise level should be at least 10 dB lower than the intrusive noise level.

(iv) Physical conditions for measurement

(a) The wind speed and direction relative to the microphone must be noted.

(b) Valid measurements can be taken in wind speeds up to 5 m/s. (acceptable with wind speeds upto 10 m/s).

(c) Windshields should always be used for outdoor measurements.

(d) Humidity levels upto about 90% and pressure variations of ± 10% have negligible effects on noise measurements.

(e) Unless the temperature is below -10°C or above 50°C measurements are usually not affected.

(f) Sound level meters are relatively insensitive to vibration while magnetic or electrostatic fields have negligible effects.

(g) Wet road surfaces give increased noise levels (so traffic noise measurements are normally taken when the road surface is dry).

(v) The Noise measurement report

According to Bruel and Kjaer (1992), the following items should be included in a noise report:

(a) The model type and serial number of the sound level meter used.

(b) The date of the last laboratory calibration.

(c) A statement of on-site calibration before and after the measurement (including the calibration equipment used).

(d) A description of the area and of the sound sources, including the type of sound (continuous, intermittent, impulsive, pure tones).

(e) A sketch of the measurement site showing the locations of the microphone and the sound sources.

(f) The time and date of the measurement.

(g) A description of the meteorological conditions.

(h) The background noise level if measuring noise intrusions.

(i) The names of the persons involved.

(j) A summary of the levels of the various noise criteria for the relevant time periods.

Environmental noise measurement reports typically include tabular values of the noise criteria levels for each hour of the measurement period and graphs showing the variation of these criteria throughout the measurement period.

3.6.7 Noise Control

Noise control may involve attention to one or all of the three elements involved. Noise is not merely a nuisance but a serious environmental problem and a health hazard.

Source ⟶ Transmission path ⟶ Receiver

(i) **Source:** It could be modified by the acoustic treatment to machine surfaces, design changes, etc. An offending noise source could be stopped or its operation limited to certain times of the day.

  (a) Designing, fabricating and using quieter machines to replace the noisy ones.

  (b) Proper lubrication and better maintenance of machines.

  (c) Installing noisy machines in soundproof chambers.

  (d) Covering noise producing machine parts with sound-absorbing materials.

  (e) Using silencers to control noise from automobiles, etc.

(ii) **Transmission path:** It could be modified by containing the source inside a sound insulating enclosure, by the construction of a noise barrier or by the provision of absorbing materials along the path.

(iii) **Receiver:** The protection of the receiver by altering the work schedule or by the provision of ear protection is mainly applicable to those working with noisy machinery. For people working in noisy installations, ear-protection aids like earplugs, noise helmets, headphones etc. must be provided to reduce occupational exposure.

3.6.7.1 Sound Insulation provided by buildings

(i) When sound pressure waves meet a wall or another building surface they exert a fluctuating pressure, causing it to vibrate. Sound is then radiated by the vibrating surface into the space on both sides of the surface.

(ii) Part of the sound energy is reflected, part is absorbed by the surface and the remainder is transmitted.

(iii) Vibrations will be larger at certain frequencies, the lowest of these being called the natural frequency of the element. Natural frequency depends on the weight, surface area and rigidity of the building element. Since natural frequencies of floors and walls tend to be low, it is much more difficult to insulate against noise with predominantly low frequencies.

3.6.7.2 Sound reduction index (Transmission loss)

(i) It is the amount of sound stopped by a building element such as a door.

(ii) It is dependent on the frequency of the sound but depends primarily on the weight and rigidity of the building element.

(iii) The actual sound reduction achieved may be less, due to cracks, ventilators or other leakage around the edges etc.

3.6.7.3 Acoustic zoning

(i) Increased distance between the source and receiver by zoning of noisy industrial areas, bus & railway stations, aerodromes etc. away from the residential areas would minimize noise pollution.

(ii) There should be silence zones near the residential areas, educational institutions, and hospitals.

3.6.7.4 Sound insulation at construction stages

(i) Sound travels through the cracks that get left between the door and the wall. For reducing noise this space should be packed with sound absorbing material.

(ii) Sound insulation can be done by constructing windows with double or triple pane of glass and filling the gaps with sound absorbing materials.

(iii) Acoustical tiles, perforated plywood etc. can be fixed on walls, ceilings, floors etc. to reduce noise (especially for sound proof recording rooms etc.)

3.6.7.5 Planting of trees

Noise absorbing plants such as Ashok, Banyan, Neem, Casuarina, Tamarind, Coconut etc. can be planted along road sides and in the front of buildings such as educational institutions, hospitals, libraries etc.

3.6.7.6 Legislative measures

(i) Minimum use of loudspeakers and amplifiers especially near silence zones.

(ii) Banning pressure horns in automobiles.

(iii) Framing a separate Noise Pollution Act.

3.6.8 Effects of Noise Pollution

The generation of unreasonable noise within the environment is regarded as a form of pollution because it lowers the quality of life. There have been several specific ways in which excessive noise can affect people adversely.

Noise has been found to interfere with our activities at three levels; (a) audio logical level in referring with the satisfactory performance of the hearing mechanism; (b) biological level interfering with the biological functioning of the body; and (c) behavioural level affecting the sociological behaviour of the subjects. Because of this the noise affects categorically, performance, physiology and psychology.

Noxious has been known to cause of nervous disorder, headache, high blood pressure and short memory. The various effects of noise may be as follows. Various psychological effect of noise pollution is summarized below:

(i) Depression and fatigue, which considerably reduces the efficiency of a person.

(ii) Insomnia as a result of lack of undisturbed and refreshing sleep.

(iii) Straining of senses and annoyance as a result of slow but persistent noise from motorcycles, alarm clocks, call bells, telephone rings etc.

(iv) Affecting of psychomotor performance of a person by a sudden loud noise (sound)

(v) It is a cause of frustration and is associated with difficulty in concentration, disturbance of rest, physical and mental fatigue. Low frequency noise of 50 to 60 dB affects the higher centre of brain and causes an alternation in the normal sleep pattern and prevents sound sleep.

(vi) Noise, which is an annoyance also causes irritation dis-satisfaction, dis-interest and affects work "performance. Noise has been reported both to improve and to decrease work efficiency, depending on its intensity, duration, frequency distribution etc.

3.6.8.1 Various physiological as well as pathologic effects of noise pollution

Noise pollution affects human health, comfort and efficiency. It cause contraction of blood vessels, makes, the skin pale, leads to excessive secretion of adrenalin hormone into blood stream with is responsible for high blood pressure.

(i) It causes muscles to contract leading to nervous breakdown, tension and even insanity.

(ii) Noise effects are anxiety, stress reaction and fright. These adverse reactions are coupled with a change in hormone content of blood, which in turn produces increased rate of heart beat, constriction of blood vessels, digestive sperms and dilation of pupil of eye.

(iii) The most immediate and acute effect of noise is the impairment of hearing, which diminishes by the damage of some part of auditory system. When exposed to very loud and sudden noise acute damage occurs to the eardrum. Prolonged exposure to noise of certain frequency pattern will lead to chronic damage to the hair cells in the inner ear.

(iv) Auditory fatigue appears in the 90 dB associated with whistling and buzzing in ears. Temporary deafness occurs at 4000-6000 Hz, and this effect is known as temporary threshold shift (tts), permanent loss of hearing occurs at 100 dB due to continuous noise exposure. Under such conditions, the auditory threshold shift is called Permanent Threshold Shift (PTS). Besides chronic hearing loss, there may be instantaneous damage or acoustic trauma, which may be caused by very high intensity impulsive noise resulting from an explosion or sudden excessive noise of more than 150 dB. Physiological effects of noise pollution include neurosis, hypertension, increase in sweating, hepatic diseases, giddiness, peptic ulcers, undesirable change in gastro intestinal activities behavioural and emotional stress.

(v) Noise mainly interferes with man's communication. It is easily visualized that a conversation can be carried on in whisper in a still place, while one has to shout to make sense in a noisy factory.

(vi) Blood gets thickened by excessive noise. Changes in breathing amplitude have also been reported due to impulsive noise.

(vii) Noise causes cosinophilia, hyperglycarmia, hypokalaemia and hypoglycarmia by a change in blood and other body fluids.

(viii) Noise causes chronic headache and irritability; work, which needs a high degree of skill, is considerably affected. The overall working efficiency goes down when noise level goes up.

(ix) Loud and sudden noise such as sonic boom produces a startle effect, which may damage the brain. Sonic booms can also cause physical damage to property i.e. windows may break due to it. Sudden noise can be much more harmful than a continuous noise.

REVIEW QUESTIONS

1. **What is humus? What is its significance?**

   Ans. Humus is the organic matter present in the soil formed by decomposition of dead plant and animal remains. It is an important element of the top soil that determines the soil fertility.

2. **What are the effects of noise on human health?**

   Ans. Noise has direct physiological effects such as hearing damage (including hearing loss and tinnitus, or ringing in the ears), as well as cardiovascular and hormonal disturbances. Indirect effects include sleep loss, interference with concentration and learning, mood changes and aggression, and social isolation.

3. **What is 'Eutrophication'?**

   Ans. 'Eutrophication' is a natural process which literally means 'well-nourished or enriched'. It is a natural state in many lakes and ponds which have a rich supply of nutrients from sewage, animal wastes and detergents.

Chapter 4

Water Treatment

4.1 INTRODUCTION

All naturally occurring water supplies are recharged by precipitation. Although generally considered pure, rain water may, and often does, contain impurities. These impurities, however, are much lower in concentration than that found in water once it reaches the earth's surface. Surface water presents many opportunities for the introduction of mineral and organic substances, microorganisms and other forms of pollution (contamination). Pollution as used in this context endures the presence of any foreign substances (organic, inorganic, radiological or biological) that tend to lower its quality to such a point that it constitutes a health hazard or impairs the quality of the water. When water runs over or through the ground surface, it may pick up soil particles which may be seen as cloudiness or turbidity. It also picks up particles of organic matter and bacteria. As surface water seeps downward into the soil and through the underlying material to the water table, most of the suspended particles are filtered out. This natural filtration may be partially effective in removing bacteria and other particulate materials. However, the chemical characteristics of the water may change and vary widely when it comes in contact with mineral deposits. As surface water seeps down to the water table, it dissolves some of the minerals contained in the soil and rocks. Ground water, therefore, often contains more dissolved minerals than surface water. The following four categories are used to describe drinking-water quality.

(i) **Physical:** Physical characteristics relate to the quality of water for domestic use and are usually associated with the appearance of water, its color or turbidity, temperature and in particular, taste and odour.

(ii) **Chemical:** The chemical characterization of drinking water includes the identification of the components and their concentrations.

(iii) **Microbiological:** Microbiological agents are important to public health and many also are significant in modifying the physical and chemical characteristics of water.

(iv) **Radiological:** Radiological factors must be considered in areas where the water may have come in contact with radioactive substances. The radioactivity of the water is of public health concern.

4.2 WATER QUALITY STANDARDS

There are two sets of water quality standards such as quality of raw water and quality of treated potable water:

(i) The quality of raw water is governed in the European Union (EU) by two directives.

(a) For **raw surface water**, Directive 75/440/EEC is known as the surface water directive intended for abstraction of drinking water.

(b) For **raw ground water**, Directive 80/68/EEC is known as the ground water protection directive against pollution by dangerous substances.

(ii) The drinking water directive 80/778/EEC relates to the quality of treated water intended for human consumption.

4.2.1. Water Quality Parameters

The EU water quality parameters are discussed as follows.

4.2.1.1 Organoleptic parameters

The parameters which can be sensed by human organs, i.e. eyes, nose and tongue are called **organoleptic parameters** like colour, turbidity, odour and taste. Colour in water is caused by dissolved minerals, dyes or humic acid from plants. Colour is measured in units of mg/l on the platinum cobalt (Pt/Co) scale. Raw water is of very good colour quality at values less than 10 mg/l, is acceptable at 100 mg/l. and is unacceptable at levels greater than 200 mg/l.

Turbidity is due to the presence of particulate matter and is a measure of the ability of water to scatter light. It is caused by the presence of very fine suspended or clay particles. Turbidity is measured in nephelopmetric turbidity units (NTU). Water is of very good turbidity quality if turbidity is less than 0.1 NTU. It is of acceptable turbidity if levels are less than 1 NTU and considered unacceptable if the value is greater than 5 NTU.

Odour and taste in water are caused by the presence of the by –products of plant and animal microorganisms, especially hydrogen sulphide.

4.2.1.2 Physicochemical parameters

Maximum admissible concentrations (MAC) for drinking water are:

Temperature should be 25°C

		$6.5 < pH < 8.5$
Conductivity	–	400 μs/cm
Chlorides	-	250 mg/l
Sulphates	-	250 mg/l
Calcium	-	100 mg/l
Magnesium	-	50 mg/l
Sodium	-	150 mg/l
Potassium	-	12 mg/l
Aluminum	-	0.2 mg/l
Total dry residues	-	1500 mg/l

Depending on its source, raw water may need particular treatment to satisfy eleven parameters listed above.

Table 4.1 EU drinking water directive parameters

	Group description	Parameter	Maximum admissible concentration (MAC)	Comment
A	Organoleptic parameters	Colour	20mg/l Pt/Co scale	
		Turbidity	10mg/l SiO_2	
		Odour	Dilution of 2 at 12°C	
		Taste	Dilution of 2 at 12°C	
B	Physicochemical parameters	Temperature	25 °C	
		pH	6.5< pH<8.5	GV
		Conductivity	400 μS/cm	GV
		Chlorides	250 mg/l CL	GV
		Sulphates	250 mg/l SO_4	
		Calcium	100 mg/l Ca	GV
		Magnesium	50 mg/l Mg	
		Sodium	150 mg/l Na	
		Potassium	0.2 mg/l k	
		Aluminium	0.2 mg/l Al	
		Total dry residues	1500 mg/l	
C	Parameters concerning substances undesirable in excessive amounts	Nitrates		
		Nitrites	50 mg/l NO_3	
		Ammonium	0.1 mg/l NO_2	mg/l GV
		Kjeldahl N	0.5 mg/l NH_4	
		Oxidizability	1 mg/l N	
		Hydrogen sulphide	5 mg/l O_2	
		Substances extractable in chloroform	Undetectable μg/l No increase above background	mg/l dry residue in GV
		Hydrocarbons	10 μg/l	
		Phenols	0.5 μg/l C_6H_5OH	GV
		Boron	1000 μg/L B	
		Surfactants	200 μg/l (lacryl sulphate)	GV
		Organochlorines	1 μg/l	
		Iron	200 μg/l Fe	
		Manganese	50 μg/l Mn	GV
		Copper	100 μg/l Cu	GV
		Zinc	100 μg/l Zn	
		Phosphorous	5000 μg/l P_2O_5/l	
		Fluoride	1000 μg/l F	

		Suspended solids	0	
		Barium	100 μg/l Ba	GV
D	Parameters concerning toxic substances	Arsenic	50 μg/l As	
		Cadmium	5 ug/l Cd	
		Cyanides	50 μg/l Cn	
		Chromium	50 μg/l Cr	(Total)
		Mercury	1 ug/l Hg	
		Nickel	50 μg/l Ni	
		Lead	50 μg/l Pb	
		Antimony	10 mg/l Sb	
		Selenium	10 mg/l Se	
		Pesticides	0.1 μg/l	
		PAHs	0.2 μg/l	
E	Microbial parameters	Total coliforms	MPN $\leq$ 1/100 ml	
		Faecal coliforms	MPN $\leq$ 1/100 ml	
		Faecal streptococci	MPN $\leq$ 1/100 ml	
		Sulphite reducing clostridia	MPN $\leq$ 1/20 ml	
F	Minimum required for softened water	Total hardness	>60 mg/l as $CaCo_3$	
		Alkalinity	>30 mg/l HCO_3	

GV=guide value

Example 4.1 Calculate the hardness in mg/l $CaCO_3$ of following water sample.

		Equivalent weight
Na^+	35	23
Mg^{2+}	9	12.2
Ca^{2+}	48	20
K^+	1	39

Hardness computed in the presence of Mg^{2+} and Ca^{2+}

$$\text{Hardness, mg/l } CaCO_3 = \frac{Mg^{2+}\ (\text{mg/L}) \times 50\ \text{mg/meq}}{\text{eq wt } Mg^{2+}} + \frac{Ca^{2+}\ (\text{mg/L}) \times 50\ \text{mg/meq}}{\text{eq wt } Ca^{2+}}$$

$$\text{Hardness} = 156.9\ \text{mg/l } CaCO_3$$

4.3 WATER TREATMENT PROCESSES

Objectives of Water Treatment

The water from the surface sources may have some characteristics which are unsuitable for human consumption, industrial use, commercial use etc. The following are some of those characteristics:

(i) It may be turbid.

(ii) It may contain colour.

(iii) It may contain acids, salts and gases which have corrosive action and may impart hardness to water.

(iv) It may contain bacteria which may cause water borne disease.

Therefore, the objective of water treatment is to remove those impurities and to make the water suitable for domestic, industrial and commercial uses (Table 4.2).

Table 4.2 Classes of Water Treatment

There are four classes of water treatment Table 4.2 and 4.3.

Class	Description	Source
A	No treatment	Borehole water, occasional upland water
B	Disinfection only	Borehole water, occasional upland water
C	Standard water treatment	Lowland rivers and reservoirs
D	Special water treatment	Some rural supplies (Fe, Mn), industrial water

(i) In **Class A,** there is a very short time of supply from source to user.

(ii) In **Class B**, disinfection is practical to maintain purity along the supply pipeline. Also aeration is sometimes used to eliminate hydrogen sulphide odours and tastes and increase the oxygen levels in the water.

(iii) In **Class C**, the standard water treatment includes pre-treatment, sedimentation and filtration, disinfection and possibly fluoridation.

(iv) In **Class D**, the special water treatment include additional processes like membrane technology, iron and manganese removal, chemical oxidation, carbon adsorption etc.

Table 4.3 Flow Chart Outline of unit process in different raw water classes

Treatment Categorization	Groundwater supply Class B	Standard water treatment Class B	Special water treatment Class B
	Intake	Intake	Intake
Pre-treatment	Aeration	Coarse screening Fine screening Pumping Storage Equalization Neutralization Aeraton Chemical pre-treatment	Screening Pumping Storage Equalization Neutralization Aeration Softening Algae removal
Primary treatment		Coagulation & mixing Flocculation Sedimentation	Coagulation & mixing Flocculation Sedimentation
Secondary treatment		Filtration	Filtration
Disinfection	Disinfection	Disinfection	Disinfection
Advanced treatment			Adsorption Activated carbon Membrane process Fe & Mn removal Halogenated compounds removal
Fluoridation	Fluoridation	Fluoridation	Fluoridation
Distribution network	Yes	Yes	Yes

4.4 PRE-TREATMENT OF WATER

They may include screening, Storage: equalization and neutralization, aeration and chemical pre-treatment, softening, algae removal and pre-chlorination as discussed below:

4.4.1 Screening

Types of screens

(i) ***Coarse screens:*** They are typically inclined bars of 25 mm diameter and 100 mm spacing to prevent large floating materials from entering the treatment plant.

(ii) ***Fine screens:*** They are typically with mesh openings of about 6 mm diameter.

(iii) ***Micro screens:*** Here the mesh openings range from 20 to 40 μm. Such screens are used only as the main (physical) treatment process for relatively uncontaminated waters and moderately coloured waters.

4.4.2 Storage

Storage may be an open impoundment on a fast stream or a smaller reservoir to balance the flow going to the treatment plant. Storage tanks also assist in the treatment process as water allowed to sit in tanks will permit some of its suspended settleable matter to begin to settle. As such, storage tanks may act as initial settling tanks. Storage should be equivalent to 7 to 10 days of the average water demand. This period may be adequate to reduce most pathogens by exposure to daylight. The period should not be long, so as not to encourage the growth of other organisms, including undesirable algae. Water is allowed to settle for some days in a storage tank will be cheaper to treat at the plant. However, the cost of building storage tanks is expensive and their maintenance is also expensive. Another type of storage system is when tanks provide a storage time of about 12hrs. This is commonly used to reduce pumping costs and of using pumping at more economical night time electricity rates.

4.4.3 Aeration

Aeration is the supply of oxygen from the atmosphere to water to effect beneficial changes in the quality of water. It is commonly used process for groundwater and less common for surface waters. Aeration is used:

(i) To release excess H_2S gas which may cause undesirable tastes and odours

(ii) To release excess CO_2 this may have corrosive tendencies on concrete materials.

(iii) To increase the O_2 content of water in the presence of undesirable tastes due to photosynthetic algae (fishy smell), which release volatile oils on decomposition.

(iv) To increase the O_2 content of water which may have negative taste, colour and stain properties due to the presence of iron and manganese in solution. The addition of oxygen assists the precipitation of iron and manganese.

4.4.4 Chemical pre–treatment

In pre-treatment, greater amounts of chemicals are required to affect the same result as some of the chemical is masked and absorbed by turbidity in the water. Only two chemical pre-treatment processes such as pre-chlorination and activated carbons are discussed here:

4.4.4.1 *Pre – chlorination*

This process is used on low turbidity water with a high *Coliform* count. Chlorine is injected into the water stream and stays in the settling tanks for over a period, oxidizes and precipitates iron & manganese. It also pathogens and reduces colour. Doses of 5mg/l should be used (distinct from 1mg/l in post-chlorination) and this is expensive. Pre-chlorination is also beneficial in the reduction of ammonia from both surface and groundwater supplies.

4.4.4.2 *Activated carbon*

The addition of activated carbon as an adsorbent is used for many purposes including:

(i) Removal of photosynthetic algae.

(ii) Improvement of colour and odour.

(iii) Removal of selective organic compounds.

Activated carbon can be used either as powdered activated carbon (PAC) or granular activated carbon (GAC). PAC has a lower capital cost but also a lower efficiency than GAC. Doses may vary from 3 to 20 mg/l. The PAC gets deposited in the sand filters and the water impurities causing the undesirable tastes and odours are adsorbed to the PAC during filtration. PAC is normally used for intermittent control of taste and odour problems only when problem seems to be persistent, GAC is used. GAC consists of a specifically designed GAC filter bed which is used in either the up flow or down flow modes. However, the use of GAC beds is more applicable to advanced water treatment processes in the production of very high quality water.

4.5 CONVENTIONAL METHODS

4.5.1 Sedimentation

Sedimentation is the solid-liquid separation using gravity settling to remove suspended solids (Reynolds, 1982).

4.5.1.1 Theory of sedimentation

The particles which do not change their shape, size and weight while settling down in a fluid are known as **discrete particles**. The suspended impurities in water consist of discrete particles such as inorganic and organic solids having specific gravity of about 2.65 and 1.04 respectively. The particles having specific gravity more than 1.20 readily settle down at the bottom of the tank due to the force of gravity. This phenomenon of settlement is known as **hydraulic subsidence**. But the lighter particles cannot settle down by force of gravity. Such particles are converted to settle able size by the application of some coagulant in water. The process of settlement depends upon the following factors:

(i) Velocity of flow

(ii) Size & shape of particle

(iii) Viscosity of water

After long investigation with the nature of discrete particles, Mr. G.G. Stokes developed an expression for the velocity of settlement. The expression is as follows:

$$V = 418(S - S_1)\, d^2 \left[\frac{3T + 70}{100}\right]$$

Where, V = Velocity of flow in mm/sec

S = Specific gravity of particle

S_1 = Specific gravity of water

d = Diameter of particle

T = Temperature in °C

The particles are given maximum opportunity to settle down by decreasing the velocity of flow and increasing the length of travel (i.e. by increasing the length of sedimentation tank).

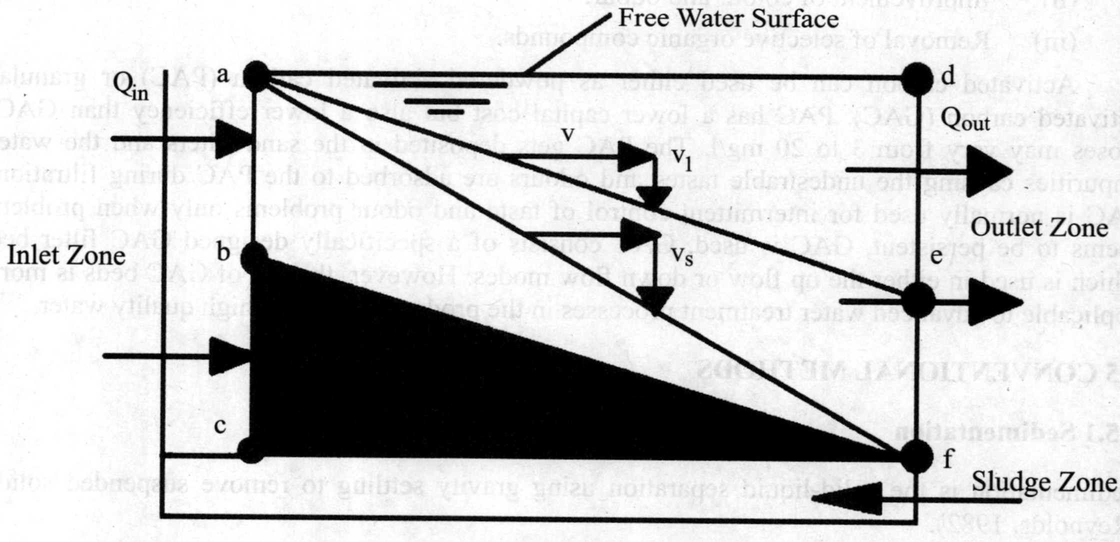

V = Horizontal velocity V_s = Stokes settling velocity W = Width

L = Tank length H = Tank Height-Sludge Depth

Fig. 4.1 Elevation of Type 1 Settling Tank

4.5.1.2 Purpose of sedimentation

The following are the purposes of sedimentation:

(i) To reduce the heavy sediment load before the water enters the coagulation tank.

(ii) To make the process of coagulation more easy.

(iii) To reduce the amount of coagulant.

(iv) To reduce the cost of coagulation process.

(v) To make the working of other treatment processes more efficient.

In water treatment, sedimentation processes used are:

(i) ***Type-I:*** To settle out discrete non-flocculent particles in a dilute suspension. This may arise due to the plain settling of surface waters prior to treatment by sand filtration (Fig. 4.1).

(ii) ***Type-II:*** To settle out flocculent particles in a dilute suspension. This may arise after chemical coagulation and flocculation where the non-discrete particles are chemically assisted to coagulate.

Other types of settling are combinations of Type I & II.

In type-I settling, the particles settle out individually and it is assumed that there is no flocculation or coagulation between the particles. It is typically the first physical process of surface waters where discrete organic and grit particles are given adequate time to settle. In type-I settling, a particle will accelerate vertically downwards until the drag force F_D equals the impelling force F_1 and thereafter the particle settles at a constant velocity known as **Stokes' velocity** Vs the impelling force:

$$F_1 = (Y_s - Y_w)V_p \qquad (1)$$

Where, Ys = weight density of solid particles $\rho_s g$

Yw = weight density of water $\rho_w g$

V_p = volume of particle

The drag force

$$F_D = C_D A_S \rho_w \left(\frac{V_s^2}{2}\right) \qquad (2)$$

Where, C_D = drag coefficient ≈ 0.4 for spheres

$$C_D = \frac{24v}{V_s d} \text{ for laminar flow for Re <100} \qquad (3)$$

As = sphere section area orthogonal to velocity vector

Vs = settling or Stokes' settling velocity

v = kinematic viscosity = μ / ρ

Equating, $F_1 = F_D$ – (4)

$$(Ys - Yw)\frac{\lambda}{6}d^3 = \frac{24v}{V_s d}\frac{\lambda}{4}d^2\, Pw\left(\frac{Vs^2}{2}\right) \qquad (5)$$

Solving, $$Vs = \frac{g}{18\mu}(Sp-1)d^2 \quad (6)$$

Where, S_p = Specific gravity of particles

Or, $$V_s = \frac{g}{18\mu}(\rho_s - \rho_w)d^2 \quad (7)$$

This is known as **Stokes' law** for settling velocity of discrete particles and applies for $R_e < 0.5$. A tank for preliminary settling of raw water is assumed to behave as follows:

(i) Type I settling applies.

(ii) The flow entering and leaving the tank is uniform.

(iii) There are three zones within the tank:

a. An inlet zone

b. An outlet zone

c. A sludge zone

(iv) The particle distribution throughout is uniform.

(v) The particles on entering the sludge zone stay there until scraped off the bottom.

4.5.1.3 Sedimentation of flocculent particles – Type –II

Flocculent particles are those particles that are chemically assisted to come together and produce large particles, and thus settle. Coagulation is the first process of adding the coagulated chemical which changes the particles electric charge and is then amenable to aggregation. Flocculation is the second process of getting the coagulated mix to form larger flocks. While the particles are settling, they are also flocculating and so increase in size and mass during the settling process (Fig. 4.2). This phenomenon occurs in settling of chemically coagulated potable water and wastewater.

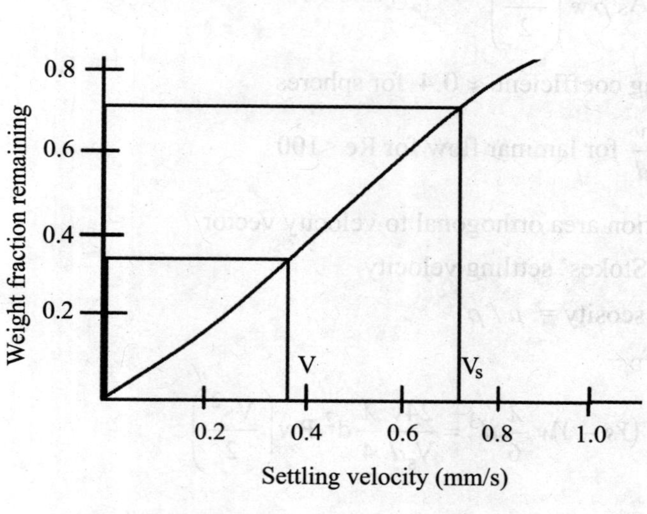

Fig. 4.2 Typical Discrete Particle Settling Curve

4.5.2 Coagulation and flocculation

4.5.2.1 Coagulation

The theory of coagulation can be explained by the following phenomenon:

(i) **Floc formation:** When some chemicals (known as coagulants) are mixed with water thoroughly a thick gelatinous precipitate is formed which is known as **floc** due to destabilization of the initially stable suspended impurities (followed by inter particle bridging). This floc has got the property of attracting the suspended impurities in water and settle down towards the bottom of tank.

Before identifying the ideal coagulant, specific properties of the suspended particles (impurities) are identified. These are: its classification and electric change.

(a) **Classification** determines whether a particulate in suspension has an affinity for water adsorption are hydrophilic and those that do not adsorb water are termed hydrophobic.

(b) **Electric charge:** It is observed that the ions of floc possess positive electric charge and the colloidal particles possess negative electric charge. So, the floc attracts the colloidal particles while it travels towards the bottom of tank.

(ii) **Coagulants:** The three most common coagulants are- aluminum sulphate (alum), ferrous sulphate (ferric) and ferric chloride,

As aluminum sulphate is corrosive in nature, it is packaged with H_2O. When alum is added to raw water, the following reactions take place:

$$Al_2\,(SO_4)_3\,.14\,H_2O + 3Ca\,(HCO_3)_2 \rightarrow 2Al\,(OH)_3 + 3Ca\,(SO_4) + 14\,H_2O + 6CO_2 \qquad -(1)$$

The object is the production of a floc, and in the above case this is an aluminum hydroxide floc. However, if there is insufficient alkalinity in the water, alkalinity is added by means of lime (calcium hydroxide) addition.

This reaction is all follows:

$$Al_2\,(SO_4)_3\,.14H_2O + 3Ca\,(OH)_2 \rightarrow 2Al\,(OH)_3 + 3Ca\,(SO_4) + 14H_2O \qquad -(2)$$

Alum calcium hydroxide aluminum hydroxide calcium sulphate

The optimum pH range for aluminum hydroxide floc formation is 4.8 to 7.8, as within this range this floc is insoluble.

4.5.2.1.1 *Coagulant aids*

Depending on the raw water quality, microfloc promotion may be inadequate by chemical coagulants alone and may require the addition of coagulant aids (Fig. 4.3 a and b). They essentially are:

(i) Polyelectrolytes

(ii) Lime addition

(iii) pH correction: lime, sulphuric acid

Polyelectrolytes are long – chain synthetic organic chemicals (SOC) which are used to optimize coagulation. They can be cationic, anionic or polyampholites and the huge range and strengths available allow almost every conceivable coagulant requirement to be satisfied.

High turbidity and high alkalinity water are easy to treat, as flocs form readily. High turbidity and low alkalinity water may need lime to improve the alkalinity and optimize coagulation. Low turbidly and low alkaline water is difficult to treat, requiring alkalinity correction with lime and high doses of high molecular weight polyelectrolyte.

4.5.2.2 Flocculation

The phenomenon of the formation of floc is termed as **flocculation**. The efficiency of flocculation depends on the following factors:

(i) **Dose of coagulant:** The dose of coagulant should be ascertained by laboratory test for the sample of water.

(ii) **Mixing:** The mixing of coagulant should be perfectly done for the quick formation of floc.

(iii) **pH value:** The pH value of water should be determined in laboratory to select the type of coagulant.

While settling down, the floc attracts more and more suspended impurities and thus the size of floc goes on increasing and ultimately the surface area of floc becomes sufficiently wide to arrest colloidal matters, organic matters and some amount of bacteria (Fig. 4.3 and 4.4).

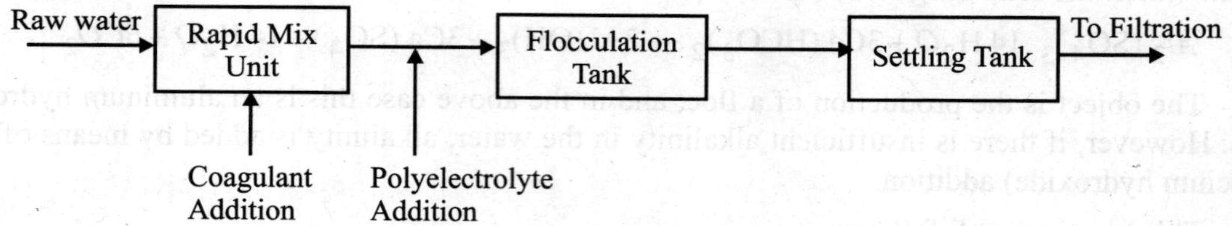

(a)

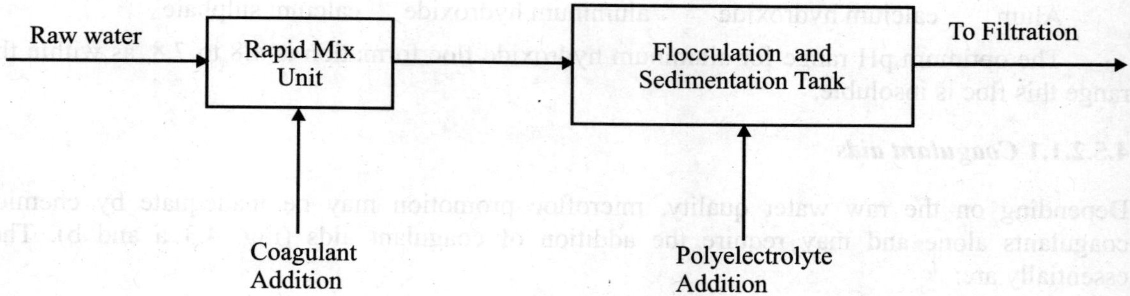

(b)

Fig. 4.3 Primary Water Treatment: [a] US Practice [b] UK Practice

4.5.3 Filtration

Filtration is the process of passing water through a porous medium to get better quality water than the influent. The theory of filtration is explained by the following phenomenon as discussed below:

(i) **Mechanical straining:** The voids of sand grains behave like a fine sieve. So, when the water is allowed to pass through the sand bed, the suspended particles are arrested by the voids and the clear water is allowed to flow towards the bottom.

(ii) **Sedimentation:** The voids of the sand grains act like small sedimentation tanks. The fine suspended particles are arrested in these tanks by the gelatinous film developed during the process of filtration.

(iii) **Biological action:** When the water is allowed to pass through the sand bed, a film of zoological jelly is formed around the sand grains. The bacteria are caught in these voids. The zoological film helps to develop bacterial colonies. The bacteria consume the organic impurities for their survival. Thus the harmful organic matters are converted to harmless compounds by the biological action.

(iv) **Electrolytic action:** The process of filtration may be explained by the ionic theory, where two opposite electric charges are found to be neutralizing each other when they come in contact. It is found by observation that the sand grains of filter and the colloidal suspended particles possess electrical charge of opposite polarity. Hence, they attract one another and neutralize themselves. Ultimately, the chemical characteristics of water are changed while passing through the bed of sand.

4.5.3.1 Classification of filtration systems (Table 4.4)

FILTRATION

- Gravity Filtration
  - Slow Sand Filtration
  - Rapid gravity Filtration
- Pressure Filtration
- Cake or depth Filtration

(i) Gravity filtration

Here the water percolates through the filters by the force of gravity.

4.5.3.1.1 *Slow sand filtration*

A slow sand filter (SSF) is a rectangular concrete open box structure containing (Fig. 4.4):

(i) Supernatant layer of raw water.

(ii) Bed of fine sand, supported on a thin layer of gravel.

(iii) System of under drain.

(iv) Inlet and outlet structures.

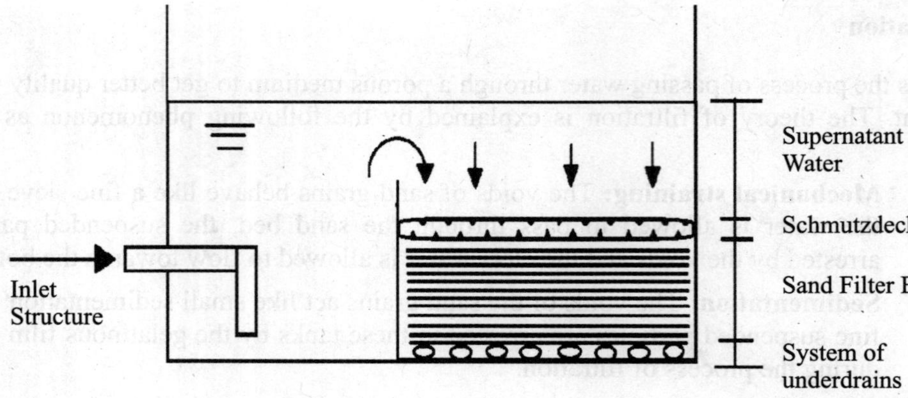

Fig. 4.4 Slow Sand Filter

(a) Supernatant layer: The inlet structure allows the water to flow on to the schmutzdecke layer without damaging it. This layer is at the top of the sand bed and is composed of living and dead microrganisms. The inlet structure also facilitates the drainage of the supernatant water during the cleaning process. The supernatant water provides a head of water sufficient to drive the water through the sand filter while creating a detention period of several hours for the raw water.

(b) Filter media of sand: The fine size of 0.15 to 0.3 mm with a uniformity coefficient no. greater than 2 is generally used as the filtering media. The depth of sand layer varies from 60 – 100 cm.

(c) Base materials: Clean gravels of different sizes are used as base materials. It is placed over the under drain system.

(d) Appurtenance: The following appurtenances are provided:

- A vertical air pipe is passed through the filter media for proper functioning of filter.
- A device for measuring loss of head.
- An adjustable telescopic tube to maintain constant discharge.

4.5.3.2 Filtration mechanism

The removal of impurities is thought to be primarily in the Schmutzdecke layer. The mechanisms are thought to be both physical and biological, the latter contributing to the fact that many micro-biological parameters improve significantly on flowing through the slow sand filter. Water is allowed to stand over the sand bed to depth equal to the depth of filter media. This depth is known as filtering head. Water percolates through the sand bed and gets collected in the outlet chamber through the under drain system. The water from the outlet chamber is then taken to the next unit for further treatment.

4.5.3.3 Cleaning operation

As the process of filtration goes on, the filtering media gets clogged due to the impurities and the loss of head goes on increasing gradually. When the loss of head reaches a certain limit, the working of filter is stopped. At that time the filter requires cleaning. For the purpose of cleaning, the top layer of sand is scrapped to a depth of about 25 mm and replaced by clean sand before the filter is started again about two days later.

4.5.3.4 Rate of filtration

The rate of filtration in a slow sand filter is very low. It is generally 100 to 200 liters per hour per square meter of surface area of filter.

4.5.3.5 Efficiency

(i) Turbidity: It can remove turbidity to the extent of 0.5 to 0.1 NTU

(ii) Colour: It can remove colour to the extent of about 25%.

(iii) Bacteria: It can remove bacteria to the extent of about 95%. Slow sand filters have filter runs varying from 2 to 6 months.

4.5.3.1.2 *Rapid gravity filters*

Rapid gravity filtration (RGF) is used to filter chemically coagulated water and so produce a high quality drinking water.

4.5.3.1.2.1 *Theory*

It is observed that the rate of filtration is more in coarse sand than that in fine land. So, the theory of rapid gravity filter is based on the principle of increasing the rate of filtration by providing coarse sand as filter media. The filtration head is also increased to increase the pressure head and the rate of filtration.

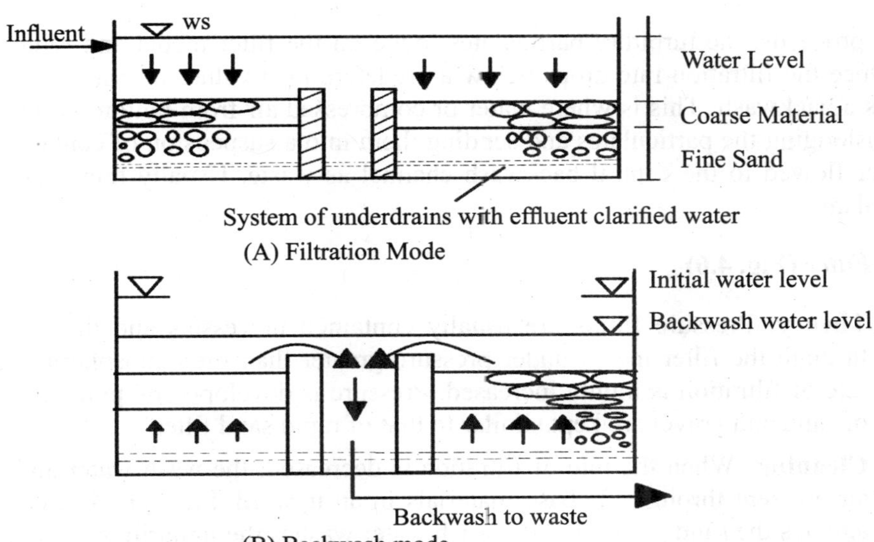

Fig. 4.5 Schematic of Rapid Gravity Dual Media Filter : (a) down flow filtration mode; (b) up flow backwash mode

4.5.3.1.2.2 *Filtration mechanism*

Filtration removes suspended particles (turbidity) by the simple physical method of filtration. Some biological activity may occur in the breakdown of ammonia to nitrate as in nitrification:

$$NH_4^+ + 2O_2 \xrightarrow[\textit{Nitrobacter}\text{ bacteria}]{\textit{Nitrosomans}\text{ and}} \underset{\text{Nitrate}}{NO_3^-} + 2H^+ + H_2O$$

The solid removal mechanisms in filtration are a combination of settlement, straining, adhesion and attraction. Thus particles, much smaller than the interstitial spaces between sand grains is removed. Both RGF and SSF operate in a mode where the water filters are vertically down through the media under gravity.

Because of the increased filtration rate, filter runs on rapid filters are 20 to 60 hours. Cleaning is achieved by agitating the bed either mechanically or with compressed air and washing water upwards through the bed to the surface, from where it is decanted as wastewater: This backwash water is then wasted or returned to the beginning of the plant for treatment.

4.5.3.1.2.3 *Types of rapid gravity filters*

(i) Single medium, usually sand or anthracite.

(ii) Dual medium, usually sand and anthracite,

(iii) Multimedia, usually garnet, sand and anthracite,

In the latter two, the top media is the coarse material of anthracite, followed by sand and the final material of garnet.

4.5.3.1.2.4 *Washing of filters*

As filtration proceeds, the turbidity particulates lodge on the filter media and reduce efficiency over time. Once the filtration rate drops below a predetermined value (or time) the system stops and performs a backwash. This is where water or compressed air from bottom is sent up through the media, dislodging the particulates and sending them into a suspension of contaminated water, which is over flowed to the central backwash channel as waste. Usually this water is recycled through the plant.

(i) *Pressure Filter* (Fig. 4.6)

(a) **Theory:** Pressure filters are usually contained in vessels and the water is forced through the filter media under pressure greater than the atmospheric pressure. The rate of filtration is highly increased. Pressure is developed by pumping. The quality of sand and gravel are kept similar to that of rapid sand filter.

(b) **Cleaning:** When the rate of filtration is decreased, the wash water and compressed air are sent through the base materials in an upward direction. The compressed air agitates the sand grains and the wash water washes the impurities in sand grains.

(c) **Efficiency:** The pressure filter is less efficient than rapid sand filter in removing turbidity, colour and bacterial load.

It is not suitable for public water supply projects. It is suitable for industrial plants, private estate, small colonies etc.

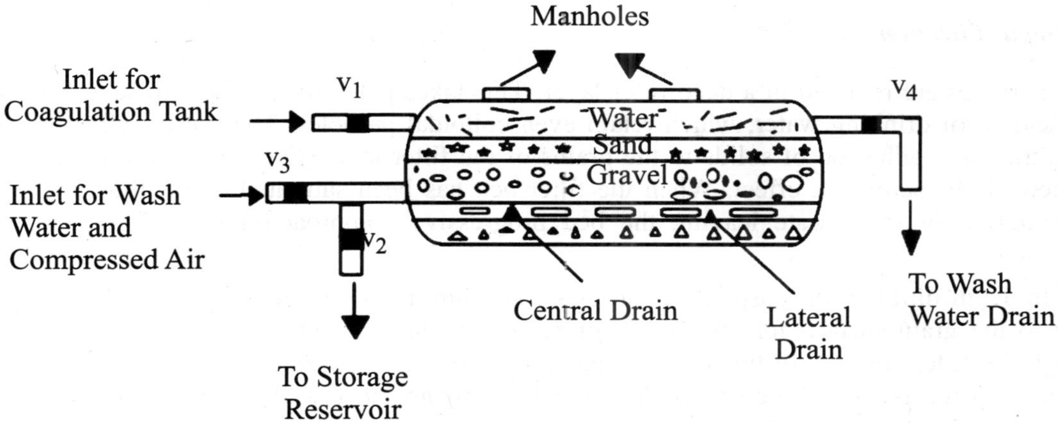

Fig. 4.6 Pressure Filter

Table 4.4 Properties of Slow Sand Filter and Rapid Sand Filter

No.	Particulars	Slow Sand Filter	Rapid Sand Filter
1.	Area	Requires large area for installation	Requires small area for installation
2.	Quality of sand	Fine sand, effective size 0.25-0.35 mm, uniformity coefficient 2 to 2.75	Coarse sand, effective size 0.35-0.65mm, uniformity coefficient 1.20 to 1.80
3.	Base materials	Gravel size varies from 3-65 mm	Gravel size varies from 3-40 mm
4.	Method of cleaning	Scraping top layer of sand to a thickness of about 25 mm	Back washing by water and compressed Air
5.	Period of cleaning	1 to 3 months	2 to 3 days
6.	Supervision	Skilled supervision not required	Skilled supervision required
7.	Rate of filtration	100 to 200 lit per hr per m^2 of surface area of filter	3000 to 6000 lit. per hr per m^2 of surface area of filter
8.	Efficiency	Highly efficient in removing bacteria	Less efficient in removing bacteria
9.	Economy	It is not economical	It is economical
10.	Sustainability	It is suitable for towns	It is suitable for large cities

(ii) *Cake filtration*

Cake filtration is the most frequently used model. Here it is assumed that the solids are deposited on the upstream side of the filter medium as a homogeneous porous layer with a constant permeability. As soon as the first layer of cake is formed, the subsequent filtration takes place at the top of the cake and the medium provides only a supporting function. Thus, if the flow rate dV/dt is constant, the pressure drop will increase linearly, proportional to the quantity of solid deposited. This model can be applied especially for all hard, particulate solids.

(iii) *Depth filtration*

Solid particles are retained in a deep filter layer. This takes place for example in sand filters for clarification of drinking water, which retain even colloidal particles. The typical effect of deep bed filtration is adhesion of solids to the grains of the filter layer. Only rather big particles are retained by the screening effect. When the filter bed has been saturated with solids, the solids concentration in the filtrate leaving the bed progressively approaches that of the incoming suspension.

In depth filtration the suspended particles enter into the porous medium (called grains) and move to the grain surface for attachment or re-entrain into the fluid, repeating this procedure through the filter column. In this case the medium provides the surface area for attachment and cake growth forms around the grains. As the cake deposits and fills the open volume inside the column, the porosity, flow rate, and pressure drop all change with time (example: sand filters).

The processes in depth filtration and cake filtration can be observed and investigated. The suspension (water and diatomite as the solid) flows from the hopper into the top of the filter element, where the solids are separated off. The filtrate flows through a flow meter into the drain. The filter element has a porous filter medium at the bottom. In cake filtration, the filter medium provides the foundation for build-up of the filter cake. In depth filtration, the filter medium supports the bulk solids (filter medium layer; gravel). Twin tube manometers measure the pressure loss over the filter element. The well-structured instructional material sets out the fundamentals and provides a step-by-step guide through the experiments.

Example 4.2 In a water treatment settling unit, water having temperature of 20°C carries solid particles with an average diameter of 0.05 mm. and specific gravity 1.2. Using stokes law, calculate settling velocity of settling particles.

Solution

For a temperature of 20°C

γ = 1.01 centi–stokes

d= 0.05 mm

s= 1.2

Substituting in stokes law, $V = \dfrac{\dfrac{g}{18(s-1)d^2}}{\gamma}$

$$= \frac{981 \times 10}{18\,(1.2-1)\,\dfrac{(0.05)^2}{1.01}} = 0.27 \text{ mm/sec.} = 0.027 \text{ cm/sec. (Ans)}$$

Example 4.3 Calculate the %age distribution of HOCL in water at 20°C and pH 8.0.

Solution. The %age distribution of HOCL is given by;

$$\frac{\text{HOCL}}{\text{HOCL} + \text{OCL}} \times 100$$

We know,

$$\frac{([\text{H}])^{+}\ (OCL)^{-}}{(\text{HOCL})} = \text{Ki}$$

$$=> \frac{\text{OCL}^{-}}{\text{HOCL}} = \frac{\text{Ki}}{\text{H}^{+}}$$

Taking $\text{Ki} = 2.5\times 10^{-8}$ moles/lit. At 20°C

$$\text{H}^{+} = 10^{-8} \text{ At pH 8.0}$$

$$\%\text{age HOCL} = \frac{100}{1+\dfrac{\text{OCL}^{-}}{\text{HOCL}}} = \frac{100}{1+\dfrac{\text{Ki}}{\text{H}^{+}}} = \frac{100}{\dfrac{1+2.5\times 10^{-8}}{10^{-8}}} = \frac{100}{3.5} = 28.6 \qquad \text{(Ans)}$$

Example 4.4 Find out the design parameters of a circular tank for daily flow of 13000 m^3. Design dimensions are diameter (d) 26 m and water depth (h) 2.1 m.

Solution. For a circular tank volume (V) is to be calculated as $V=d^2\ (0.011\ d + 0.785\ h)$

$$\text{V} = (0.11 \times 26 + 0.785 \times 2.1) \times 26^2 = 1307.7 \text{ m}^3$$

$$\text{Detention time } \phi = \frac{\text{V}}{\text{Q}} = \frac{1307.7\text{m}^3}{\dfrac{13000m^3}{24}} = 2.41 \text{ hours within 2-8}$$

$$\text{Plan area } \text{A}_\text{p} = \frac{\pi}{4}.\text{d}^2 = \frac{\pi}{4}\times 26^2 = 530.66\text{m}^2$$

$$\text{Surface Overflow Rate (SOR)} = \frac{13000\dfrac{m^3}{day}}{530.66m^2} = \frac{24.5\dfrac{\text{m}^3}{m^2}}{day}$$

Design range (20-35)

$$\text{Weir Overflow Rate (WOR)} = \frac{\dfrac{\text{Q}}{\text{perimeter (m)}}\text{m}^3}{\text{day}} = \frac{13000}{3.14\times 26} = 159.2 \text{ m}^3\text{/m/day}$$

Design range (150-300)

4.6 DISINFECTION

Disinfection refers to operations aimed at killing or rendering harmless the pathogenic micro – organisms. The process of destroying all micro – organisms (either harmful or harmless) is known as ***sterilization*** (Fig. 4.7). The requirements of a good disinfectant are that it should:

(i) Be toxic to micro-organisms at concentrations well below the toxic threshold to humans and higher animals.

(ii) Have fast rate of kill.

(iii) Be persistent enough to prevent re-growth of organisms in the distribution systems.

4.6.1 Disinfection Curve

The disinfection curve has three zones (Fig. 4.7):

(i) The initial lag phase

(ii) Exponential die off rate

(iii) Final slow phase

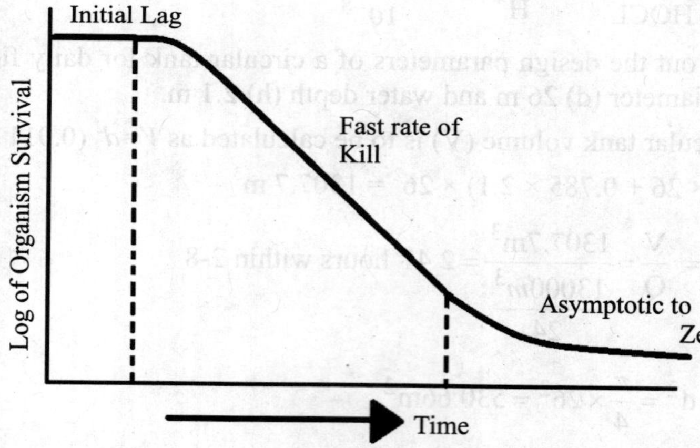

Fig.4.7 Disinfection Curve

4.6.2 Factors Resulting in Low Efficiency of Disinfection

(i) Turbidity

(ii) Resistant organisms (Giardia)

(iii) High amount of organic material

(iv) Deposits of iron and manganese

(v) Oxidizable compounds.

Viruses are more resistant to disinfectants than bacteria and need additional exposure time and higher concentrations. Turbidity producing colloids and iron and manganese deposits can shield organisms and use up the disinfectant.

4.6.3 Methods of Disinfection

4.6.3.1 Boiling

When water is heated to boiling temperature (100^0C) for at least 10 – 15 minutes, the bacteria is completely removed. Boiling also removes some of the dissolved salts. It is suitable for domestic purpose.

4.6.3.2 Iodine & bromine

Iodine & bromine also have the property of killing bacteria. So, sometimes these are used for disinfection. The dose of iodine or bromine should be 8 – 10 ppm. These chemicals are available in the form of small pellets. This method is suitable for small water supply installations like factories, institutions, military, survey camps etc.

4.6.3.3 Lime

Normally, lime is added to water to remove some of the dissolved salts. But, when excess lime is added to water, it is found to act as disinfectant. Excess lime increases the pH – value of water (i.e. increases alkalinity) which is detrimental to bacteria because the bacteria cannot resist alkalinity of water. It is found that the pH – value having 9 – 10 can remove bacteria to the extent of 99 percent. But after treatment the residual lime should be removed by the method of recarbonation.

4.6.3.4 Potassium permanganate

Potassium permanganate is a powerful oxidizing agent. It oxidizes the organic matters present in water and hence the bacteria are killed. This is mostly used for disinfecting the water of wells in village area, swimming pools, ponds etc. The dose of this chemical is about 2 – 3 ppm and the contact period is generally 2 – 3 hours.

4.6.3.5 Silver

Silver is found to be very effective in killing bacteria. Silver foils are spread over the filter media and water is passed through it. The water absorbs some portion of silver which kills bacteria. The dose of silver varies from 0.5 – 1 ppm. As silver is costly, it is not suitable for public water supply schemes. It is suitable for domestic use only.

4.6.3.6 Chlorine dioxide

At normal conditions, it is a yellowish or yellow – green or reddish gas that liquefies at approximately 10°C. The solution in water is not stable and degrades, especially when exposed to light. At higher temperatures it is explosive. Therefore chlorine dioxide must be produced on site before application.

It has been established that 50 percent of the used ClO_2 is transformed into chlorite (ClO_2^-) and chlorate (ClO_3^-). Studies of mutagenic activities indicate that at concentrations of < 1 mg chlorine dioxide per liter, no increase in the mutagenic activity in drinking water occurs, whereas there is a significant increase in mutagenic activity at higher doses.

4.6.3.7 Chloramines

When chlorine and ammonia (NH_3) are both present in water, they react to form products collectively known as **chloramines**. The inorganic chloramines consist of three species:

1. $NH_{3(aq)} + HOCl \longrightarrow NH_2Cl + H_2O$ (Monochloramine)

2. $NH_2Cl + HOCl \longrightarrow NHCl_2 + H_2O$ (Dichloramine)

3. $NHCl_2 + HOCl \longrightarrow NCl_3 + H_2O$ (Trichloramine)

The species formed as a result of the combination of chlorine and ammonia depend on a number of factors, including the ratio of chlorine to ammonia – nitrogen, chlorine dose, temperature, pH and alkalinity. Compared to chlorine dioxide or chlorine, chloramine is less effective disinfectants and has been described below:

(i) Have an algicidous effect.

(ii) Have a detrimental effect on taste and colour.

(iii) Show an efficiency that is heavily dependent on the pH value, a higher efficiency having been observed at lower pH values.

(iv) Do not react with organic material or phenols.

(v) Are persistent and provide continued protection against regrowth in the distribution system.

4.6.3.8 Ozone

Ozone, the allotropic form of oxygen, can be produced in a high – strength electrical field from oxygen in pure form or from the ionization of clean, dry air.

$$O_2 \xrightarrow{\text{High voltage}} O + O$$

$$O + O_2 \rightarrow O_3$$

Ozone is a powerful oxidant which reacts with reduced inorganic compounds and with organic material. The difference however is that an oxygen atom, instead of a chlorine atom is added to the organics, the end result being an environmentally acceptable compound. Once this ozone demand has been met, ozone reacts vigorously with bacteria and viruses. It is reported to be more effective than chlorine in inactivating resistant strains of bacteria and viruses.

(i) Since ozone is chemically unstable it must be produced on – site and used immediately. Typical dosages range from 1 to 5 g/m^3 with energy consumptions of 10 to 20 kwh per kg ozone. So costs of ozonation become 2 to 3 times higher than the costs for chlorination. Due to re-growth of micro – organisms, the treatment of water with ozone as the (last) disinfection step does not occur very often. Since no residual remains, it is necessary to use small amounts of chlorine after ozonization to provide continuous protection against re-growth in the distribution system. Properties of ozone as a disinfectant are:

(ii) Especially effective in killing viruses.

(iii) Improvement of odour and taste.

(iv) Transformation of almost non – degradable substances into easily degradable ones.

(v) Microflocculation effect.

(vi) Largely pH independent.

(vii) Re-growth of microrganisms within the water supply system due to the production of more easily degradable substances.

(viii) Formation of a number of toxic compounds.

(ix) No remaining residuals (of disinfectant).

4.6.3.9 UV Radiation

Irradiation with UV light is a promising method of disinfection. The properties of UV radiation as a disinfectant include:

(i) Necessity of having clear water (turbidity free) and thin sheets of water

(ii) No residual

(iii) Photo –oxidation of compounds may occur

(iv) No odour or taste problems

(v) No chemicals added

UV light spans the wavelength of 200 to 390 nm. The most effective band for disinfection is in the shorter range of 250 to 280 nm. This is the range where the UV light is absorbed by the DNA of the micro – organisms which then lead to a change in the genetic material so that they no longer can multiply.

4.6.3.10 Chlorination

The process of disinfection of water by the application of chlorine is known as chlorination. Chlorine is considered a very efficient oxidizing, bleaching and disinfecting agent. In water, chlorine reacts as follows:

$$Cl_2 + H_2O \longrightarrow H^+ + Cl^- + HOCl$$

HOCl is hypochloric acid. Depending on the pH value of water, HOCl can dissociate to hypochlorite:

$$HOCl \rightarrow H^+ + OCl^-$$

Hypochloric acid and hypochlorite ions represent the 'free chlorine residual' which is the primary disinfectant employed.

A recent theory states that the free chlorine partly unites with the cell of pathogen and forms some products known as chloro – products which are poisonous to the pathogen with subsequent blocking of an essential enzyme and destruction of cell wall takes place. Factors affecting the process are:

(i) Chemical form of chlorine
(ii) pH
(iii) Concentration
(iv) Contact time
(v) Type of organisms
(vi) Suspended solids
(vii) Temperature

The action of chlorine is pH dependent. At high pH (above 8.5) chlorine is practically inactive. But at low pH –value (below 7), chlorine is very effective in killing pathogens.

The dosage of chlorine should also be proper which can be determined by laboratory test (i.e. by break point chlorination).

Instead of adding Cl_2 gas to water it is also possible to use chlorine in an ionized state, e.g. as compounds $Ca\ (ClO)_2$, or $NaClO_2$. This procedure is chosen particularly when small amounts of water are to be treated with a relatively small expenditure on technical equipment. Another consideration for choosing hypochlorites is for safety reasons. Forms of chlorination has been described below:

(i) **Plane-chlorination:** When the raw water is supplied to consumers by applying chlorine treatment only, it is termed as **plain-chlorination**. In this case, the raw-water is sufficiently clear and it does not require primary treatments like primary sedimentation coagulation and filtration. Generally, in case of emergencies, for military camps, training camps, survey camps, etc. the process of plain-chlorination is adopted.

(ii) **Pre-chlorination:** In rainy season, when the water is suspected to be highly contaminated then a dose of chlorine is added to raw water before it enters the sedimentation tank. The application of chlorine is termed as **pre-chlorination**. It is an advance dose of chlorine before the actual period of chlorination. The idea of pre-chlorination is to reduces bacterial load before it enters the treatment plant. .

(iii) **Post-chlorination:** When chlorine is added to water after all the treatments are over and just before it enters the distribution system, it is termed as **post-chlorination**. This dose of chlorine is added to water to control the contamination when it flows through the distribution system.

(iv) **Double-chlorination:** When post-chlorination and pre-chlorination are both adopted in water supply scheme to safeguard the public health from epidemic, then it is termed as **double-chlorination**. This system is urgently needed in case of flood or other natural calamity when the possibility of bacterial contamination is very high.

(v) **Break Point-chlorination:** The break point-chlorination is a method of determining the chlorine demand of raw water. There is no chlorine demand in pure water (Fig. 4.8). If chlorine is added to such water, the chlorine will come as residual chlorine which is represented by a straight line R_1.

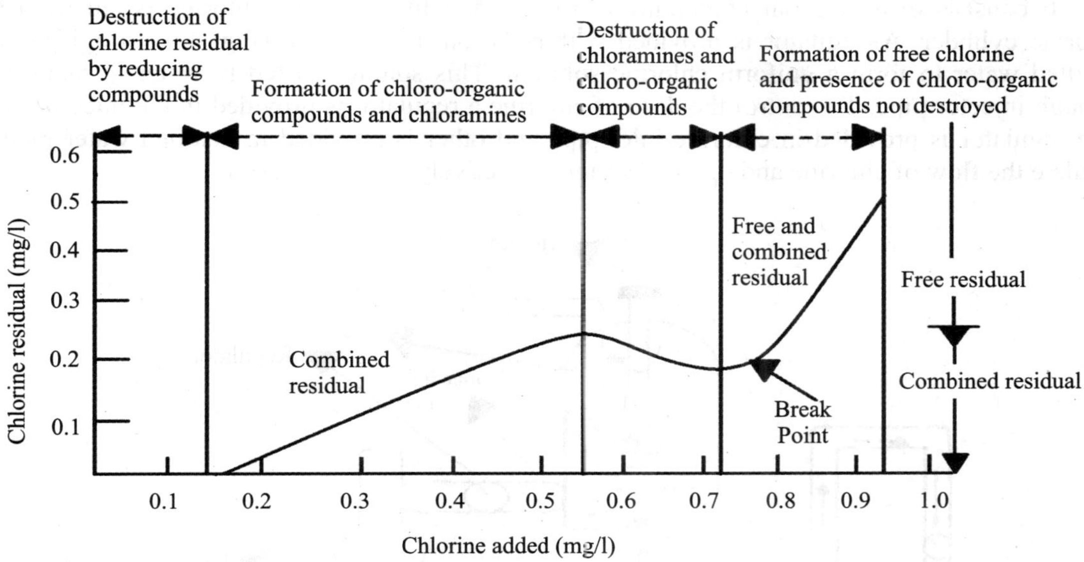

Fig. 4.8 Break Point-Chlorination

When chlorine is added to contaminated water, it will kill the bacteria and oxidize the organic maters. In the beginning, the chlorine continues to killing bacteria and then it starts accumulating up to a certain point. If dose of chlorine goes on increasing, it will start emitting bad smell but after sometime the bad smell observed to disappear. The critical point is known as **break point**. If further chlorine is added, it will come as residual chlorine which is shown by a straight line R_2. So, the actual dose of chlorine should be determined by carefully finding the break point formation.

(vi) **Super-chlorination:** The process of application of chlorine beyond the formation of break point is termed as **super-chlorination**. This treatment is necessary when there is an epidemic due to some water borne disease. But, in treatment much residual chlorine in water may appear which is unpleasant to consumers. To neutralize the residual chlorine to an allowable limit, de-chlorination should be done. The amount of residual chlorine in water may be determined by laboratory test.

(vii) **De-chlorination:** The process of removal of excess residual chlorine from water is known as **de-chlorination**. The residual chlorine should not be removed completely, but some amount of chlorine of about 0.25 to 0.50 ppm should remain in water so that it may not be polluted in distribution system. For de-chlorination, various types of chemicals may be used such as sodium sulphite, activated carbon, potassium permanganate, sulphur dioxide, etc.

4.6.3.10.1 *Mixing device of chlorine*

Chlorine is mixed with water with a device known as chlorinator. These are various types of chlorinators patented by different manufacturers. A simple type is described below:

It consists of a glass jar containing distilled water in which the chlorine is fed from the chlorine cylinder. An agitator is provided within the jar as shown in (Fig. 4.9). It agitates the distilled water to form a uniform chlorine solution. This solution is fed to water in main pipe through injector pipe. To control the dose of chlorine a regulator is provided in the injector pipe. One regulator is provided in chlorine inlet pipe and other is provided in distilled water pipe to regulate the flow of chlorine and distilled water respectively.

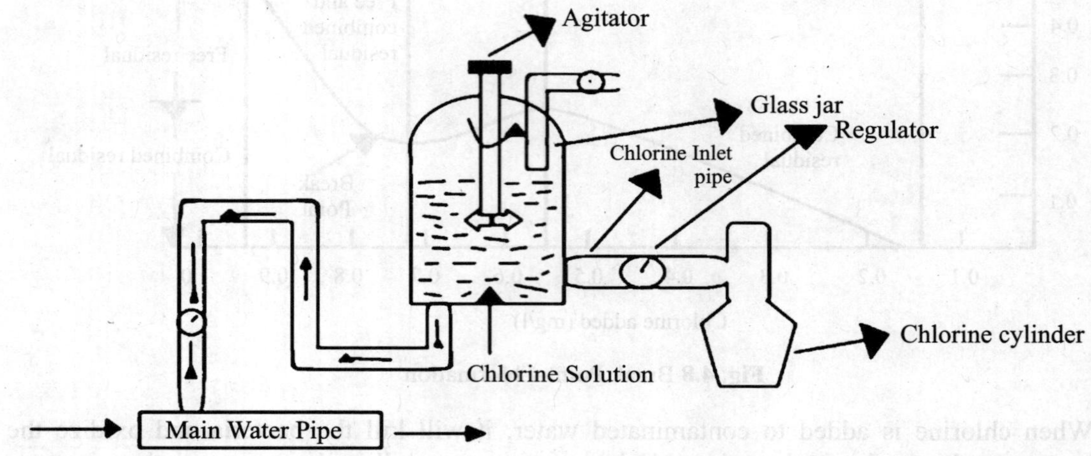

Fig. 4.9 Chlorinator

4.6.3.10.2 *Tests for residual chlorine*

The residual chlorine in water is tasted by the following tests such as orthotolidin test and starch-iodine test

(i) Orthotolidin Test: This test is carried out by the following procedure:

(a) 1 ml of orthotolidin solution is added to sample of water in a test tube.

(b) A yellow colour will be formed. The intensity of colour will indicate the amount of residual chlorine which should be ascertained by comparing with colour standard.

- Deep yellow-Indicates very high chlorine content.
- Medium yellow-Indicates moderately high chlorine content.
- Light yellow – Indicates slightly high chlorine content.
- Lemon yellow –Indicates safe for potable water.

(ii) Starch- Iodine Test: This procedure is carried out by following procedure:

(a) 10ml of potassium –iodide is mixed with 1 litre of sample of water. The mixture is agitated thoroughly.

(b) 5ml of starch solution is added to above solution and a blue colour is formed.

(c) This colour is removed by titration with sodium thiosulphate solution.

The chlorine content is ascertained as follows:

Chlorine content=0.3546 X Quantity of thiosulphate solution required. (In ppm)

4.6.3.11 Fluoridation

Fluoride is a natural trace element, found in small but widely varying amounts in waters. Ground waters are more likely to have higher fluoride content. However, most surface waters have negligible amounts.

At the optimum concentrations in potable water, fluoride reduces dental caries from 20 percent to 40 percent among children who ingest the water from birth. However, in the past decade, the relative beneficial impact of fluoridation is masked as other sources of fluoride have appeared, e.g. fluoride toothpaste, fluoride tablets etc. Fluoride has been added to water in different forms such as sodium fluoride, sodium silicofluoride and hydrofluosilic acid. The procedure in countries that have fluoride public water supplies is to add it in doses of 1 mg/L, and this is now mostly in the form of hydrofluosilicic acid.

4.7 ADVANCED WATER TREATMENT PROCESS

The purposes of advanced water treatment processes are:

(i) To take water treated by standard processes and to improve it to an exceptionally high quality as often required by particular industries eg. beverage, pharmaceutical.

(ii) To treat water containing specific chemical or microbiological contaminants to an acceptable standard e.g. the removal of iron and manganese, the removal of blue – green algae, the removal of specific organics.

4.7.1 Techniques for Advanced Water Treatment

(i) Iron and manganese removal

(ii) Ion exchange and inorganic absorption

(iii) Adsorption of organics

(iv) Oxidation including chemical oxidation

(v) Membrane processes including reverse osmosis

4.7.1.1 Iron and manganese removal

Both elements are essential for all living creatures. Iron plays an important role in the formation of the red blood cells haem (together with protein it is called haemoglobin). Manganese in animals is important for growth and healthy functioning of the nervous system. In spite of this fact, excessive quantities of both elements must be removed from water, as otherwise disturbance and damage within the distribution system would occur.

4.7.1.1.1 *Procedure by which Fe and Mn get into water*

Bacteria use organic material as food, converting it ultimately to CO_2 and H_2O. For that purpose, they need oxygen. The easiest way to obtain oxygen is to take it out of water. Therefore, a raindrop, for example gradually loses its oxygen on its way through the soil towards the groundwater table. After having used all oxygen in the water, the micro – organisms look for another source of oxygen. This is obtained by scavenging the oxygen from commonly occurring metal oxides, like Fe_2O_3 and MnO_2. Iron and manganese are therefore reduced to ferrous and manganous (Fe^{2+} and Mn^{2+}) compounds, which dissolve in water. Ground waters that do not contain any oxygen are called **reduced ground waters.**

4.7.1.1.2 *Removal procedures for iron*

If the amount of iron in water is fairly low (<1mg/L), the removal of iron is often achieved in a pressure filter containing the usual bed of sand, but sometimes incorporating a layer of proprietary material such as Polarite or an iron core which acts as a catalyst.

As an alternative to sand filtration, it is possible to use semi –calcined dolomite limestone as the filter medium (contains Ca / $MgCO_3$ and oxides). This material can be effective for the removal of soluble iron by precipitating ferric hydroxide at its alkaline surface and the material can help render the water less aggressive.

If the amount of iron is large (>10 mg/l), the traditional practice was to treat the water as a surface water (where the iron is dissolved as a colloid) providing a coagulation step before rapid sand filtration. This method requires high demand in equipment and personnel. It is now possible to decrease filtration rates to approximately 5 m/hr. or use a sedimentation tank to allow a large proportion of the ferric hydroxide to settle out. Multimedia filtration, involving the use of a filter containing several layers of different materials, has proven to be highly effective. Raw water with about 15 mg/L Fe and high concentrations of humic acids can be treated at filtration rates of 15m/h.

4.7.1.1.3 *Removal of Manganese*

Manganese occurs in natural waters as Mn^{2+}. There is no way to oxidize it by aeration in the normal pH range even if the amount is 10 mg/L or higher. The pH has to be 9 or 10 before oxidation takes place. This can be achieved only with strongly alkaline filter materials. For the removal of Mn^{2+}. It has proven favourable to add $KMnO_4$ solution as a conditioning agent prior to the filtration step without changing the pH. The stoichiometrically evaluated value of 1.9 kg $KMnO_4$ per kg Mn is often found to be too high (or too low), depending on simultaneously occurring reduction and oxidation processes. The addition of $KMnO_4$ has to be carried out for several days (3 days to 1 month) before the process works on its own.

4.7.1.2 Ion exchange

Hard water (> 50 mg/L $CaCO_3$) contains an excess of calcium and magnesium cations. In the ion exchange process, sodium ions are exchanged for calcium and magnesium ions on an equivalence basis and there is no net decrease is dissolved solids. For demineralization, however, the exchanged ions must not contribute dissolved solids to the effluent. This is accomplished by exchanging hydrogen for the dissolved cations and hydroxide for the dissolved anions. The two then combine in equal amounts to form H_2O_2 leaving no residual and not affecting the pH. The resins are regenerated with acids and bases respectively. If Na_2R is an exchange sodium resin (R is the complex base) then water softening is represented by:

$$Mg^{2+} + Na_2R \rightarrow MgR + 2Na^+$$

$$Ca^{2+} + Na_2R \rightarrow CaR + 2Na^+$$

The Ion – exchange process must be carried out in two or more steps (Fig. 4.10). Generally, the cations are removed first, followed by the anions. The process of water softening does not remove all the hardness, as minimum values are required for health purposes. The softening capacity of exchange resins varies from 100 to 1500 eq/m^3 (equivalents).

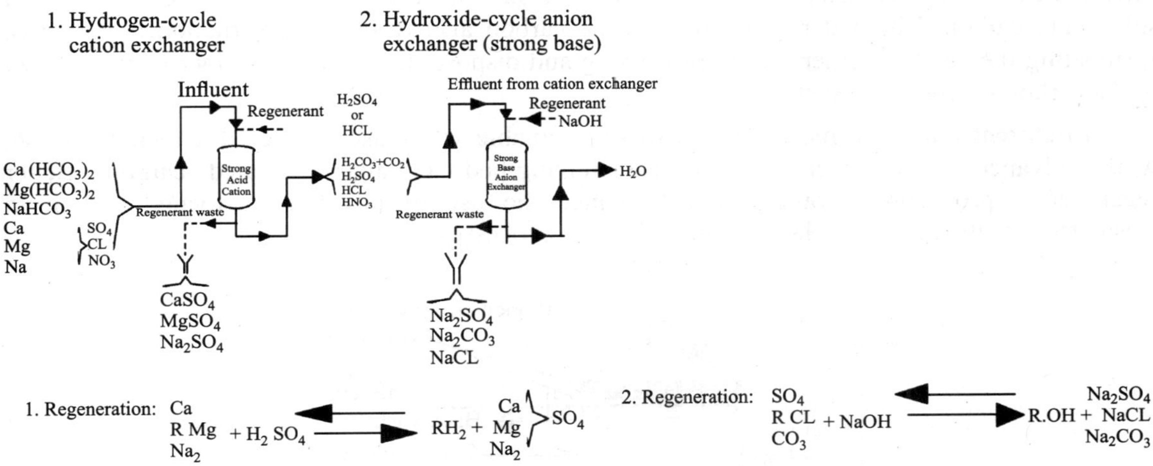

Fig. 4.10 Two-Stage-Ion-Exchange System

4.7.1.3 Adsorption

Adsorption can be defined as the physical and chemical process of accumulation of substances at an interface between the liquid and solid phases. The material removed from the liquid phase is called the ***adsorbate,*** and the material providing the solid surface is called ***adsorbent***. Adsorbents used in water treatment industry include:

(i) Activated carbon ------- PAC or GAC

(ii) Activated alumina

(iii) Clay colloids

(iv) Hydroxides

(v) Adsorbent resins

The adsorbent most commonly used in water and wastewater treatment is activated carbon. Activated carbon is crushed into granules (GAC) ranging from 0.1 to 2 mm in diameter or is pulverized to a very fine powder. Although adsorption properties and mechanisms are essentially the same, application techniques for GAC and PAC are considerably different. Dose rates range from 1 to 100 g/m^3.

The contact system for granular activated carbon (GAC) consists of a cylindrical tank which contains a bed of the material (Fig. 4.11). The water is passed through the bed with sufficient residence time allowed for completion of the adsorption process. The system may be operated in either a fixed – bed or moving – bed mode. Fixed – bed systems are batch operations that are taken off the line when the adsorptive capacity of the carbon is used up.

In wastewater treatment, powdered activated carbon can be added to the aeration basin and removed with the biological solids in the secondary clarifier. In this case, both the refractory and biodegradable organics are adsorbed. Biomass growth on the carbon surface utilizes the biodegradable fraction. Removal efficiency for biodegradable organics may be improved by this process, but usually at the expense of refractory organic removal efficiency. Use of powdered activated carbon in secondary wastewater systems results in an inseparable mixture of biological solids and carbon. Thermal regeneration of the carbon also results in destruction of biomass, eliminating the need for other sludge processing and disposal techniques, but increasing the size of the carbon regeneration system.

In current practices, most systems treating potable water use powdered activated carbon, while advanced wastewater systems use moving-bed, granular activated carbon. Better regeneration procedures would greatly enhance the use of powdered activated carbon in wastewater treatment, particularly in the secondary processes.

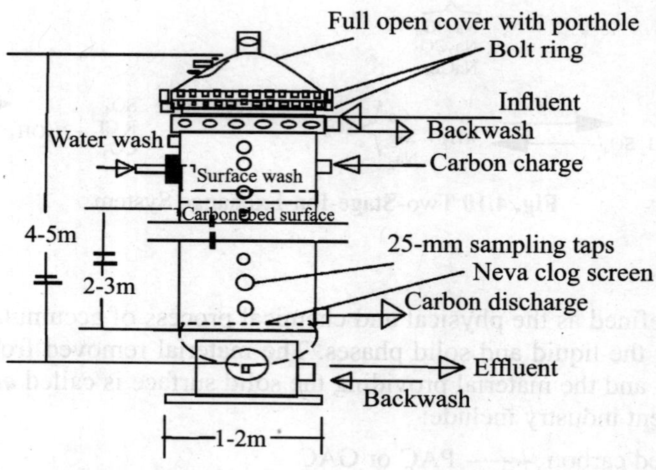

Fig. 4.11 Typical Activated Carbon Adsorption Column (adapted from Metcalf & Eddy, Inc. [4-40])

GAC is used as a final filter bed after sand filtration. It produces very high quality water and can reduce chlorine levels. Sometimes industries receiving municipal water will treat it with GAC prior to use e.g. pharmaceutical industry.

4.7.1.4 Chemical oxidation

Chemical oxidation is the resultant reaction when two or more chemical species are added with the purpose of increasing the oxidation state of one. For one chemical species to increase its oxidation state (lose electrons) the other chemical (s) in the reaction must be reduced (gain electrons). The following equation is where Fe^{2+} is oxidized by HOCl:

$$2Fe^{2+} + HOCl + 5H_2O \rightarrow 2Fe(OH_3) + Cl^- + 5H^+$$

The ferrous ion increased in oxidation state from +2 to +3, i.e. it lost one electron, chlorine was reduced from Cl^{1+} to Cl^{1-}, i.e., it gained two electrons. Large complex organic, ring - structured detergents and phenolic and humic compounds can be broken into simpler compounds by strong oxidants such as ozone or chlorine. Further oxidation by chemical or biological means may result in stable end products. Added advantages of this process may include:

(i) Ammonia removal

(ii) Oxidation of inorganic substances such as iron & manganese

(iii) Disinfection

(iv) Improvement of taste and odour

(v) Colour removal

Traditionally chlorine was used but in recent years attention is focused on chlorine alternatives, because of the production of trihalomethanes (THM) when chlorine reacts with natural organics. The alternative to chlorine as an oxidant is chloramines, ozone and potassium permanganate.

4.7.1.5 Membrane process

Membrane separation techniques include microfiltration (MF), ultrafiltration (UF), reverse osmosis (RO) and electrodialysis (ED). These are broadly sophisticated filtration techniques used to filter out minute impurities. MF and UF separate molecules according to their size and molecular mass (Fig.4.12).

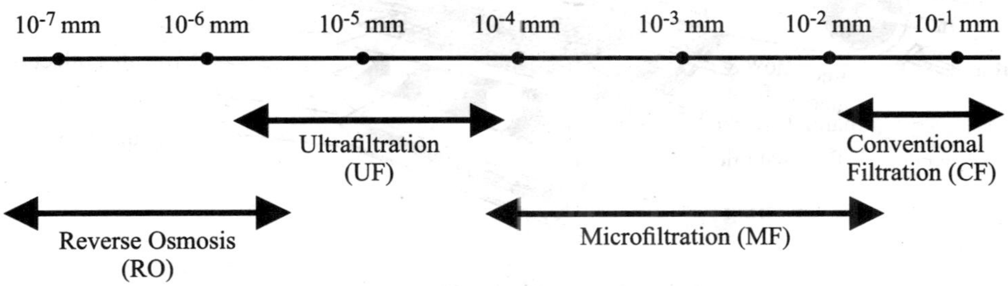

Fig. 4.12 Separation Processes and Associated Particle Size

(i) MF: MF allows macromolecules (10^{-4} to 10^{-3} mm) to flow through the membrane. Bacteria that are larger than 10^{-4} mm are prevented from passing through the membrane wall.

(ii) UF: Here the filters have cut-off threshold in the range of 10^{-6} to 10^{-4} mm pore size.

(iii) RO: This is a solubilization diffusion technique that makes use of a semi – permeable membrane which acts as a barrier to dissolved salts and inorganic molecules. This system uses pressure to drive water through the membrane against the force of osmotic pressure. The membrane commonly used in RO is composed of cellulose acetate and is about 100 μm thick. RO systems can operate at 90% efficiency or better with respect to total dissolved solids. In addition to inorganic ions, the membranes also remove residual organic molecules, turbidity, bacteria and viruses (Figs 4.13 and 4.14).

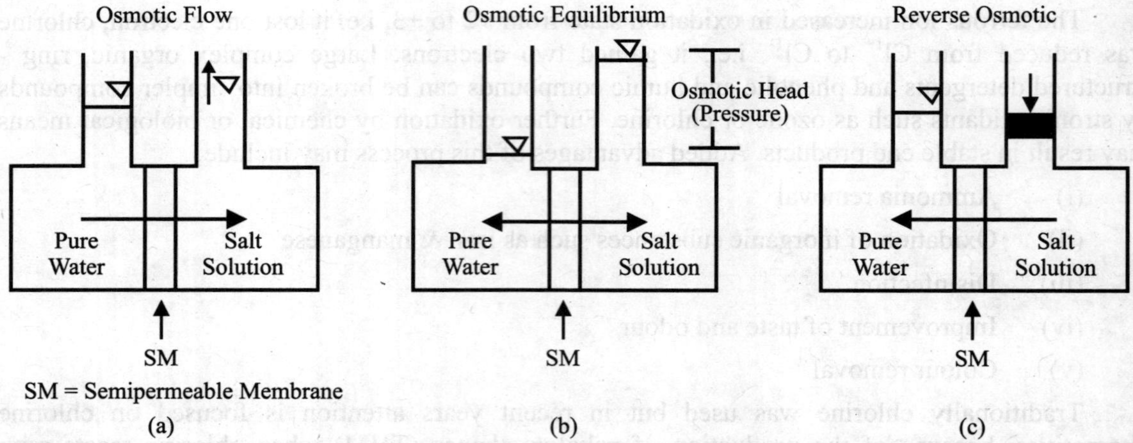

Fig. 4.13 Schematic of Reverse Osmosis

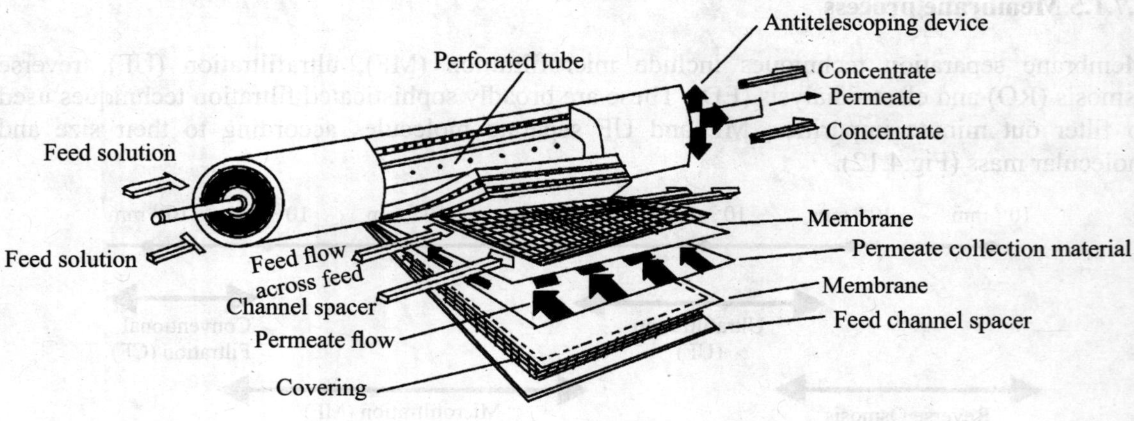

(A) Spiral-wound ultrafiltration model

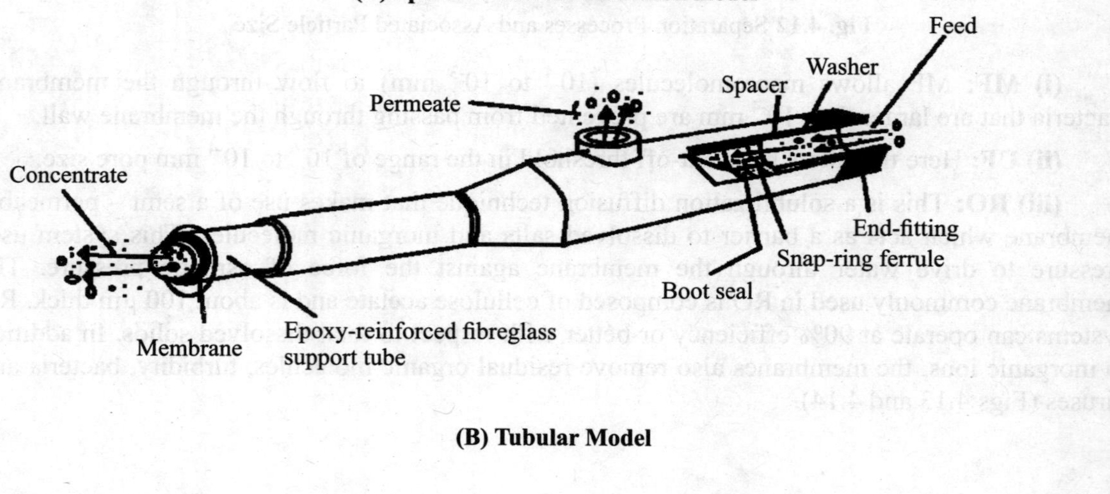

(B) Tubular Model

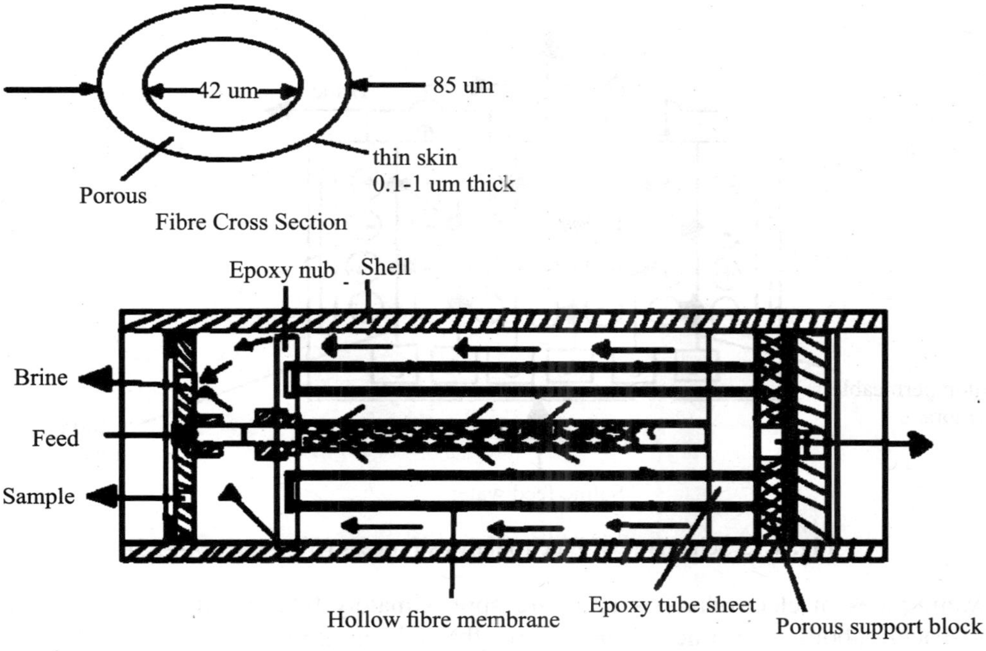

(C) Hollow-fibre Membrane

Fig. 4.14 Filtration Models

Flux rates of approximately 0.1m^3 /m^{2}d are typical for spiral-wound and tubular systems with considerably lower rates for the hollow–fiber units. However, much more membrane surface area is available in modules packed with hollow fibers than in comparably sized spiral-wound modules, and product water per module unit is approximately the same. Modules are placed in parallel to provide the necessary capacity and in series to increase efficiency.

Reverse osmosis systems can operate at 90% efficiency or better with respect to total dissolved solids. In addition to inorganic ions, the membranes also remove residual organic molecules, turbidity, bacteria, and viruses.

(iv) ED: It is an electrically charged membrane process where the ions are transferred through a membrane from a less concentrated to a more concentrated solution. The flow of pure water is tangential to the membrane while the flow of ions is perpendicular (Bilstad, 1992).

The electrodialysis process uses a series of membranes from ion-exchange resins. These membranes will selectively transfer ions. One membrane is cation-permeable, that is, it will pass cations but will reject anions, while the other membrane is anion-permeable and rejects cations. When parallel channels are constructed by alternating membranes and an electrical current is passed across them, an electrodialysis cell is formed as shown in Fig. 4.15. Cations are drawn toward the cathode, passing through the cation-selective membrane but being stopped by the anion-selective membrane. The opposite action occurs with anions, resulting in ions being removed from one channel and concentrated in the adjoining channel.

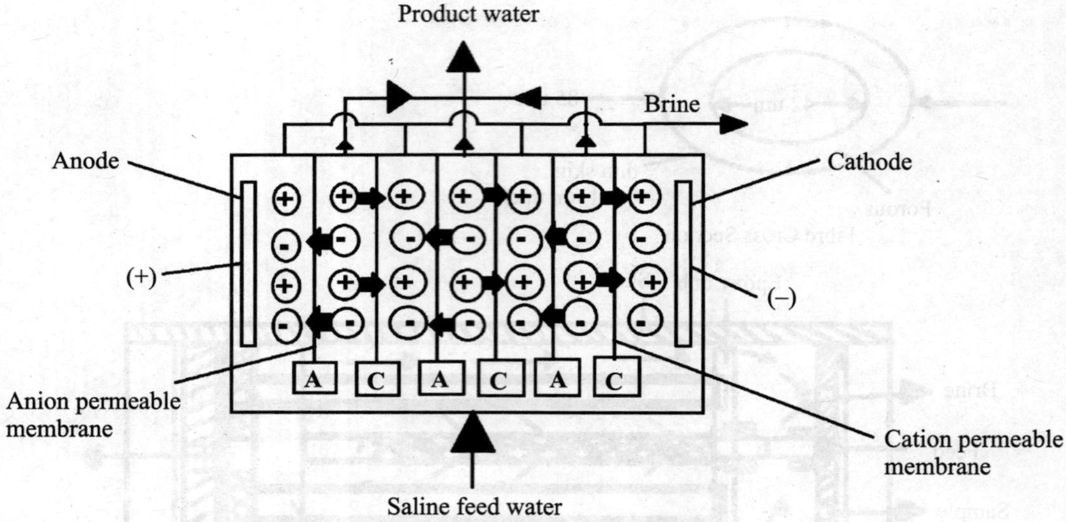

Fig. 4.15 Schematic of Electrodialysis Process (from Lacey [4-35])

Membranes in electrodialysis units are approximately 0.5mm thick and are separated by porous spacers about 1mm thick. Water flows through the porous spacers. Several membranes and spacers are sandwiched together into one electrodialysis cell.

A contact time of 10 to 20 s is required with removal efficiencies of about 25-60%. Cells are placed in series to increase efficiency and in parallel to meet flow requirements. Under ideal conditions, approximately 90% of the feed water is deionized, with the ions being concentrated in the remaining 10%.

Both reverse osmosis and electrodialysis produce a waste stream that may range from 10-25% of the feed water. In potable water supply systems, an additional volume of water and wastewater treatment systems, the concentrated wastewater streams must be disposed of properly.

4.7.1.6 Electrodialysis

Dialysis is a process in which the dissolved ionic components are separated from solution (especially colloidal solution) through the use of semipermeable ion selective membrane. In nature, kidney walls act as natural semipermeable membrane which allows dissolved impurities from blood pass through the membrane and secrete through urine. Persons with kidney failure have to undergo artificial dialysis. Dialysis is a slow process. This slow process can be expedited with application of electric field. The process known as electrodialysis finds many applications in water and wastewater treatment.

Example 4.5 Determine the daily requirement of water treatment chemicals in kg for a water treatment plant (WTP) handling a flow of 500 m^3/hour. Dosage requirement from jar test for 1 litre as:

5 ml of 10 gm/l alum solution

2.2 ml of 5 gm/l lime solution

1 ml of 1 gm/l polyelectrolyte solution

2 ml of 5 gm/l chlorine solution

Solution. Alum solution strength 10 gm/l

5 ml contains 50 mg

∴ Alum dosage is 50 mg/l. Similarly lime dosage is 2.2 x 5 = 11 *mg/l*

Polyelectrolyte dosage 1 x 1 = 1 *mg/l*. Chlorine dosage 2 x 5 = 10 *mg/l*

∴ Daily alum requirement

$$\frac{500 \times 1000 \times 24 \times 50}{10^6} = 600 \text{ kg}$$

Daily lime requirement

$$\frac{500 \times 1000 \times 24 \times 11}{10^6} = 132 \text{ kg}$$

Polyelectrolyte daily requirement

$$\frac{500 \times 1000 \times 24 \times 1}{10^6} = 12 \text{ kg}$$

Daily chlorine requirement

$$\frac{500 \times 1000 \times 24 \times 10}{10^6} = 120 \text{ kg}$$

To prevent the formation of dichloramine, ammonia should be dosed in the ratio.

$$Cl_2 : NH_3 \sim 4:1$$

∴ Ammonia requirement 120 x 0.25 = 30 *kg/day*

Example 4.6 Design a slow sand filtration (SSF) unit for a flow of 960 *m^3/day*. Assume filtration rate for SSF 0.16 m/h.

Solution. Filtration area = $\dfrac{\text{flow in } \dfrac{m^3}{h}}{\text{rate } \dfrac{m}{h}} = \dfrac{960/24}{0.16} = 250 \text{ m}^2$

Choose a tank dimension 26 x 12. Height of unit should comprise of water height 1 *m*, media bed height 0.9 *m* and system under drain 0.5 *m*. Total required height 1 + 0.9 + 0.5 = 2.4 *m*.

Choose the dimension 26 x 12 x 2.4 *m*.

Example 4.7

(a) Design a rapid gravity filtration (RGF) unit for a flow of 50,000 m^3/day. Assume filtration rate 12 m/h.

(b) Calculate the amount of sludge solid generated in a water treatment plant (WTP) for the following data water flow 50,000 m^3/day.

Inlet suspended solid (SS) concentration 420 mg/L. Alum dosage 50 mg/L. WTP efficiency 85%

$$\frac{50,000}{24} m^2/h$$

Solution (a) Filtration rate $= \dfrac{h}{12\dfrac{m}{h}} = 173.6\,m^2$

Choose the dimension 16 x 12 *m*

Height of the unit should comprise of water height 1 *m*

Bed height 0.8 *m* under drain 0.5 *m*

Required total height 0.5 +0.8+1 = 2.3 *m*

Choose the dimension 16 x 12 x 2.3 *m*

(b) Mount of suspended solid to be removed as sludge from WTP with 85% efficiency

$$= \frac{50,000 \times 420 \times 0.85 \times 1000}{1000 \times 1000} = 17850 \frac{kg}{day} = 17.85 \text{ Ton per day}$$

One mole of alum $Al_2(SO_4)_3\,14\,H_2O$ produces two moles of $Al(OH)_3$ floc

594 kg of alum → 2 x 78 kg of $Al(OH)_3$

(molecular mass of alum) [molecular mass of $Al(OH)_3 \sim 78$]

i.e., 1 kg of alum produces 0.26 kg of chemical sludge.

Alum requirement in a day $= \dfrac{50,000 \times 1000 \times 50}{1000 \times 1000} = 2500 \dfrac{kg}{day}$

Generation of chemical sludge = 2500 x 0.26 = 650 kg/day

Total sludge generation = 17850 + 650 = 18500 kg/day = 18.5 Tons/day

Example 4.8 For disinfecting water supply, it is required to treat 5,00,000 litres of daily supply with 0.5 ppm of chlorine. If the disinfectant is available in the form of bleaching powder containing 30% of available chlorine, calculate the amount of bleaching powder required to treat the daily flow of water.

Solution Dosage of chlorine $= \dfrac{5,00,000 \times 0.5}{1,00,000} = 0.25\,kg$

For the bleaching powder with available chlorine of 30%, amount of bleaching powder required

$$= 0.25 \times \frac{100}{30} = 0.833\,kg \qquad \textbf{(Ans)}$$

Example 4.9 Find the required of alum to treat 10^7 l/day of water at alum dosage of 30mg/l.

Solution: Water to be treated = 10^7 l/day.

Alum dosage = 30 mg/l = $30 \times 10^{-6} \dfrac{\text{kg}}{\text{l}}$

Therefore, Alum required per day= $10^7 \times 30 \times 10^{-6}$ = 300 kg/day

Alum requirement = 300 kg/day

Example 4.10 Determine the size of a rapid sand filter unit for filtering 0.15 ml/hr of water, assuming a rate filtration = 6000 L/hr/m^2 and a length to width ratio=1.25.

Solution Water to be filtered =0.15 ml/hr = 0.15×10^6 l/hr

Rate of filtration = 6000 l/hr/m^2

Therefore, Area of the filter required = $\dfrac{0.15 \times 10^6}{6000} = 2\,5\text{m}^2$

Length (L) to width (W) ratio = (L/W) = 1.25

Thus, we have $L \times W = 25,$ ....(i)

$L/W = 1.25$ .....(ii)

Multiplying both sides of equation (i) and (ii) we have

$L^2 = 25 \times 1.25 = 31.25$, or L=5.59m

Substituting this value in eq. we have 5.59 × W = 25,

Or $W = \dfrac{25}{5.59} = 4.47$ m

Rounding these values, we get L= 6m, W = 4.5m.

The size of the filter = 6m × 4.5m (Ans)

Example 4.11 Calculate the amount of bleaching powder required to treat 10^7 l/day of water with chlorine dosage 0.8 mg/L, assuming that the available chlorine in the bleaching powder is 40%.

Solution Water to be treated = 10^7 L/day

Chlorine dosage = 0.8 mg/L = $0.8 \times \dfrac{10^{-6}\ \text{kg}}{\text{L}}$

Available chlorine = 40% = 0.40 of bleaching powder

Thus, bleaching powder dosage = $\dfrac{\text{chlorine dosage}}{0.40} = \dfrac{0.8 \times 10^{-6}}{0.40} \dfrac{\text{kg}}{\text{L}}$

Thus, bleaching powder required = $\dfrac{0.8 \times 10^{-6} \times 10^7}{0.40} = 20 \dfrac{\text{kg}}{\text{day}}$

Bleaching powder required = $20 \dfrac{kg}{day}$

Example 4.12 From the following analysis report, suggest the treatment methods, with the help of a flow sheet (Fig. 4.16)

Suspended solid (SS)	120 mg/l
Iron as Fe^{2+}	2 mg/l
Hardness as $CaCO_3$	72 mg/l
Fluoride as F^-	3.4 mg/l

Solution The high SS concentration can be removed by sedimentation and coagulation operation.

(i) Iron as Fe^{2+} is appreciable high and hence requires to be removed by aeration method

(ii) Fluoride concentration also exceeds the acceptable limit and hence excess fluoride is to be removed by lime precipitation method.

(iii) The total hardness of 72 mg/l is soft water and hence requires no treatment.

(iv) The special treatment methods are to be incorporated along with the standard conventional treatment.

Suggested flow sheet

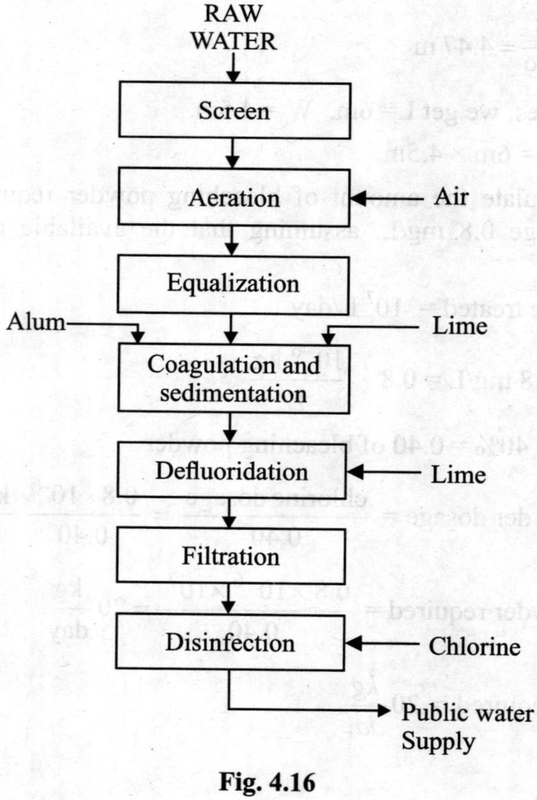

Fig. 4.16

Review Questions

1. **What do you mean by water quality standards and parameters?**

   Ans. The quality of water depends upon the purpose of water consumption (human consumption) and non-consumptive use (agriculture, industry and irrigation). Depending upon the use of water, certain quality criteria are established and based on these criteria; quality standards are specified by health and other regulating agencies to ensure that the water quality in available resource is suitable for the consumption.

2. **How can the quality of water be assessed?**

   Ans. The quality of water depends upon the purpose and permissible limits of the physical and chemical parameter.

3. **What is the difference between softening water and filtering water?**

   Ans. Filtering water involves separating mineral particles, like particulates, iron, hydrogen sulfide or other organic matter, from H_2O. By passing water through a filter bed, or media bed, these granular particles are trapped - and clean water passes through the bed. Softening water involves ion exchange to remove dissolved minerals - like calcium, magnesium, iron and manganese - that can't be trapped in a filter bed. Softeners use fresh resin beads with sodium attached to the resin. As water enters the tank, dissolved calcium and magnesium are attracted to the resin. The resin passes up the sodium in exchange for the dissolved chemicals and the water is then rid of these impurities.

4. **What are drinking water disinfectants?**

   Ans. Chlorine and chloramine are common disinfectants used for killing potential bacteria in water systems and added at the drinking water treatment plant.

5. **Write the formula for settling velocity in sedimentation as per Stock's law.**

   Ans : $$v = \frac{d^2 (p - L) \times g}{18 \mu}$$

   v = sedimentation rate or velocity of the sphere

   d = diameter of the sphere

   p = particle density

   L = medium density

   μ = viscosity of medium

   g = gravitational force

6. **What is flue gas desulphurization?**

   Ans. Flue gas is bad or highly polluted gases and flue gas desulphurization (FGD) stands for removal oxides of sulphur from emissions. FGD process may be of two types:

   1. Generative process

   2. Non-generative process

Generative process mainly based on reactions which results in some marketable products such as calcium sulphate (Gypsum), H_2SO_4.

$$SO_2 + CaSO_3 \rightarrow CaSO_3 + CO_2$$

$$2CaSO_3 + O2 \rightarrow 2CaSO_4 \text{ (Calcium sulphate/Gypsum)}$$

$$2SO^3 + \text{Excess } O_2 \rightarrow 2SO_3$$

$$2SO_3 + 2H_2O \rightarrow 2H_2SO_4$$

Non-generative process generates residue which must be disposed and the reaction based on the reaction of SO2 with compounds like:

a) Lime (CaO) b) Lime stone ($CaCO_3$)

c) Soda ash (Na_2CO_3) d) Caustic soda (NaOH).

Some of the reactions of non-regenerative process are-

$SO_2 + CaO \rightarrow CaSO_3$ (calcium sulphite)

$2CaCO_3 + SO_2 \rightarrow CaSO_3 + CO_2$

7. What is the difference between softening water and filtering water?

Ans. Filtering water involves separating mineral particles, like particulates, iron, hydrogen sulfide or other organic matter, from H_2O. By passing water through a filter bed, or media bed, these granular particles are trapped - and clean water passes through the bed. Softening water involves ion exchange to remove dissolved minerals - like calcium, magnesium, iron and manganese - that can't be trapped in a filter bed. Softeners use fresh resin beads with sodium attached to the resin. As water enters the tank, dissolved calcium and magnesium are attracted to the resin. The resin passes up the sodium in exchange for the dissolved chemicals and the water is then rid of these impurities.

8. What are drinking water disinfectants?

Ans. Chlorine and chloramine are common disinfectants used for killing potential bacteria in water systems and added at the drinking water treatment plant.

9. What do you understand by the term polyelectrolyte?

Ans. Polyelectrolytes are long-chain molecules used in the conditioning of sludge. The neutralize surface charges and thus cause bridging across fine particles (or floes) to form larger particles.

10. How is the rapid gravity filters classified with respect to the medium they use?

Ans. With respect to the medium used, the rapid gravity filters are classified as follows:

(i) Single-medium filter (usually sand or anthracite)

(ii) Dual-medium filter (usually sand or anthracite)

(iii) Multi-medium filter (usually garnet, sand or anthracite)

11. What is the mechanism of pathogen kill by chlorination?

Ans. The mechanism of pathogen kill by chlorination is considered to be as follows:

(i) A penetration by chlorine molecules into the cell of micro-organisms

(ii) Subsequent blocking of an essential enzyme

(iii) Destruction of the cell membrane and cells

12. What are the factors that affect the process of chlorination?

Ans. The process of chlorination is affected by the following factors:

(i) Chemical form of chlorine

(ii) The pH of water

(iii) Chlorine concentration

(iv) Contact time

(v) Type of organisms

(vi) Suspended solids

(vii) Temperature

13. What is reduced ground water? Enlist its properties.

Ans. Ground water that does not contain any oxygen is called reduced ground-water. Its properties are as follows:

(i) It is clear and colourless.

(ii) It has metallic smell and/or metallic taste.

(iii) It is free of oxygen.

(iv) Sometimes it has a smell of H_2S.

14. What is chlorine demand?

Ans. Chlorine demand is the amount of chlorine required, with a contact time of 30 min, to kill all micro-organisms in water and to oxidize or react with the chemical compounds present in it. Chemical demand also includes the residual amount of chlorine desired (0.2 mg/L) to prevent further contamination of water in the distributed system.

15. What are the various forms of raw water impurities?

Ans. Raw water may contain impurities in several forms which include the following:

(i) Particulate matter (size $> 10^{-1}$ mm)

(ii) Suspended matter (10^{-3} mm $<$ size $< 10^{-1}$ mm)

(iii) Colloidal particles (10^{-6} mm $<$ size $< 10^{-3}$ mm)

(iv) Dissolved matter (size $< 10^{-6}$ mm), e.g., humic acid, tannic acid, colour.

It is the objective of the water treatment industry to reduce the above impurities in raw water to acceptable levels. The forms of impurities present will help to decide the type and level of treatment to be used.

16. Chennai city has a water supply problem with no practical surface water supply option. Suggest a practical solution to solve the water problem of the city.

Ans. Chennai has no feasible fresh water source nearby. The city is situated on sea shore. So the desalination process based on the principle of reverse osmosis is one promising feasible option.

17. Suggest a suitable treatment method for each

i. Colloidal solids
ii. Dissolved solids
iii. Suspended solids
iv. Ammonia
v. Odour
vi. Excess hardness
vii. Pathogenic bacteria
viii. Colour
ix. Boiled feed water

Ans.

Colloidal solids	Coagulation and sedimentation
Dissolved solids	Ion exchange
Suspended solids	Sedimentation
Ammonia	Air stripping and adsorption
Odour	Aeration
Excess hardness	Lime-soda process
Pathogenic bacteria	Chlorine disinfection
Colour	Adsorption
Boiled feed water	Ion exchange

Self-assessment Questions

1. Explain the action of chlorine, when mixed with water.
2. How is chlorine applied in water?
3. Explain different forms of chlorination.
4. How is orthotolidin test carried out?
5. State the procedure of starch-iodine test.
6. Distinguish between the following:
   (a) Pre-chlorination and post-chlorination
   (b) Super- chlorination and de-chlorination
   (c) Plain- chlorination and break point -chlorinations

Chapter 5

Waste Water Treatment

5.1 INTRODUCTION

The study of waste water plays an important role in determining the process of treatment. Again, the treatment should be such that the effluent can be disposed of to the natural water sources safely. The characteristics are divided into three categories.

(i) Physical characteristics

(ii) Chemical characteristics

(iii) Biological characteristics

Before carrying out the testing, the sampling should be done properly. The testing involves the physical, chemical and biological tests as discussed below:

5.1.1 Physical Tests

The physical tests include the following observations and has been discussed below:

(i) Colour

The colour depends on the discharge of industrial wastes. However, to designate the colour, the platinum cobalt scale may be used.

(ii) Turbidity

The turbidity of water is due to the suspension of fine particles. It is measured by turbidimeter. Turbidity is expressed as ppm or mg/l

(iii) Temperature

Temperature cannot be controlled by any treatment. But temperature of the water sample should be noted, as the biological activities depend on temperature. At greater temperature biological activities are high and viscosity is low. When viscosity is low, the efficiency of treatment plant is more.

5.1.2 Chemical Tests

The chemical tests involve the determination of factors such as total solids, chlorine demand, dissolved oxygen, biochemical oxygen demand, chemical oxygen demand, pH, nutrients (N and P), VOCs as discussed briefly below:

(i) Total solids

It is very essential to know the quantity of total solids in water, so that it helps to know the rate of deposition of sludge in the primary sedimentation tank and at the same time it is possible to indicate the organic and inorganic substances in water which help in water treatment. The amount of total solids is found out as follows:

A known amount of water is taken and the water is evaporated at 100°C. The residue is dried properly and weighed. The weight of the dry residue represents the total solids. Again, the total solids may be of two types volatile solids and suspended solid discussed below:

(a) ***Volatile solids***: The dried total solids are heated or ignited in electric furnace. After ignition the remaining solids are weighed. The loss of weight will indicate the volatile solids present which are due to the presence of organic matters.

(b) ***Suspended solids:*** The solids which settle down due to the phenomenon of hydraulic subsidence are known as suspended or settleable solids.

(ii) Chlorine demand

A requisite amount of chlorine is required to break up the organic matters present in water and for killing the pathogenic bacteria. The necessity of chlorine is known as **chlorine demand**. It is found out by the process of break – point chlorination.

(iii) Dissolved oxygen

One of the most important water quality parameters is the amount of dissolved oxygen (DO) present. The saturated value of DO in water is modest of the order of 8 to 15 mg/l, depending on temperature and salinity.

Oxygen – demanding wastes are substances that oxidize in the receiving body of water, reducing the amount of DO available. As DO drops, fish and other aquatic life are threatened and in the extreme case, killed. In addition, as DO level fall, undesirable odors, tastes and colors reduce the acceptability of water as a domestic supply and reduce its attractiveness for recreational uses. Oxygen demanding wastes are usually biodegradable organic substances contained in municipal wastewaters or in effluents from certain industries, such as food processing and paper production. In addition, the oxidation of certain inorganic compounds may also contribute to the oxygen demand. Even naturally occurring organic matter, such as leaves and animal droppings that find their way into surface water add to the DO depletion.

There are several measures of oxygen demand commonly used. The chemical oxygen demand or COD is the amount of dissolved oxygen needed to chemically oxidize the wastes, while the biochemical oxygen demand or BOD is the amount of DO required by microorganisms to biologically degrade the wastes.

(iv) Biochemical oxygen demand (BOD)

When biodegradable organic matter is released into a water body, microorganisms, especially bacteria, feed on the wastes, breaking it down into simpler organic and inorganic substances. When this decomposition takes place in an aerobic environment, that is, in presence of oxygen, the process produces non objectionable, stable end products such as carbon dioxide (CO_2), sulfate (SO_4), orthophosphate (PO_4) and nitrate (NO_3). A simplified representation of aerobic decomposition is given by the following equation:

$$\text{Organic matter} + O_2 \xrightarrow{\text{Microorganisms}} CO_2 + H_2O + \text{new cells} + \text{stable products}$$

When insufficient oxygen is available the resulting anaerobic decomposition is performed by completely different microorganisms. They produce end products that can be highly objectionable, including hydrogen sulfide (H_2S), ammonia (NH_3) and methane (CH_4). Anaerobic decomposition can be represented by the following equation:

$$\text{Organic matter} + H_2O \xrightarrow{\text{Microorganisms}} + CH_4 + CO_2 + H_2O + \text{new cells} + NH_3 + H_2S + \text{heat}$$

The methane produced is physically stable, biologically degradable and a potent greenhouse gas. When emitted from water bodies are often called **swamp gas.** It is also generated in the anaerobic environment of landfills, where it is sometimes collected and used as an energy source.

BOD is defined as the amount of oxygen required by living organisms in the stabilization of the organic matter of a water / wastewater body. BOD may have various units, but most often it is expressed in milligrams of oxygen required per liter of wastes (mg/l) or the equivalent g/m^3.

(v) Chemical oxygen demand (COD)

COD is a measure of the total organic carbon with the exception of certain aromatics such as benzene which are not completely oxidized in the reaction. Some organic materials such as cellulose, phenols etc., resist biodegradation along with pesticides and polychlorinated biphenyls (PCBs) which are toxic to microorganisms and these materials are oxidized in the COD test. A chemical oxidizing agent is used instead of microorganisms as is the case in the BOD test. The result is a COD that is always higher than the associated BOD in municipal wastewater.

$$\text{COD} \approx 1.6\ \text{BOD}_5$$

COD is a most used parameter, particularly as it can be completed in about 2 hours (using the titrimetric or colorimetric method) and also because it takes into account those organics that are not readily biodegradable.

(vi) pH

Often the chemical adjustment of pH is necessary to effect the desired precipitation. Oxidizing agents may be used if reduced compounds are to be removed. For example, potassium permanganate may be added to oxidize soluble forms of iron and manganese to precipitate out. Chlorine is sometimes used as an oxidizing agent as well as a disinfectant in both water and wastewater treatment.

(vii) Nutrients

Nutrients are chemicals, such as nitrogen, phosphorus, carbon, sulfur, calcium, potassium, iron, manganese, boron and cobalt that are essential to the growth of living organisms. In terms of water quality, nutrients can be considered as pollutants when their concentrations are sufficient to allow excessive growth of aquatic plants, particularly algae.

Nutrient enrichment can lead to blooms of algae which eventually die and decompose. Their decomposition removes oxygen from the water, potentially leading to levels of DO that are insufficient to sustain normal life forms. Algae and decaying organic matter add colour, turbidity, odours and objectionable tastes to water that are difficult to remove and that may greatly reduce its acceptability as a domestic water source. The process of nutrient enrichment, called **eutrophication** is an especially important one in lakes.

(viii) Volatile organic compounds (VOCs)

The term VOC means evaporated gasoline from petrol filling stations, vapours from refineries, organic solvents from paint and dry cleaning, and many others. These organic species are major contributors to smog and may be toxic or carcinogenic for humans. Their emissions are strictly regulated. Studies have found that glues and paints in our houses may have serious adverse health effects. There is a large effort today to switch to water-based paint, so that emissions of VOCs are decreased. VOCs include benzene, chloroform, methanol, carbon tetrachloride and formaldehyde among hundreds of other compounds. Gasoline is a mixture of many such compounds.

Volatile organic compounds are released into the atmosphere by anthropogenic and natural emissions which are important because of their involvement in photochemical pollution. They are hydrocarbon compounds that have low boiling points, usually less than 100°C, and therefore evaporate readily. Some are gases at room temperature. Propane, benzene, and other components of gasoline are all VOCs. Many VOCs are human-made chemicals that are used and produced in the manufacture of paints, adhesives, petroleum products, pharmaceuticals, and refrigerants. They often are compounds of fuels, solvents, hydraulic fluids, paint thinners, and dry-cleaning agents commonly used in urban settings. VOC contamination of drinking water supplies is a human-health concern because many are toxic and are known or suspected human carcinogens. VOCs can be toxic to either fishes or humans as can excess metals or pesticides.

Important parameters in determining the water quality

Quality	Turbidity (mg/L)	DO (mg/L)	pH	BOD_5 (mg/L)	NH_3 (mg/L)	Remarks
Very good	<20	>8	6.5-8.5	<1	<0.1	Class I: Saturated with O_2; low in nutrients
Good	20-40	>6	6.0-9.0	2-6	<0.3	Class II: Low nutrients
Contaminated	40-100	<4	5.0-9.0	7-13	0.3-1.0	Class III: O_2- depleting pollutants
Highly polluted	100-275	<3	4.0-10.0	>15	1.0-2.7	Class IV: Putrefying; only bacteria (no higher organisms)

5.1.3 Biological Tests

The biological tests are carried out to know the presence of bacteria, fungi, algae, protozoa etc. The presence of some bacteria are essential for the decomposition of sewage. But the presence of pathogenic bacteria indicates the degree of pollution in water.

However, the biological processes are used extensively in wastewater treatment to convert biodegradable organics and other nutrients into a more manageable form. Biological processes form the basis for secondary treatment in which dissolved and colloidal organics are converted into biomass that is subsequently separated from the liquid stream. Secondary treatment systems are designed to optimize contact between microorganisms and organics under the most favourable environmental conditions. Once separated, the biomass becomes a concentrated waste stream that must be dealt with promptly. Biological treatment of this and other organic wastewater sludges, called **sludge digestion**, is one of the most important and most difficult processes in wastewater treatment.

5.2 BOD OF WASTE WATER

The treatment of waste water involves the determination of following factors as discussed below:

5.2.1 Five – Day BOD Test (BOD_5)

If BOD of wastewater is > 3 mg/l, dilution of a wastewater sample with BOD free, oxygen saturated water is necessary (Fig. 5.1). A sample of wastewater is diluted with seeded water and placed in an airtight bottle, measuring the concentration of DO at day 0 and again at day 5. The difference in DO is the BOD_5. The standard test uses a 300 ml BOD bottle and incubation is performed at 20ºC over 5 days in light excluded environment so as to stop the photosynthesis (addition of O_2) by algae. The various dilutions of 1, 5, 10, 20, 50, 100 etc. are now readily available as given in standard methods (1992). The 5- day BOD of a diluted sample is given by:

$$BOD_5 = p\,(DO_i - DO_f)$$

Where, DO_i = the initial dissolved oxygen of the diluted wastewater

DO_f = the final DO of the diluted wastewater

$$p = \text{the dilution factor} = \frac{\text{Volume of waste water}}{\text{Volume of waste water plus dilution water}}$$

In addition to the oxygen demand by the wastewater sample, the seeded dilution water is likely to exert some oxygen demand of its own, however small. Therefore, if the above method was used, uncorrected, then BOD_5 would be too high. The corrected BOD_5 is defined as follows:

$$BOD_5 = p\,[(DO_i - DO_f) - (B_i - B_f)\,f]$$

Where, p = dilution factor

B_i, B_f = initial and final DO concentrations of the seeded diluted water (blanks)

f = ratio of seed DO in sample to seed in blanks

$$= \frac{\%\ \text{seed in } DO_i}{\%\ \text{seed in } B_i}$$

In discussing water quality, we consider the rate of decomposition of organic matter to be proportional to the amount of organic matter available. This relationship is formulated as continuous first – order reaction (Fig. 5.2).

$$\frac{dL_t}{dt} = -K_1 L_t \tag{5.1}$$

Where, L_t = BOD_r, i.e. BOD remaining, mg/l

K_1 = BOD deoxygenation rate coefficient, day^{-1}

Integrating eqn. 1, between L_0 and L_t, we get

$$L_t = L_o\, e^{-K_1 t} \tag{5.2}$$

Where, L_o = BOD_u, i.e. ultimate BOD (or ultimate carbonaceous BOD or initial BOD of effluent at point of discharge to a stream)

With reference to Fig. 5.1, we find

$$BOD_t = BOD_u - BOD_r \tag{5.3}$$

When BOD_t = BOD exerted at time *t* (or oxygen demand)

BOD_r = BOD remaining at time *t*

$BOD_t = L_o - L_t$

$= L_o - L_o\, e^{-K_1 t}$

Therefore $BOD_t = L_o(1 - e^{-K_1 t})$

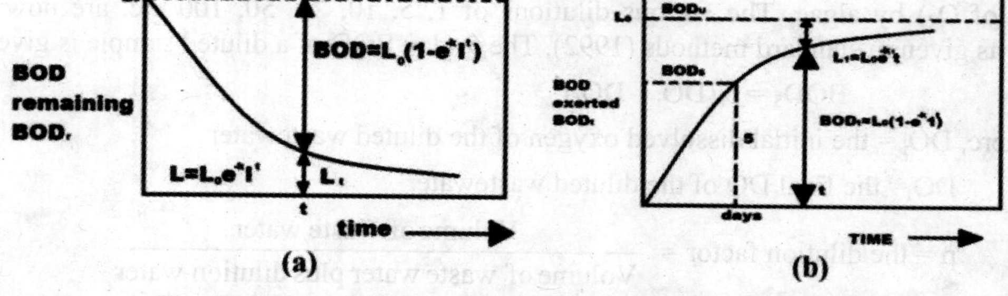

Fig. 5.1 Idealized BOD: a. Idealized BOD_r b. Idealized BOD_t

Temperature has an effect on K_1 and is related by

$$K_T = K_{20}\, \theta^{(T-20)}$$

Where, K_T = the rate at temperature T (°C)

K_{20} = the rate at 20°C (known)

$$\theta = \text{coefficient} \approx \begin{Bmatrix} 1.047 \text{ for } 20^\circ C < T < 30^\circ C \\ 1.35 \text{ for } 4^\circ C < T < 20^\circ C \end{Bmatrix}$$

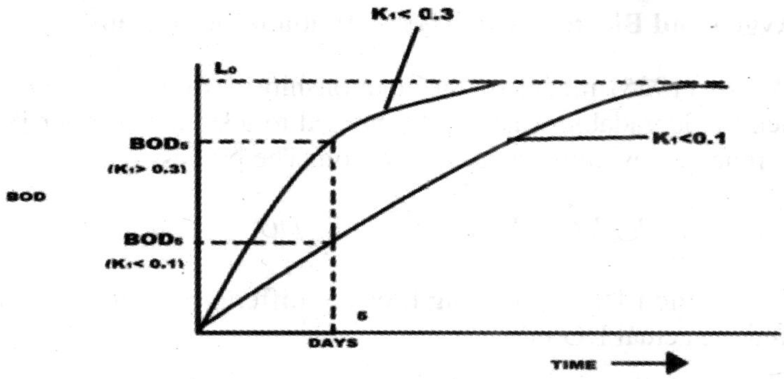

Fig. 5.2 Influence of K_1 Values on BOD

5.2.1.1 *CBOD*

When a carbonaceous waste biodegrades, it exert an oxygen demand. Note that BOD is often synonymous with CBOD (i.e. the carbonaceous biochemical oxygen demand) (Fig. 5.3).

5.2.1.2 *NBOD*

In addition there can also be can oxygen demand exerted by the oxidation of nitrogenous compounds. Ammonia is oxidized to nitrite and nitrite is oxidized to nitrate in the following microbial assisted reactions

$$2NH_3 + 3O_2 \xrightarrow{\textit{Nitrosomonas}} 2NO_2^- + 2H^+ + 2H_2O$$

$$2NO_2^- + O_2 \xrightarrow{\textit{Nitrobacter}} 2NO^{3-}$$

The above two reactions are termed **nitrification reactions**. The oxygen demand associated with the oxidation of ammonia to nitrate is called the **nitrogenous biochemical oxygen demand** (NBOD). Typically the NBOD occurs later than 5 days and does not show up in the CBOD test.

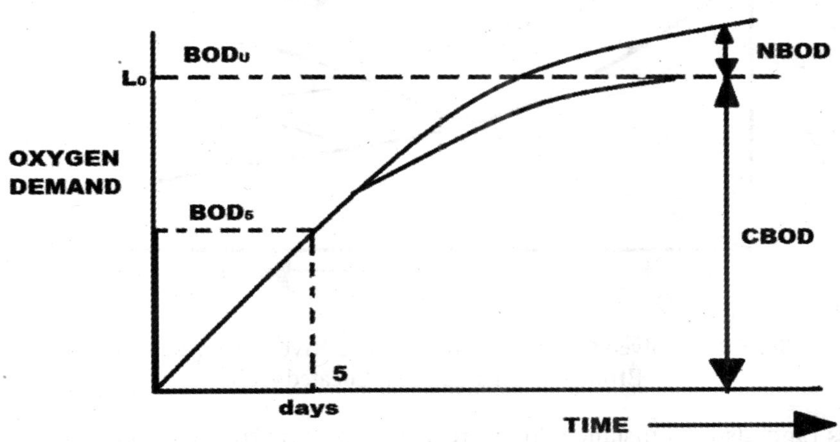

Fig. 5.3 Relationship of CBOD and NBOD

5.2.2 Dissolved Oxygen and Biochemical Oxygen Demand in Streams

The Streeter and Phelps (1925) model of the relationship of BOD and DO is still valid. They considered that when biodegradable waste is discharged to a stream or river it consumes oxygen and the later is only renewed by atmospheric reaeration. The process is

$$\frac{dDO}{dt} = K_1L_t - K_2DO = K_1\,Lo e^{-k_1 t} - K_2 DO \tag{5.4}$$

Where, DO = saturation DO deficit, mg/l. or the difference between saturation (maximum DO) and the actual DO in mg/l

L_t = CBOD, mg/l

K_1 = deoxygenation rate, day $^{-1}$

K_2 = reaeration rate, day $^{-1}$

The solution of eqn. 5.4 is

$$DO\,(t) = \frac{K_1 L_o}{K_2 - K_1}\,(e^{-K_1 t} - e^{-K_2 t}) + DO_o\, e^{-k_2 t} \tag{5.5}$$

Where, L_O = oxygen demand at t = t_o (or the BOD)

DO_0 = dissolved oxygen deficit at t = t_o

DO (t) = dissolved oxygen saturation deficit at any time t

Equation (5.5) is the Streeter – Phelps oxygen sag formula and is sketched in Fig. 5.4

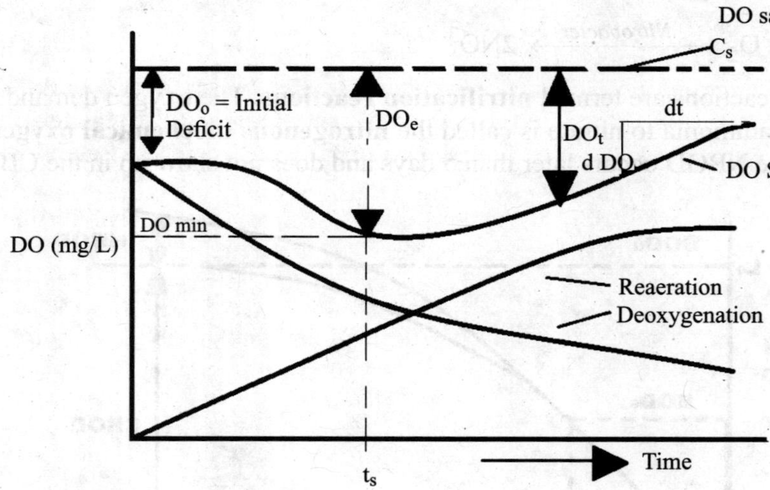

Fig. 5.4 Dissolved Oxygen SAG Curve in a River Downstream of an Effluent Discharge Point Located At X

The times (and also the distance downstream) of the occurrence of the minimum DO or the maximum dissolved oxygen deficit DO is obtained by differentiating equation – (5.4) with respect to t and setting it equal to zero.

Then, $$t_0 = \frac{1}{K_2 - K_1} \ln \left\{ \frac{K_2}{K_1} \left[1 - \frac{DO_o (K_2 - K_1)}{K_1 L_o} \right] \right\} \tag{5.6}$$

and $$DO_c = \frac{K_1}{K_2} L_o \, e^{-K_1 t} \tag{5.7}$$

5.2.3 Oxygen Diffusion into the Benthic Zone – SOD

This is often referred to as the sediment oxygen demand (SOD) and occurs when wastewater organics settle on the river bottom and biodegrade, thereby representing dissolved oxygen sink. The zero – order rate model of SOD is:

$$\pi_{SOD} = \frac{K_{SOD}}{H}$$

Where, π_{SOD} = rate of O_2 consumption, mg/l day

K_{SOD} = Oxygen uptake in sediment, mg/m^2 day

H = depth, m

In areas of significant sediments, the SOD may be a significant sink in the oxygen mass balance.

5.2.4 Modified Streeter – Phelps oxygen sag model

The previous section discussed the original oxygen sag curve from Streeter and Phelps (1925) based on the two processes of BOD oxidation and surface re-aeration. This section extends that development to include for SOD and other parameters of significance. The Streeter and Phelps model assumes a constant pollution load discharged at a single point in a river with constant flow rate and uniform cross-section. This of course is not very real as typical wastewater BOD loads of different magnitude. Also, the flow rates of rivers vary with time and cross-sections are rarely uniform or constant. In the Streeter and Phelps model, the lateral and vertical DO and BOD concentrations are assumed uniform throughout each section, i.e. essentially a plug flow assumption. Also the de-oxygenation and re-aeration rates are assumed first order with constant rates. This may be too simplistic for delicate water bodies. Recall that Eq. (5.4) for dissolved oxygen is

$$\frac{dDO}{dt} = \underbrace{K_1 L}_{deoxygenation} - \underbrace{K_2 DO}_{reaeration} \tag{5.8}$$

Considering that some of the BOD may be removed by sedimentation, Orlob (1983) writes for BOD,

$$\frac{dL}{dt} = -(K_1 + K_S) L \tag{5.9}$$

where, K_S = rate constant for BOD removal by sedimentation, d^{-1}

Thus

$$L_t = L_o\, e^{-(K_1+K_s)t} \tag{5.10}$$

and
$$DO(t) = \frac{K_2 L_o}{K_2 - (K_1 + K_s)}[e^{-(K_1+K_s)t} - e^{-K_s t}] + DO_o\, e^{-K_2 t} \tag{5.11}$$

In the above two equations which are the modifications to Eqs. (5.8) and (5.9), it is assumed that no DO is drawn from the sediment and that there is no photosynthesis (Orlob, 1983). Further modifications by Camp (1963) for the BOD and DO profile equations include:

$$\frac{dL}{dt} = -(K_1 + K_s)L + \underset{\text{BOD from sediment}}{B} \tag{5.12}$$

$$\frac{dDO}{dt} = \underset{= \text{Deoxygenation - reaeration - Photosynthesis}}{K_1 L - K_2 DO - P} \tag{5.13}$$

Where B = the addition of BOD to the water body from the benthic deposits, gm/l day

$$L(t) = \left(L_o - \frac{B}{K_1 + K_s}\right) e^{-(K_1+K_s)t} + \frac{B}{K_1 + K_s} \tag{5.14}$$

$$\text{and } DO(t) = \frac{K_1}{K_2 - (K_1 + K_s)}\left(L_o - \frac{B}{K_1 + K_s}\right)[e^{-(K_1+K_2)t} - e^{-K_2 t}]$$

$$+ \frac{K_1}{K_2}\left(\frac{B}{K_1 + K_s} - \frac{P}{K_1}\right)(1 - e^{-K_2 t}) + DO_o\, e^{-K_2 t} \tag{5.15}$$

If the BOD is exerted in the water body by the benthic deposits but there is no BOD reduction by settling, (i.e. $K_s = 0$), Eqs (5.14) and (5.15) reduce to

$$L(t) = \left(L_o - \frac{B}{K_1}\right) e^{-K_1 t} + \frac{B}{K_1} \tag{5.16}$$

$$DO(t) = \frac{K_1}{K_2 - K_1}\left(L_o - \frac{B}{K_1}\right)\left(e^{-K_1 t} - e^{-K_2 t}\right)$$

$$+ \frac{B - P}{K_2}(1 - e^{-K_2 t}) + DO_o\, e^{-K_2 t} \tag{5.17}$$

An almost endless serried of modifications can be accounted for, even including the mixing process. The reader is referred to the concise descriptions by Orlob (1983).

5.2.5 Total Organic Carbon (TOC)

The amount of carbon used for oxidation of carbon present in the organic matter and it measures all carbon as CO_2 and is expressed in mg/l or ppm. The stoichiometric requirement of oxygen for the complete oxidation of organic matter is known as ThOD. It is otherwise known as

stoichiometric oxygen demand (*SOD)* or total oxygen demand *(TOD)* which is determined from balanced chemical oxidation equation (Fig. 5.5)

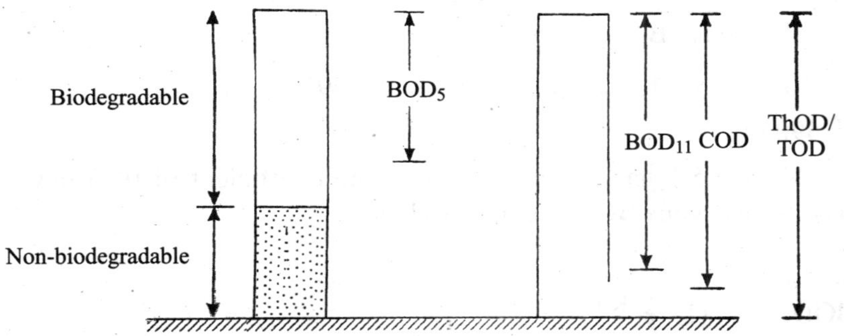

Fig. 5.5 Relationship between Oxygen Demanding Parameters

Example 5.1 The results from a BOD test, diluted by 100, are given in columns 1 and 2 in the table below compute the BOD

Time (days)	Diluted Sample DO (mg/l)	BOD (calculated) (mg/l)
0	7.95	-
1	3.75	420
2	3.45	450
3	2.75	520
4	2.15	580
5	1.80	615

Solution

Column 3 contains the BOD volumes for the seeded sample

$$BOD_5 = P\,(DO_1 - DO_5)$$

$$= 100\,(7.95 - 1.8) = 615 \text{ mg/l}$$

Example 5.2 Use e.g. 5.1 with corresponding DO values for the blank to determine the corrected BOD_5

Time (days)	Diluted sample DO (mg/l)	Blank Seeded Sample DO (mg/l)	BOD (Problem 1) (mg/l)	BOD corrected (mg/l)
0	7.95	8.15	-	-
1	3.75	8.10	420	415
2	3.45	8.05	450	440
3	2.75	8.00	520	505
4	2.15	7.95	580	560
5	1.80	7.90	615	590

Column 3 contains the given values for the DO of the blank and column 5 the corrected BOD values

$$f = \frac{\%\text{ seed in DO}_i}{\%\text{ seed in B}_i} = \frac{99\%}{100\%} = 0.99$$

$$BOD_5 = 100[(7.95 - 1.80) - (8.15 - 7.90) \text{ X } 0.99]$$
$$= 590 \text{ mg/l}$$

Example 5.3 If e.g.5.1 has a deoxygenation rate coefficient of 0.15 day^{-1}, determine the ultimate CBOD (Carbonaceous BOD) i.e. BOD_u (L_o)

Solution

$$BOD_t = L_o(1 - e^{-K_1 t})$$

$$L_o = \frac{BOD_5}{1 - e^{-0.15 \times 5}} = \frac{615}{0.527}$$

$$L_o = 1165 \text{ mg/l}$$

Example 5.4 A wastewater with a BOD of 25 mg/l is discharged through an outfall to a freshwater stream of mean velocity 0.1 m/s. The DO upstream of the outfall is 8.5 mg/l. Assuring the deoxygenation rate $K_1 = 0.25$ d^{-1} and the reaction rate $K_2 = 0.4$ d^{-1}, determine.

(a) The time and distance downstream (X_c) where DO is a minimum

(b) The minimum DO

Solution

Saturation DO = 9.2 mg/l

Initial DO = 8.5 mg/l

Initial deficit $DO_0 = 9.2 - 8.5 = 0.7$ mg/l

(a) The time at which DO_0 is minimum from equation is

$$t_c = \frac{1}{K_2 - K_1} \ln\left\{\frac{K_2}{K_1}\left[1 - \frac{DO_o(K_2 - K_1)}{K_1 L_o}\right]\right\}$$

$$= \frac{1}{0.4 - 0.25} \ln\left\{\frac{0.4}{0.25}\left[1 - \frac{0.7(0.4 - 0.25)}{0.25 \times 25}\right]\right\}$$

$$t_c = 3.02 \text{ day}$$

$$x_c = \overline{U}\, t_c = 0.1 \times 3.02 \times 24 \times 3600 \times 10^{-3}$$

$$x_c = 26.1 \text{ Km}$$

(b) The maximum deficit DO_C is

$$DO_c = \frac{K_1}{K_2} L_o e^{-K_1 t}$$

$$= \frac{0.25}{0.4} \times 25 \times e^{-0.25 \times 3.02}$$

$$DO_c = 7.3 \text{ mg/l}$$

Thus the minimum value of DO is

$$DO_{min} = DO_{sat} - DO_c$$

$$= 9.2 - 7.3 = 19 \text{ mg/l}$$

Example 5.5 From the data of e.g. 5.1, determine the total oxygen demand if the sample contains 25 mg/L of nitrogen.

Solution

From e.g. 5.1 $BOD_5 = 615 \text{ mg/l}$

e.g. 5.3 $BOD_U = 1165 \text{ mg/l}$

Nitrification: $\underset{\substack{1\text{ mole}\\17\text{ g/mole}}}{NH_3} + \underset{\substack{2\text{ moles}\\16\text{ g/mole}}}{2O_2} \rightarrow \underset{1\text{ mole}}{NO_3^-} + \underset{1\text{ mole}}{H^+} + \underset{1\text{ mole}}{H_2O}$

1 mole of NH_3 requires 2 moles of O_2

$$NBOD = 25 \times \frac{2 \times 32}{14} = 114 \text{ mg/l}$$

Thus, total oxygen demand = CBOD+NBOD

= 1165+114 = 1279 mg/l

If the concentration of nitrogen as organic nitrogen and ammonia nitrogen is known then it is possible to directly determine the NBOD. It is

$$NBOD = TKN \times \frac{2 \times 32}{14}$$

$$NBOD = TKN \text{ X } 4.57$$

Where TKN = total Kjeldahl nitrogen

i.e each 10 mg N/l exerts a 45.7 mg O_2/l oxygen demand.

Example 5.6 BOD of an effluent sample incubated for one day at 30°C was found to be 100 mg/l. What would be the 5 day BOD at 20°C ? K=0.12 at 20°C

Solution:

$$L_t = L_o (1 - e^{-Kt})$$

BOD_1 At 30°C = 100 mg/l

When t=1, L_t = 100 mg/l

K_{20} = 0.12 day^{-1}

Therefore, $K_{30} = K_{20}(e)^{T-20} = 0.12 \times (1.047)^{30-20} = 0.12 \times (1.047)^{10} = 0.19\ \text{day}^{-1}$ (Ans)

$$L_o = \frac{L_t}{1-e^{-Kt}} = \frac{100}{1-e^{-0.19\times1}} = \frac{100}{1-0.827} = \frac{100}{0.173} = 578\ \text{mg/l}$$

$$BOD_s \text{ At } 20^\circ C = L_o(1-e^{-Kt}) = 578(1-e^{-0.12\times5})$$

$$= 578(1-0.549) = 578 \times 0.451 = 260.678\ \text{mg/l}$$

Example 5.7 Compute the ThOD of 108.75 mg/l of Glucose.

Solution

The balanced equation of the reaction is:

$$C_6H_{12}O_6 + 6O_2 \rightarrow 6CO_2 + 6H_2O$$

Molecular weight of Glucose = 72 + 12 + 96 = 180

$6O_2 \rightarrow 6 \times 32 = 192$

192g of Oxygen required to oxidize 180g of Glucose

The ThOD of 108.75 mg/l of glucose is

$$108.75 \times \frac{192}{180} = 116\ \text{mg/l of Oxygen}$$

Example 5.8 Waste water is analyzed and shown to contain 120mg/l of ethylene glycol and ($C_2H_6O_2$) 150 mg/l of phenol (C_6H_6O). Determine COD and TOC.

Solution

$$C_2H_6O_2 + 2.5\,O_2 \rightarrow 2CO_2 + 3H_2O$$

$$\frac{C_2H_6O_2}{62} + \frac{2.5O_2}{2.5\times32} \rightarrow 2CO_2 + 3H_2O$$

$$COD = \frac{2.5\times32}{62} \times 120 = 154.84\ \text{mg/l}$$

$$TOC = \frac{2\times12}{62} \times 120 = 46.45\ \text{mg/l}$$

In phenol $C_6H_6O + 7O_2 \rightarrow 6CO_2 + 3H_2O$

$$COD = \frac{7\times32}{94} \times 150 = 357.45\ \text{mg/l}$$

$$TOC = \frac{6\times 12}{94} \times 150 = 114.9\ \text{mg/l}$$

Total COD = (154.84 mg/l + 357.45 mg/l) = 512.28 mg/l

Total TOC = (46.45 mg/l + 114.9 mg/l)= 161.34 mg/l

5.3 WASTE WATER TREATMENT PROCESS

Wastewaters are usually classified as industrial wastewater or municipal wastewater (Table 5.1). Industrial wastewater with characteristics compatible with municipal wastewater is often discharged to the municipal sewers. Many industrial wastewaters require pre-treatment to remove non-compatible substances prior to discharge into the municipal system. Characteristics of industrial wastewater vary greatly from industry to industry, and consequently, treatment processes for industrial wastewater also vary, although many of the processes used to treat municipal wastewater are also used in industrial wastewater treatment. Municipal wastewater is primarily organic in content and a significant number of industries including chemical, pharmaceutical and food industries have high organic waste loads. This means that the main treatment processes are geared towards organics removal (Table 5.2). The treatment processes are classified as:

(i) Pre – treatment - Physical or chemical

(ii) Primary treatment - Physical

(iii) Secondary treatment - Biological

(iv) Advanced treatment - Physical &/or chemical and / or biological

Table 5.1 Flow chart of possible unit operations (unit processes) for three wastewater types: Domestic, Chemical and Food

Treatment Categorization	Municipal wastewater	Chemical industry wastewater	Milk processing wastewater
Physical	Equalization	Equalization	
Pre-treatment	Coarse screening		Coarse screening
	Fine screening	Air stripping	Fine screening
	Grit removal	Oxidation/reduction	Grit removal
		Flotation	Flotation
Chemical		Neutralization	Neutralization
Pre-treatment			
Primary treatment	Primary treatment	Primary clarification	
Secondary treatment	Biological treatment	Biological	Bio-towers
	Activated sludge	Activated sludge	Activated sludge
	or Trickling filters		
	or Aerated lagoons		
	or RBC		
	Secondary clarification		
Nutrient removal	Biological (N)		

	Chemical (P)		
	Biological (P)		
Tertiary treatment	Sand filters	Sand filters Adsorption Chemical oxidation Oxidation	Sand filters
Advance wastewater Treatment		Ion exchange	
Sludge treatment and disposal	Yes	Yes	Yes

Table 5.2 Flow chart outline of unit processes and operation in wastewater treatment

	Unit process	*Operations*
1.	Pre-treatment	- Equalization, screening, neutralization of pH etc.
2.	Primary treatment	- Sedimentation, Chemical treatment coagulation
3.	Secondary treatment	-
(a)	Activated sludge Treatment, sludge digesters	- Activated sludge treatment, oxidation and decomposition, sludge drying and sludge disposal
(b)	Biological filters (Trickling filters) Rotating Biological contactor	- Biological oxidation, aerobic & anaerobic treatment
(c)	Stabilization Ponds (i) Facultative ponds (ii) Aerobic lagoons (iii) Anaerobic lagoons	- **Sedimentation**
4.	Tertiary treatment Electrodialysis, chemical oxidation or nutrient removal	- Ion exchange, reverse osmosis, (advanced)
5.	Disinfection	- Chlorine disinfection, U.V. radiation
6.	Release of effluent	- To agricultural fields, water bodies

Substances that impair water quality

Substance	**Undesirable effect**	**Maximum allowed level (mg/l)**
Chloride	Taste and smell	200
Fluoride	Fluorosis	1
Detergents	Taste and foaming	0.2
Hydrogen sulfide	Taste and odor	0.05
Calcium	Water hardness	50
Magnesium	Water hardness	30

Copper	Taste	0.05
Iron	Taste and color	0.1
Zinc	Taste and color	5
Ammonia	Growth of organism and pH	0.05

5.4 PRE – TREATMENT OF WASTE WATER

It is the process or processes that prepare wastewater to a condition that it can be further treated in conventional secondary treatment biological processes. In municipal wastewater, pre-treatment means the removal of floating debris, grit and oily scums. These pollutants would inhibit the biological process and possibly damage mechanical equipment.

Industrial influents may additionally require chemical pre–treatment in the form of air stripping (ammonia removal), oxidation, reduction (heavy metal removal) and air floatation (oil removal) (Fig. 5.6 and 5.7).

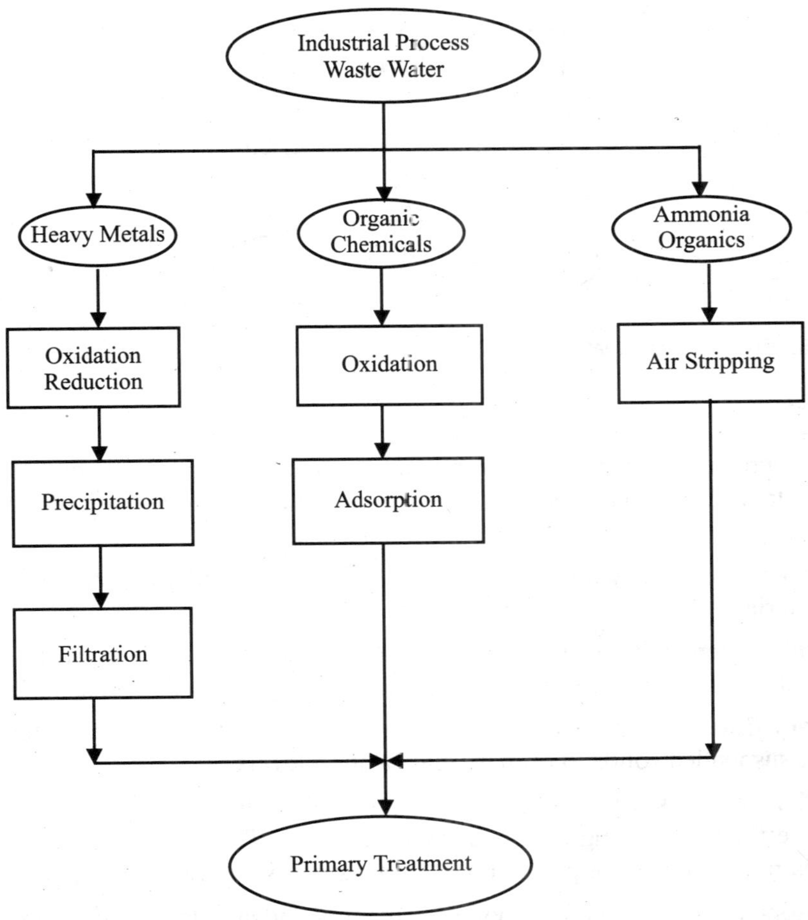

Fig. 5.6 Typical Chemical Pre-Treatment of Industrial Effluents

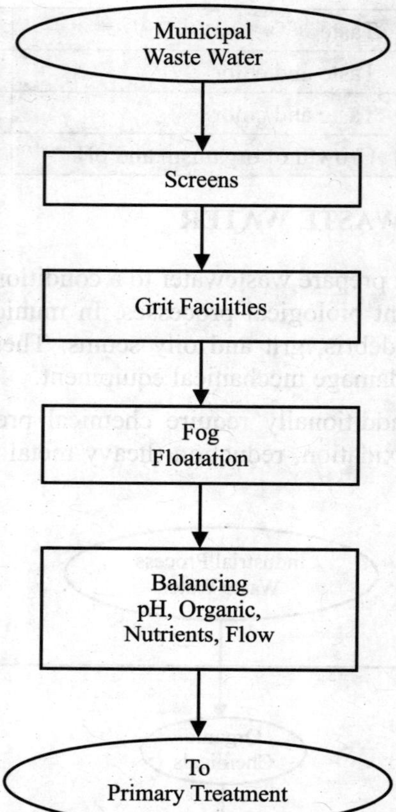

Fig. 5.7 Typical Physical Pre-Treatment of Municipal and Combined Municipal/Industrial Waste Water

5.4.1 Screenings

The objective of screening is to remove large floating materials (e.g. plastic bottles, rags etc.) and so protect downstream mechanical equipment (pumps). There are four types of screens in normal use:

- **(i)** **Coarse screens** with openings greater than 6 mm. diameter that remove large materials.
- **(ii)** **Fine screens** with openings in the range 1.5 to 6 mm diameter, which are sometimes used as a substitute for primary clarification (e.g. when activated sludge is used).
- **(iii)** **Very fine screens** with openings in the range 0.2 to 1.5mm diameter, which reduce the suspended solids to primary clarification levels.
- **(iv)** **Micro screens** with openings in the range 0.001 to 0.3 mm diameter which are used for effluent polishing as a final treatment step. These are used as a single one step treatment process for predominantly inorganic wastewater, e.g. quarry washings.

Comminutors are a traditional method of screening and shredding the retained material and then allowing it back into the flow. They are no longer recommended since items like plastic pieces can find their way to the biological plant creating inhibition conditions for the microbial population. Screens are designed to accommodate through flow velocities greater than 0.5 m/s and less than 1.2 m/s with maximum head losses of about 0.7 m.

5.4.2 Grit Channels

Municipal wastewater contains a wide assortment of inorganic solids of size about 1 mm. such as pebbles, sand, silt, egg shells, glass and metal fragments. Operations to remove these inorganic will also remove some of the larger, heavier organics such as seeds, bone chips, etc. Together, these compose the material known as grit in wastewater. Grit usually does not exist in industrial wastewater. Most of the substances in grit are abrasive in nature and will cause accelerated wear on pumps and sludge – handling equipment with which it comes in contact. Additionally these materials are not biodegradable and occupy valuable space in sludge digesters. It is therefore desirable to separate them from the organic suspended solids.

The two common types of grit collection devices are:

(i) *Helical flow Aerated Grit Chamber*

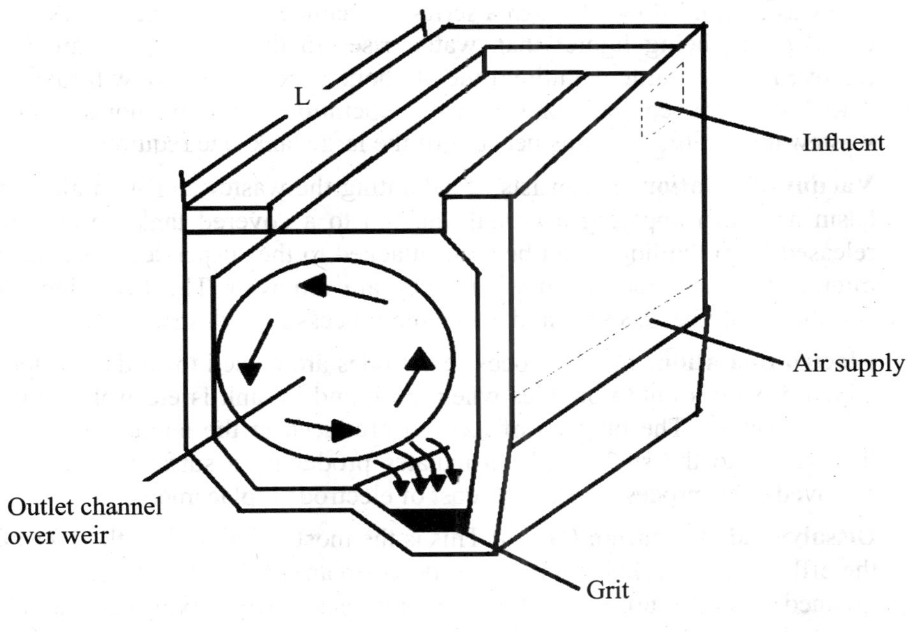

Fig. 5.8 Helical Flow Pattern in Aerated Grit Chamber

Air is introduced along one side of the channel near the bottom and this causes a spiral motion perpendicular to the main flow direction. The heavier grit particles settle while the lighter organic matter remains in suspension and is carried on to primary clarification (Fig. 5.8). The design philosophy is based on an adequate retention time in the basin of about 3 min. Aerated grit chambers have proved more efficient than the horizontal flow type and the grit tends to be cleaner.

(ii) *Horizontal Flow Grit Channel*

It is a more traditional grit chamber. The design philosophy is based on a minimum through flow velocity of about 0.3 m/s and a detention time of about 1 min. They are designed to remove grit particles greater than 0.15 mm with associated settling velocities of about 0.01 m/s.

5.4.3 Floatation

Floatation is the buoyancy unit process of separating solid particles from a liquid phase. This is not an essential unit process as the amount of solid is insignificant. The solid may include fats, oils and greases (FOGs) in municipal works and waste oil products in industrial plants. The process of separation involves introducing air bubbles at the bottom of a flotation tank. These air bubbles attach themselves to the particulate matter and their combined buoyancy encourages the particles to rise to the surface where they can be removed by skimming. Floatation is used when suspended particles have a settling velocity so low that they are not settleable in sedimentation tanks. The particulates may settle if chemically assisted by a coagulant. Floatation system includes the following as discussed:

(i) **Gravity floatation:** It is common in very small industries and automobile garages. The waste liquid flows through a series of chambers and in the process the grease or oil particles, being lighter than water, rises to the surface and are mechanically removed. A hydraulic retention time of about 30 min. and a flow through velocity of 4 to 6 m/h are required for successful operation. They are not suitable for urban wastewater treatment plants because of the large tank size required.

(ii) **Vacuum floatation:** It consists of saturating the wastewater with air in an aeration basin and then applying a partial vacuum to a covered tank. Minute bubbles are released from the liquid and become attached to the suspended particles which then migrate to the surface from where they are removed. The floatation procedure is common with the U.S fruit and vegetable processing industry.

(iii) **Electro floatation:** In this process electrodes are placed towards the floor of a tank which produces micro-bubbles when the liquid in tank is electrolyzed by means of current supply. The bubbles of oxygen produced at the anode end rise and attach themselves to the suspended particulates producing a surface scum which can be removed. This process has a high cost of electrode replacement.

(iv) **Dissolved air floatation (DAF):** This is the most successful method. Usually part of the effluent is recycled from a point downstream of the DAF unit. Recycled flow is retained in a pressure vessel for a few minutes where mixing and saturation of the flow with air occurs (Fig. 5.9). This recycled effluent is added to the DAF unit where it is mixed with the incoming raw effluent. As the pressure returns to atmospheric conditions, the dissolved air comes out of solution, forming fine bubbles, which rise to the surface bringing grease matter with them, where they are removed. The design upward flow velocity is in the range of 4 to 6 m/h and the air to solids ratio by weight is in the range of 1 to 5 percent.

(v) **Air Floatation:** It is a variation of DAF where air is directly introduced to the flotation tank by means of an impeller.

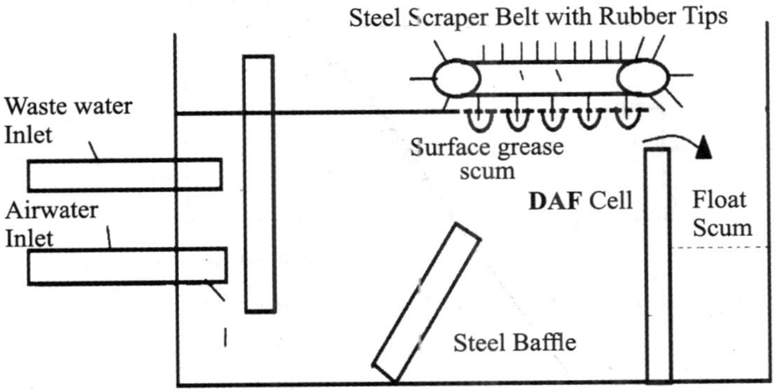

Fig. 5.9 Dissolved Air Floatation (DAF) Unit

5.4.4 Equalization

To ensure that a wastewater treatment plant receives an effluent that it is capable of handling without distress, equalization (balancing) may be required. This may include one or more of the following:

(i) Flow equalization
(ii) Organic equalization
(iii) Nutrient equalization
(iv) pH equalization (Neutralization or pH correction)

In **flow *equalization,*** the flow is balanced or spread out equally over 7 days so that the flow arriving into the plant is the same for each of the 7 days. Similarly, in case of organic or pollutant load balancing, industry may at diffluent times during the week have a high COD effluent, lasting only a few hours. If this were sent directly through the treatment plant it may cause a shock load with consequent problems. It is therefore, usual to balance the high load such that a more even load is sent to the plant for treatment. This is done by retaining the pollutant load in a balance or equalization bank, prior to treatment.

Nutrient balancing is where nutrients may be added to the influent wastewater in case the wastewater is deficient in nutrients. pH balancing may be required should an influent to a wastewater treatment plant be too high or too low in pH for optimum secondary treatment of that waste. It is desirable that the pH be in the range of 6.5 to 8.5 for activated sludge treatment systems. The objectives of balancing may be summarized as:

(i) Equalization of flows to minimize flow surges.
(ii) Equalization of organic loads to dampen fluctuations.
(iii) Neutralization of pH variations to bring it to the range 6.5 to 8.5.
(iv) Provision of continuous influent to the plant.
(v) Provision of continuous effluent from the plant to the receiving water body.
(vi) Control of high toxicity loads.

Two layouts are common such as in-line system (Fig. 5.10) and side-line system (Fig. 5.11)

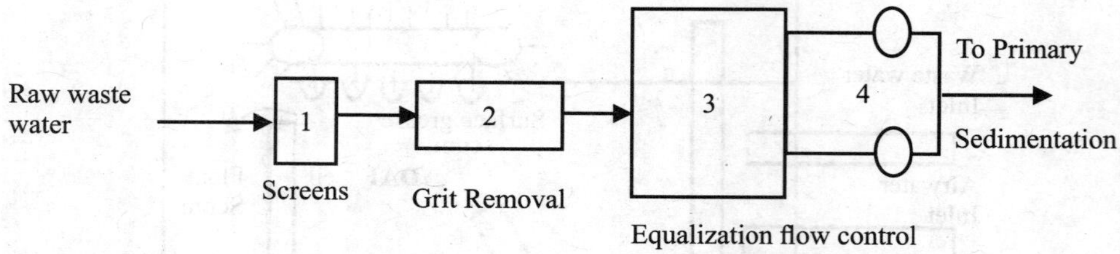

Fig. 5.10 In-line equalization (Adapted from ASCE and WEF, 1991)

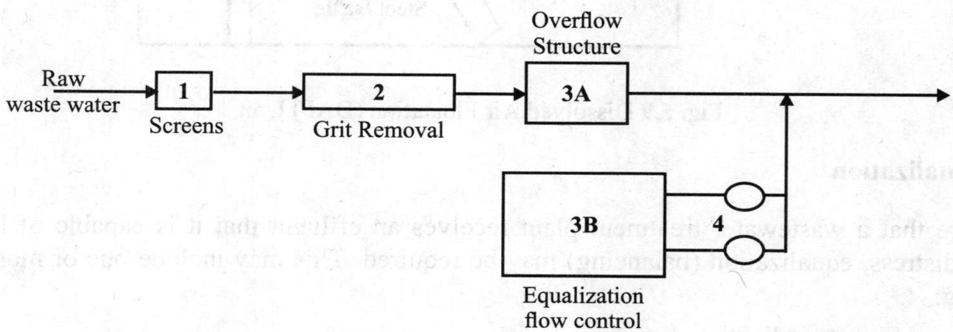

Fig. 5.11 Side-Line Equalization (Adapted from ASCE and WEF, 1991)

Equalization facilities generally may be of three types:

(i) Constant flow, variable waste strength

(ii) Constant waste strength, variable flow

(iii) Constant flow, constant waste strength

Pollutants in sewage discharges

Constituents	Sources	Effects on water quality
Oxygen-demanding substances	Organic compound; feces	Oxygen demand (BOD)
Household products	Domestic and industrial	Toxic to aquatic life
Food, beverages and oil	Domestic	Algal nutrients; harmful to aquatic life
Salts	Domestic and industrial	Water salinity
Heavy metals	Industrial	Toxicity
Particulate matter	All sources	Harmful to aquatic life
Pathogens		

5.5 PRIMARY TREATMENT OF WASTE WATER

Primary treatment is often called **clarification**, **sedimentation** or **settling**. This is the unit process where the wastewater is allowed to settle for a period ($\approx$ 2h) in a settling tank and so produce a clarified liquid effluent in one stream and a liquid – solid sludge (called **primary sludge**) in a second stream. The benefits of primary treatment include:

(i) Reduction in suspended solids.

(ii) Reduction in BOD_5.

(iii) Reduction in the amount of waste activated sludge (WAS) in the activated sludge plant.

(iv) Removal of floating material.

(v) Partial equalization of flow rates and organic load.

Primary treatment is quiescent sedimentation with surface skimming of floating matter and grease, and bed level collection and removal of settled sludge. Sedimentation is carried out in a variety of tank configurations including circular (most common), rectangular and square.

They may be flat bottomed or hopper bottomed. Wastewater enters the tank, usually at the centre, through a well or diffusion box. The tank is sized so that retention time is about 2 hrs. to 3 hrs. In this quiescent period, the suspended particles settle to the bottom as sludge and are raked towards a central hopper from where the sludge is withdrawn (Fig. 5.12). The clarified water is discharged over a perimeter weir at the surface of the tank, at a rate known as the **basin overflow rate** or **surface overflow rate** (SOR). The units of SOR are $m^3/day/m^2$. The last m^2 is the plan area of the tank. SOR is different to the weir overflow rate (WOR). The units of the WOR are $m^2/day/m$. The last m is the length (perimeter) of the weir.

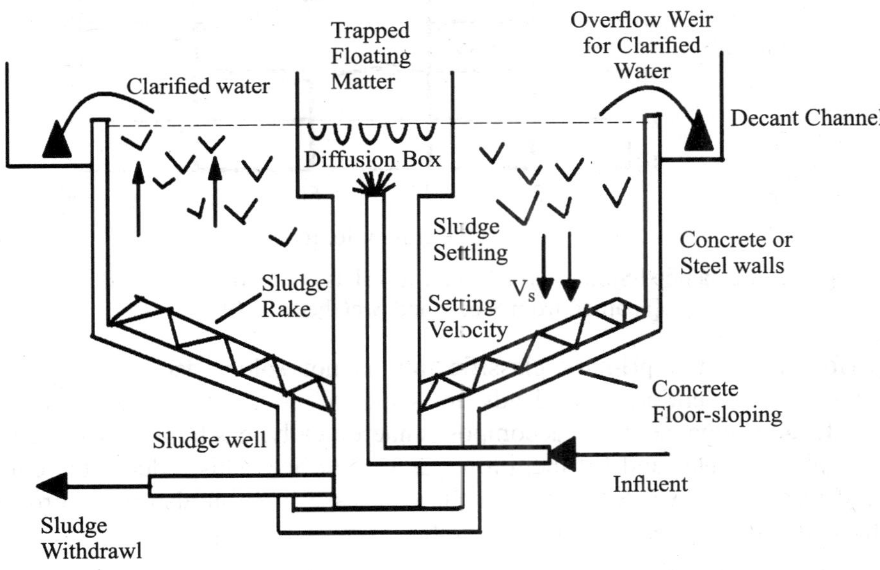

Fig. 5.12 Typical Circular Primary Settling Tank

Traditionally, the design criteria are:

(i) Basin overflow rate (Fig. 5.13):

(a) Depth

(b) Surface geometry

(c) Hydraulic retention time

(d) Weir rate (m^2/day/m)

These criteria may be adequate for the design of the tank. Performance criteria were established to monitor and improve the day to day performance. These criteria include:

(i) Influent flow rates and their variation (daily variation)

(ii) Influent waste strength rates and its variation

(iii) Recycle influent streams:

(a) From activated sludge or septage

(b) Supernatants from sludge dewatering

(c) Washings from tertiary filter processes.

The above parameters may vary from hour to hour or from day to day.

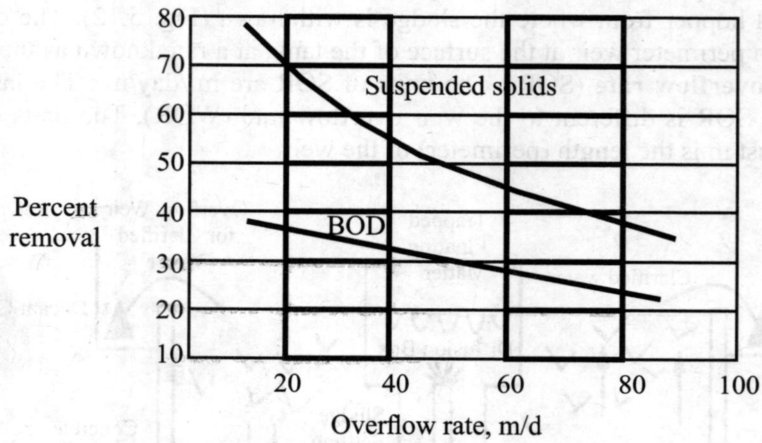

Fig. 5.13 Suspended Solids and BOD Removal as a Function of Overflow Rate [adapted from Steele and McGhee (5-50)]

The performance of a primary classification is not solely dependent on influent flow variations. Excessive retention time leads to septicity as there is no mixing in primary sedimentation. If the sludge begins to decompose anaerobically, gas bubbles will be produced and will adhere to solid particles and lift them towards the surface. This reduces the compactness of the sludge and makes removal less efficient. In such case there is also no improvement in the quality of clarified water on the influent wastewater.

The quality of sludge removed in primary sedimentation may depend on several variables, including the strength of the incoming waste, the efficiency of the clarifier and the conditions of the sludge (i.e. specific gravity, water content, etc.). With good performance management, removal rates of 50 to 70 percent for suspended solids and 25 to 40 percent for BOD_5 can be achieved.

5.5.1 Chemically Enhanced Primary Sedimentation

The addition of coagulant chemicals (iron salts, lime and alum) before sedimentation promotes flocculation of fine suspended matter into more readily settleable flocs. This increases the efficiency very substantially of SS and BOD_5 removal rates. A disadvantage of chemical coagulants is an increase in primary sludge which is a chemical – type sludge quite different to the biological sludge, from primary sedimentation.

Water Quality Index

Water quality is influenced by various geographical, natural and anthropological contaminants. Water quality index is based on some very important parameters that can provide sample indicators of water quality. The WQI translates a wide variety of environmental indicators into a simple system. Easy-to-understand information helps general public to comply with environmental regulations.

The WQI is evaluated taking into account various parameters (DO, fecal coliform, pH BOD5, temperature change, total phosphorus, nitrate, turbidity, and total dissolved solids) that influence the water quality.

5.5.2 Sludge Quantities from Primary Settling

The amount of sludge produced during primary settling will depend on the through flow, the total suspended solids and the efficiency of solid removal. This quantity can be readily determined from:

$$Sm = Q \times TSS \times E \; Kg/day$$

Where, Q = through flow, m^3/day

TSS = total suspended solids, mg/l

E = removal efficiency

Sm = sludge quantity, Kg / day

5.6 SECONDARY TREATMENT OF WASTE WATER

The effluent from primary treatment still contains 40 to 50 percent of the original suspended solids and 70 to 80 percent of the dissolved organics and inorganics. To meet the minimum Environmental Protection Act (EPA) standards for discharge, the organic fraction, both suspended and dissolved must be significantly reduced. This organic removal, referred to as **secondary treatment** may consist of chemical – physical processes or biological processes. The main purpose of secondary treatment is to reduce the BOD_5 value which does not benefit as much as SS from primary settling.

In other words, secondary treatment should be a process that is capable of biodegrading the organic matter into non – polluting end products, e.g. H_2O, CO_2 and biomass (sludge). The end product liquid effluent should be well stabilized or well oxidized so that it does not provide a food source for aerobic bacteria in the receiving water body.

5.6.1 Principles of Biological Oxidation

The mechanisms of removal of organic matter are discussed below:

5.6.1.1 Air stripping

Air stripping of VOC occurs in aerobic systems. The breakdown of carbonaceous material by aerobic degradation emits CO_2 and other VOCs into the atmosphere.

5.6.1.2 Adsorption

Adsorption of non-degradable organics onto biological solids is not significant, but it does occur for particular organics including the pesticides, e.g. lindane. Heavy metals will adsorb onto biomass and bioaccumulate, resulting in end product sludges containing heavy metals.

5.6.1.3 Biodegradation

It is the dominant mechanism of organics removal for municipal and most industrial wastewaters. Most treatment plants now use the activated sludge system for this purpose (Fig. 5.14). The understanding of the activated sludge process was first introduced in Manchester by Arden and Lockett (1914), has made significant advances.

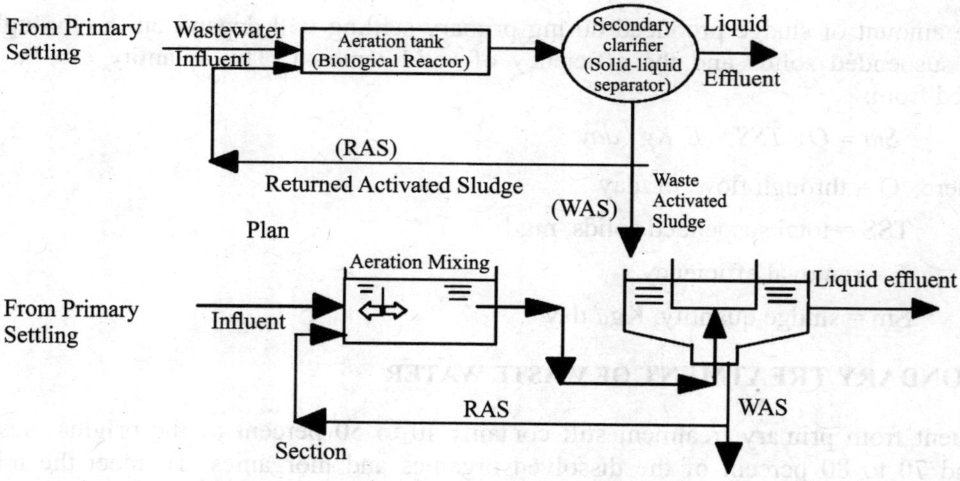

Fig. 5.14 Typical Layout of Activated Sludge System

It comprises of two box structures, the aeration tank and the secondary clarification tank. The **aeration tank**, while having many possible configurations, basically retains the influent wastewater for a number of hours (or days) in a well – mixed / aerated environment, before forwarding the effluent for further settling to the **secondary clarification tank**. The end products of the clarification tank are clarified liquid effluent, ready for discharge to open water bodies and a liquid – solid sludge.

A fraction (about 29 percent) of the sludge is returned to the aeration tank and is called **returned activated sludge** (RAS). The sludge contains a high density of live microbial biomass and in returning part of it; an active population of microbes are always maintained in the aeration tank. The influent wastewater is the food source for the resident microbes in the aeration tank. The microbes biodegrade this feedstock into new microbial cells in the presence of aerated water.

Other end products include CO_2, NO_2^- and SO_4^-. In the clarifier, the excess biomass settles out as sludge and about 80 percent of this is removed for further treatment and subsequent disposal.

The mixed liquor suspension in the aeration tank contains wastewater, living and dead micro-organisms and inert biodegradable and non – biodegradable suspended and colloidal matter. The particulate fraction of the 'mixed liquor is called the **mixed liquor suspended solids** (MLSS). This value is generally of the order of 2000 to 4000 mg/l for a healthy microbial suspension in the aeration tank. The MLSS of RAS is of the order of 10, 000 to 20, 000 mg/l. This is a measure of the microbial population and essential to keep the MLSS in the aeration tank > 2000 mg/l, otherwise the microbial population is not large enough to biodegrade the incoming organics.

5.6.1.3.1 *Bacterial Growth in Pure Cultures*

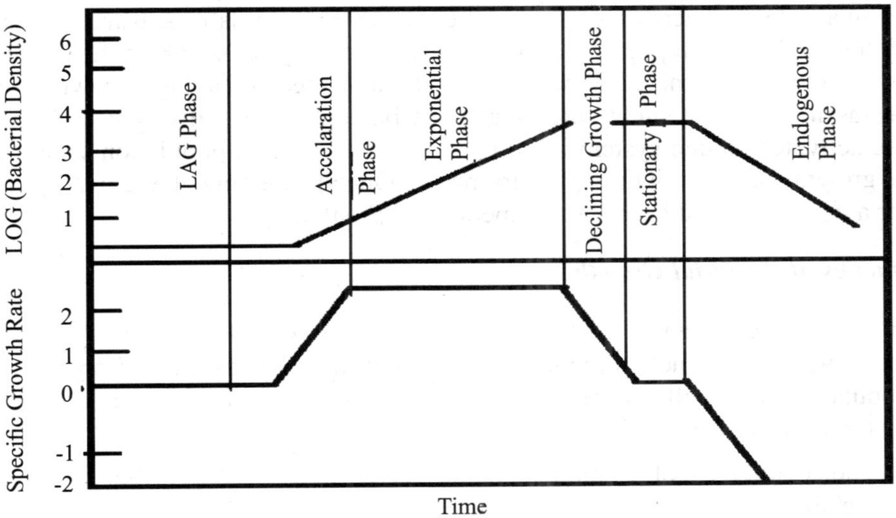

Fig. 5.15 Characteristic Growth Curves of Cultured Microorganisms

Fig. 5.15 is a schematic growth curve for typical bacteria. At different points on the growth curve, the cells may have different size and a different mass of sobbed substrata. When a bacterial inoculum is placed in a closed vessel with ideal conditions and excess food, the growth and decay of the bacterial cell number follows the curve of Fig. 5.15. The micro-organisms must first become acclimated to their surrounding environment and to the food provided. The acclimation period, called the **lag phase**, is represented by segment ⊥ on the curve and will vary in length, depending on the history of the seed organisms. If the organisms have been accustomed to a similar environment and similar food, the lag phase will be very brief or may be up to six weeks for some complex industrial wastes. Once growth has been initiated, it will proceed quite rapidly. Bacterial cells reproduce by binary fission; that is, cells divide into segments that separate to become two new independent cells. A significant parameter called the regeneration or doubling time is the time required for a cell to mature and separate, depends on environmental factors and food supply and may be as short as 20 to 60 min, although it may take a few days in exceptional environments. When maximum growth is occurring, the rate of reproduction is exponential according to the equation:

$$N = 2^{n-1}$$

Where N is the number of organisms produced from one individual after n regeneration times. Maximum growth thus occurs at a logarithmic rate and represented by segment 2 on the growth curve called the **log phase**.

Maximum growth cannot continue indefinitely. The food supply may become limiting, environmental conditions may change (i.e. overcrowding, waste product buildup etc.). Segment 3 of the curve **stationary phase** represents that the population remains constant with the rate new cell synthesis equal to the death rate of old cells.

The following phase, called the **endogenous phase**, is where the organisms are surviving of their own stored energy and consuming the dead cells. The number of cells begin to decline and there is then an accelerated death phase when there is no biodegradable substrata. This schematic is typical of the real microbial population of activated sludge systems, where the population tends to be complex, interrelated mixed populations with each species having its own growth curve. Most organisms in activated sludge are single cell bacteria with soil origins. The objective in operating the activated sludge system is to try to retain the mixed population within a desirable range on the growth curve for optimum performance. Outside the range there may be an excess of organisms or a deficiency and both lead to operational problems.

5.6.1.3.2 *Kinetics of Bacterial Growth*

When a biodegradable organic food source is supplied to heterotrophic (utilizing organic materials for energy, as distinct from autotrophic which use CO_2 as the carbon source) micro – organism population in a well – aerated environment, the response is as follows, adapted from Ekama and Marais (1984):

(i) The readily soluble biodegradable COD goes through the cell wall and is metabolized quickly.

(ii) The slowly biodegradable particulate COD is adsorbed on to the organisms and stored. This quick reaction removes all the particulate and colloidal COD. Over time the stored COD is broken down by extracellular enzymes, transferred through the cell wall and metabolized. The rate limiting step in the overall synthesis is the enzymatic breakdown at rates about 10 percent of the readily biodegradable COD rate.

(iii) Some of the COD metabolized is converted to new cells, while the remainder is lost as heat in the energy process required for the new cell synthesis. Oxygen externally supplied is used up in this energy process, such that the amount of oxygen utilized is proportional to the COD lost.

(iv) At the same time, there is a net loss of live biomass, termed **endogenous mass loss**, where some of the organisms utilize as food their own stored food materials and dead cells. The endogenous degradation is continuous and relatively constant at about 10 to 20 percent per day.

The biochemical equation for bacterial cell respiration and synthesis in using organic matter as substrate in the activated sludge process is:

$$\text{Organic matter} + O_2 + \text{Nutrients} \xrightarrow{\textit{Bacteria}} CO_2 + NH_3 + \text{New biomass} + \text{other end products}$$

For mixed cultures, as in activated sludge, biomass is the parameter of interest, rather than the number of organisms. In the log growth phase, the rate of biomass increase is proportional to the initial biomass concentration and is represented by the first order equation:

$$\frac{dx}{dt} = \mu x \tag{5.18}$$

Where,

$\frac{dx}{dt}$ = growth rate of biomass, mg/l/day

x = concentration of biomass, mg/l

μ = specific growth rate constant, d^{-1} (mass of cells produced/ mass of cells present per unit time)

Let Xo represent the biomass at time $t = 0$, integration of equation (5.18) is

$$\text{In X} = \text{InXo} + \mu t$$

$$\text{In}\left(\frac{X}{Xo}\right) = \mu t$$

$$X = Xo \, . \exp(\mu t) \tag{5.19}$$

The growth rate that follows equation (5.19) is called the **exponential growth rate**. It is assumed that equation (5.19) holds as long as there is no change in the biomass composition or environmental conditions (Fig. 5.16).

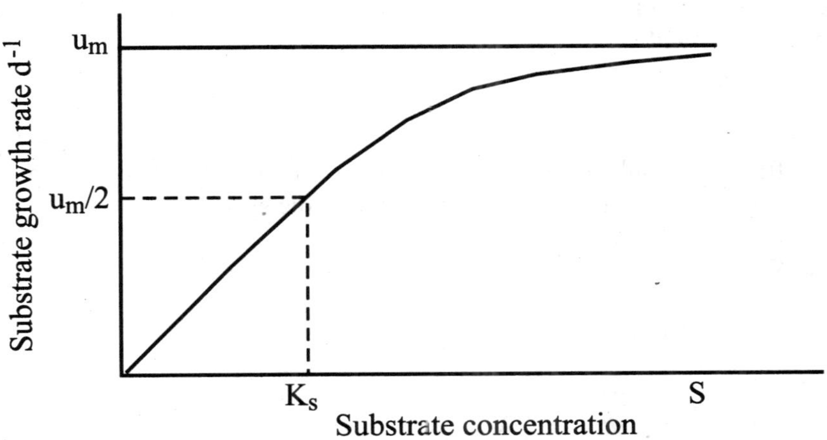

Fig. 5.16 Substrate Concentration versus Specific Growth Rate

In mixed cultures, it is not yet possible to measure. Typically, there are essential requirements for microbial growth including availability of substrate or nutrients. If an essential requirement is limited, growth slows or ceases. Monod (1949), noting that the growth rate (dx/dt) was a function of organism concentration (X) and also of a limiting nutrient concentration, showed experimentally that

$$\mu = \mu_m \frac{S}{K_s + S} \tag{5.20}$$

Where,

S = concentration of limiting substrate, mg/l.

μ_m= maximum growth rate, d^{-1}

K_s = half – saturation constant, i.e. the concentration of S when $\mu = \mu_m/2$ (mg/l) as shown in Fig. 5.16

Equation (5.19) assumes only growth of micro-organisms. However, there is growth of one set of organisms and die –off of another set, occurring simultaneously. To take account of die-off, an endogenous decay (rate = K_d) is used, so equation (5.18) becomes

$$\frac{dx}{dt} = \mu x - k_d X$$

$$\text{Biomass production} = \frac{dx}{dt} = \left(\frac{\mu_m S}{Ks + S}\right) X - K_d X \tag{5.21}$$

If all substrate (S) could be converted to biomass (X) then the rate of substrate utilization is:

$$\text{Ideal:} \quad -\frac{ds}{dt} = \frac{dx}{dt} \tag{5.22}$$

(Similar to the energy analogy, dissipation = production).

However, such an idealization cannot occur due to inefficiencies in the conversion process and a yield coefficient (Y<1) is introduced such that the rate of substrate utilization is in excess of the rate of biomass generated:

$$\text{Real:} \quad -\frac{ds}{dt} = \frac{1}{Y}\frac{dx}{dt} \tag{5.23}$$

Where, Y =fraction of substrate converted to biomass, mg/l of substrate typically 0.4 to 0.8 in aerobic systems and 0.08 to 0.2 for anaerobic systems

$$\text{Anaerobic systems} \quad = \frac{dx}{ds} \tag{5.24}$$

Substitution into equation (5.23) becomes:

$$\text{Substrate utilization:} \quad -\frac{ds}{dt} = \frac{1}{Y}\left(\frac{\mu_m S X}{K_s + S}\right) \tag{5.25}$$

Equation (5.22) for biomass production and equation (5.25) for substrate utilization are the fundamental biological process design equations for different reaction configurations. They are used later in the mass balance equations for complete mix and plug flow activated sludge systems. They are also the principal equations in anaerobic digestion.

5.6.1.3.3 *The F/M Ratio*

The F/M (food to microbes) ratio is the most useful design and operational parameter of activated sludge systems. The activated sludge system is a continuous process with growth and decay of micro – organisms. A system achieves equilibrium when the food substrate and the micro – organisms consuming it are in balance. The equilibrium parameter is known as the **F/M ratio** or **the food to microbes ratio**. This ratio controls the rate of biological oxidation and the mass of organisms, by maintaining microbial growth in either the log, declining or endogenous phase (Gray, 1990).

The type of activated sludge can be defined by its F/M ratio (Fig. 5.17)

(i) Extended aeration 0.03 < F/M <0.8

(ii) Conventional 0.8 < F/M < 2

(iii) High rate FM > 2

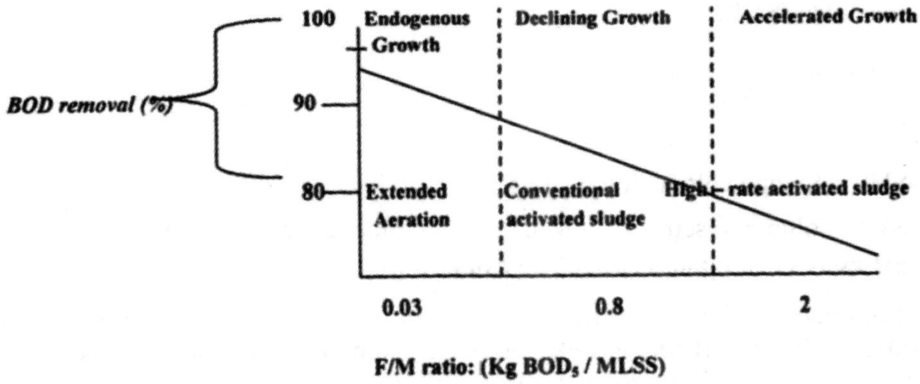

$$F/M = \left(\frac{KgBOD_5 / m^3}{KgMLSS / m^3} \right) / day$$

Fig. 5.17 Schematic of F/M Ratio

The F/M ratio is defined as:

$$F/M = \frac{BOD\ of\ sewage\ (kg/m^3)\ \times\ influent\ flow\ (m^3/d)}{Reactor\ solids\ (kg/m^3)\ \times\ Reactor\ volume\ (m^3)}$$

$$= \frac{SoQ_o}{XV}$$

$$\therefore F/M = \frac{S_o}{QX}$$

Where,

S_o = Concentration of influent BOD (Kg/m^3)

Q_o = Influent flow rate (m^3/day)

X = Concentration of reactor solids i.e. MLSS (Kg/m^3)

V = Reactor volume (m^3)

Q = Hydraulic retention time (days)

In the log or accelerated growth phase there is an excess of substrate, characterizing a high F/M ratio (>1.0). In the endogenous phase the F/M ratio is low at values generally less than 0.4 and ideally at around 0.2 for plug and 0.1 for complete mix systems. Removal rates of BOD are then highest and this is conventionally called **extended aeration**.

5.6.1.3.4 *Sludge Settlement Parameters*

In the process design of activated sludge, a principal feature is the use of recycle sludge. If the aeration basin did not receive this recycle sludge, it would ultimately become deficient in micro – organism population. One parameter used in the control of the amount of recycle sludge is the sludge volume index (SVI).

$$SVI = \frac{SV}{MLSS} \times 1000$$

Where,

SVI = Sludge volume index, ml/g

SV = Volume of settled solids in 1 liter cylinder after 30 min, ml/l

MLSS = Mixed liquor suspended solids, mg/l

Sludge has good settling characteristics if the SVI range is 80 to 120 for an MLSS range of 2000 to 3500 mg/l. As the MLSS increase, the solid loading is higher, producing a lower SVI. An alternative to a lower SVI is to increase the aeration basin size.

5.6.1.3.5 *Nitrification & Denitrification*

The nutrients, nitrogen and phosphorus contribute to eutrophication of lakes and slow moving waters. Therefore, most countries legislate for the removal of nitrogen and phosphorus.

$$N_2 \rightarrow \underbrace{\overset{\text{Ammonia N}}{(NH_3^+ - N)} \rightarrow \overset{\text{Nitrite N}}{(NO_2^- - N)}}_{\textit{Nitrification}} \rightarrow \underbrace{(NO_3^- - N)}_{\text{Denitrification}} \rightarrow N_2$$

Organic (N) and Ammonia (N) are undesirable in wastewater effluents since they both have a nitrogenous oxygen demand and Ammonia (N) is toxic to fish life. Nitrification is the biological process, using nitrifying bacteria (*Nitrosomonas* and *Nitrobacter*) to convert ammonia N to nitrate N in two process steps as follows:

$$NH_4^+ + \frac{3}{2}O_2 \xrightarrow{\textit{Nitrosomonas}} NO_2^- + 2H^+ + H_2O$$

$$NO_2^- + \frac{1}{2}O_2 \xrightarrow{\text{Nitrobacter}} NO_3^-$$

The end product of nitrification, nitrate N, leads to the stimulation of algae growth and is also associated with the infant disease of methanoglobinaemia. **Denitrification** is the biological process to convert nitrate N to inoffensive nitrogen gas (N_2). Conventional activated sludge processes did little for nitrification and usually nothing for denitrification. Extended aeration, however, because of its long hydraulic retention time (HRT) and low oxygen levels (near to anoxic state) towards the downstream end of the oxidation ditch achieves a level of nitrification.

5.7 ACTIVATED SLUDGE SYSTEMS

The more common activated sludge systems such as complete mix, plug flow, oxidation ditch, contact stabilization and sequencing batch reactors are discussed as below:

5.7.1 Complete Mix Reactors

The complete mix reactors have uniform characteristics throughout the entire reactor. Aeration can be provided by surface aerators with adjustable outflow weirs or by submerged bubble diffuser aeration systems. The operating food substrate to microbe's ratio (F/M) ranges from 0.04 to 0.07. DO levels are maintained throughout at not less than 2 mg/l. Generally the RAS from the clarifier is fed directly to the aeration basin where it is completely mixed with the existing contents (Fig. 5.18). A less desirable route for the RAS is to join with the influent stream and mix in the influent pipe before entry to the aeration basin.

Advantages of the complete mix system include the ability to withstand shock loads, due to the low F/M ratio, and good flexibility in being able to utilize a wide range of loads. It is particularly suitable for high organic industrial wastewater.

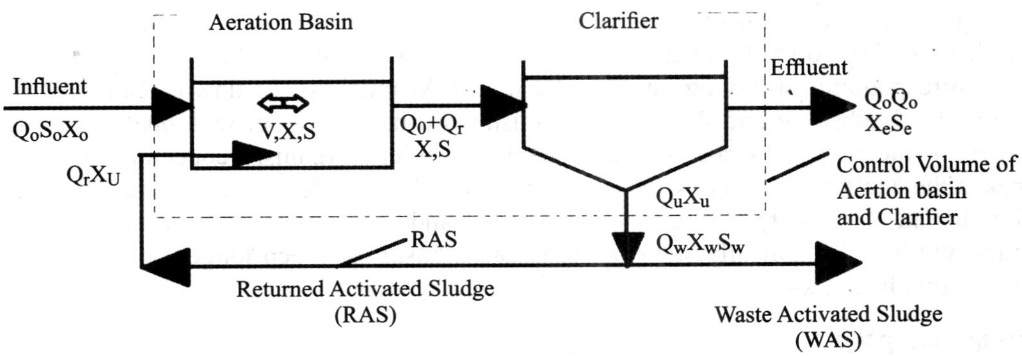

Fig. 5.18 Typical Complete Mix Activated Sludge System

For Fig 5.18, mass balance equations for biomass production (Eq.5.22) and for substrate utilization (Eq. 5.25) are as follows:

(i) **Mass balance of biomass production** – This uses the control volume of the aeration tank and clarifier (shown by dashed lines in Fig 5.18).

Influent Biomass+Biomass Production = Effluent Biomass + Sludge Wasted Biomass

(ii) **Mass balance of food substrate** - This uses the control volume of the aeration tank and clarifier.

Influent Substrate+Substrate Consumed=Effluent Substrate+Sludge Wasted Substrate

5.7.2 Plug Flow Reactors

Plug flow means that a plug of substrate influent to an aeration basin is moved forward without too much interaction with the plug that went before or after it. This means that satisfactory mixing occurs in the lateral direction, but none in the longitudinal direction (Fig. 5.19).

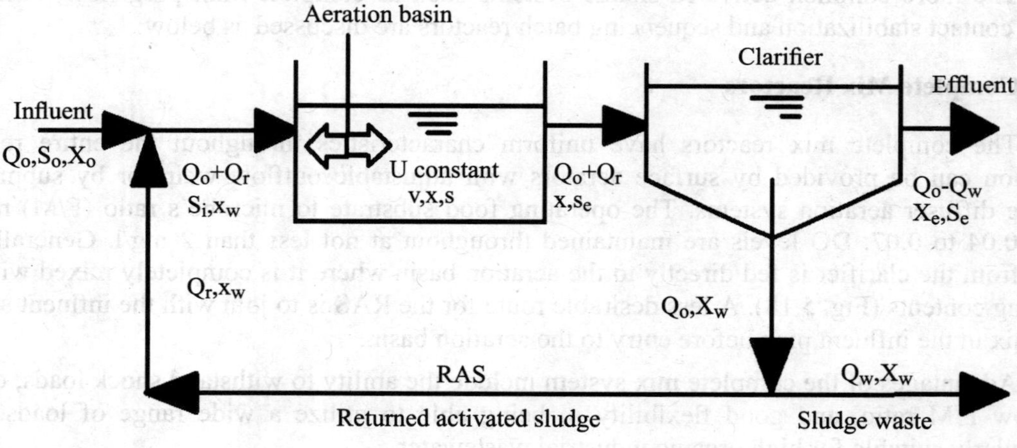

Fig. 5.19 Typical Plug Flow Activated Sludge System

The plug flow system is characterized by a high organic loading at the influent end of the basin, reducing as the outfall weir is approached. At the influent end there is an excess of food substrate corresponding to the log phase and a high F/M ratio. At the downstream end, there is shortage of food substrate and the micro-organisms are in the endogenous phase. Through the aeration basin the food substrate decreases while the micro–organism concentration increases. This makes the analysis of kinetics more complex than the complete mix system. Advantages of plug flow include the ability to treat fully all influent and allow no plugs to go untreated. This is so because the influent typically stays in the aeration basin for much longer periods than in the case of the complete mix system.

5.7.3 Oxidation Ditch

The aeration basin is usually a racetrack configuration with cage or brush aerators at one or more locations. The influent enters the basin just upstream of the aerator and moves forward as a plug from a location of high oxygen levels to a zone of low oxygen levels (Fig. 5.20). The effluent leaves the basin upstream of the influent. Sludge is recycled, as is typical with activated sludge. Oxidation ditches are characterized by long HRTs (~ 24h) and long SRTs or MCRTs (20 to 30 days). High – quality liquid effluents are also characteristic of oxidation ditches, but a disadvantage is a thin sludge which is difficult to thicken.

Extended aeration is more common with oxidation ditch systems. Extended aeration means long HRTs (16 to 24 h), enabling variable flow rates and flow strengths to be handled. The endogenous phase is the operational phase of the micro–organisms and while nitrification is common, it is also possible to achieve denitrification. Primary clarification is omitted as the key objective is to minimize the amount of sludge generated. Ideally, all substrate removal is converted to energy and then oxidized so that no excess biomass is generated and so no sludge handling is required.

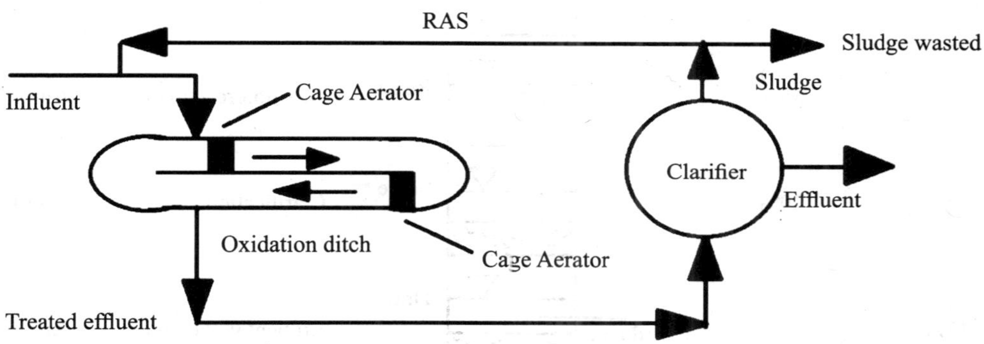

Fig. 5.20 Typical Layout of an Oxidation Ditch System

5.7.4 Contact Stabilization

In this system aeration is carried out in two phases in two different tanks. In the **contact tank** the suspended organic matter is adsorbed by the microbial mass and the dissolved organic matter is absorbed by the biomass (Fig. 2.10). The retention time is 30 to 60 min. The second tank, called the **stabilization tank,** with a 2 to 3 h retention period, is where the solids have been removed in a settlement zone and are further stabilized by re-aeration before being combined with the influent wastewater. The MLSS tends to be about 2000 mg/l in the contact tank and up to 20, 000 mg/l in the aeration basin.

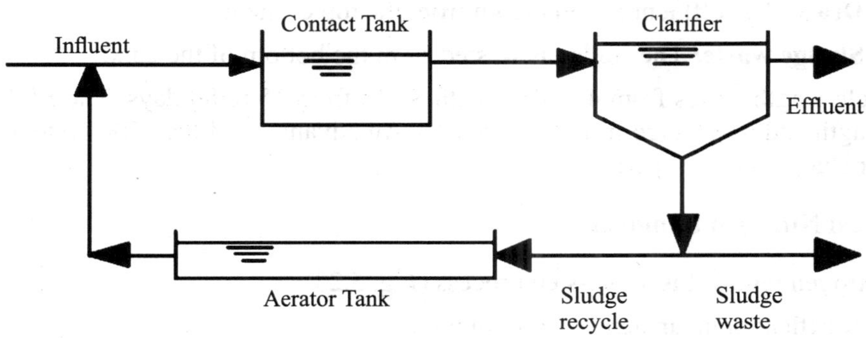

Fig. 5.21 Layout of Contact Stabilization System

The BOD removal efficiency is in the range 80 to 90 percent. This system is used for expansion of existing systems (where space is a limitation) and also in package plants.

5.7.5 Sequencing Batch Reactors (SBRs)

A sequencing batch reactor is a complete mix activated sludge system without a secondary clarifier. Within the single aeration basin, five different sequences are followed as shown in Fig. 5.22.

Volume	Cycle time	Purpose	Operation
25-100%	25%	Add substrate	Air on/off
100%	35%	Aeration/reaction time	Air on
100%	20%	Clarification	Air off
100->35%	15%	Effluent withdrawl	Air off
35->25%	5%	Sludge wasting	Air on/off

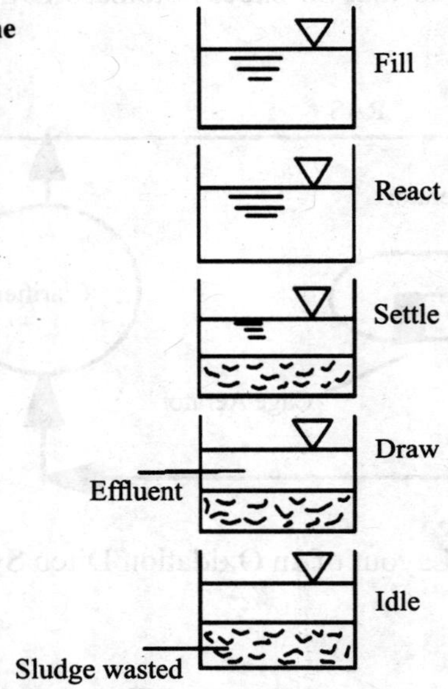

Fig. 5.22 Typical configuration for an SBR (adapted from USEPA, 1986)

Aeration and clarification are carried out in one tank. The sequences are as follows:

- **(i)** **Fill:** The basin fills with influent.
- **(ii)** **React**: The basin is aerated when 100 percent full.
- **(iii)** **Settle:** The basin is allowed to settle where sedimentation and clarification occurs.
- **(iv)** **Draw:** The effluent is withdrawn from the top of the tank.
- **(v)** **Sludge waste***:* The sludge is wasted from the bottom of the tank.

The cycle length varies from 4 to 48 h with SRTs from 15 to 80 days. The F/M ratio varies with cycle length and may range from 0.03 to 0.18. An advantage of the SBR system is that there is no need for sludge recycle period.

5.7.6 Biological Nitrogen Removal

Biological nitrogen removal is a two-step process (Fig. 5.23):

- (i) Nitrification in an aerobic environment.
- (ii) Denitrification in an anoxic environment.

5.7.6.1 Nitrification

Nitrification or ammonia oxidation depends on the population of nitrifying bacteria, which depend on temperature, ammonia concentration, organic substrate, pH, DO concentration, system mean cell residence time and the presence of nitrifying inhibitors:

$$NH_4^+ \xrightarrow[O_2]{\textit{Nitrosomonas}} NO_2^- \xrightarrow{\textit{Nitrobacter}} NO_3^-$$

The complete conversion from ammonia to nitrate is accomplished by two groups of aerobic microbes: *Nitrosomonas* for ammonia to nitrite and *Nitrobacter* responsible for conversion of nitrite to nitrate.

Complete reaction of ammonia to nitrate via nitrite is as follows:

$$2NH_4^+ + 3O_2 \xrightarrow{\textit{Nitrosomonas}} 2NO_2^- + 4H^+ + 2H_2O$$

$$2NO_2^+ + O_2 \xrightarrow{\textit{Nitrobacter}} NO_3^- + 2H^+ + H_2O$$

Total oxidation reaction is:

$$NH_4^+ + 2O_2 \longrightarrow NO_3^- + 2H^+ + H_2O$$

Based on this stoichiometric consideration, the oxygen requirement for complete conversion of ammonia to nitrate is 4.57 gm of O_2 per gm of nitrogen.

5.7.6.2 Denitrification

Denitrification follows, on completion of nitrification. It is achieved biologically under anoxic conditions, i.e. in the absence of oxygen. The energy required for denitrification comes from the organic content of the wastewater. This is the process where nitrite / nitrate is reduced to stable nitrogen. The denitrifying organisms are primarily facultative heterotrophs (use organic carbon for all cell tissue).

$$NO_3^- + N \rightarrow N_2$$

5.7.6.3 Combined Nitrification – Denitrification System

Rather than having to use an external carbon source for denitrification and to exploit the oxygen-saving advantages of denitrification, several nitrification – denitrification systems have evolved. In Fig. 5.23, the raw wastewater or primary clarified effluent is first passed through an anoxic tank where there is no aeration (no mixing). Here no carbon oxidation takes place but denitrification does take place. In the **first tank** the characteristic of the influent is high BOD, high ammonia and high nitrate levels. Without oxygen, the bacteria for carbon oxidation and ammonia reduction to nitrate do not proliferate. The denitrifying bacteria proliferate in an environment rich in carbon and low in oxygen. In the **anoxic tank**, there will be a minimum of mixing to obviate septicity and to maintain a forward flow. The effluent from the anoxic tank is fed to the aerobic tank where levels of oxygen are maintained in excess of 2 mg/l throughout. Here the environment is suitable for carbon oxidation with a plentiful substrate supply, a high oxygen level and dense biomass of heterotrophic bacteria. With an adequate hydraulic retention time, the same aerobic environment is suitable for nitrifying bacteria where nitrification of ammonia to nitrate is completed. The well – nitrified, well – oxidized effluent from the aerobic tank is fed to the secondary clarifier for final settlement. An essential return cycle of nitrified effluent is returned to the anoxic tank for denitrification.

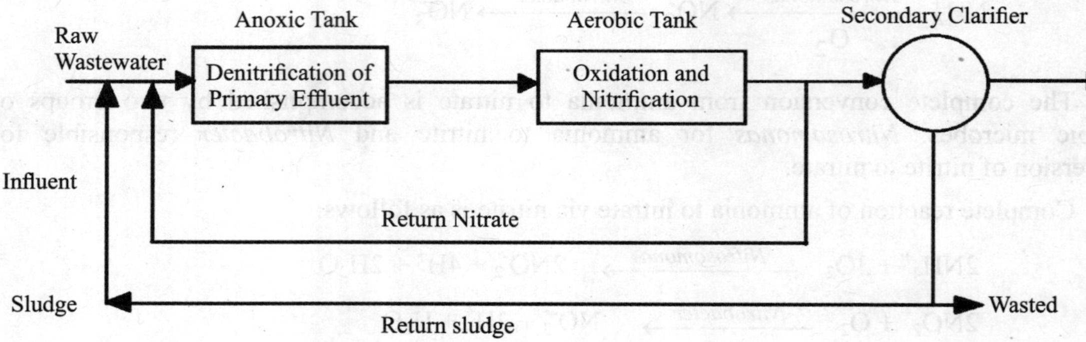

Fig. 5.23 Two – State biological Nitrogen Removal

> **Integrated Wastewater Management**
>
> Louisina, USA, coastal wetlands effluents are rich in nitrates and phosphates after secondary treatment. It increases the production of wetlands plants which improves water quality as these nutrients are used by the plants. When the plants die, their organic material (stems, leaves, roots) causes the wetlands to grow vertically, partially offsetting wetland loss due to sea level rise. Besides, there are also significant economic savings in the application of treated wastewater to wetlands, because the financial investment needed is small compared to tertiary treatment. The utilisation of isolated wetlands can serve as a practical solution to improving water quality in small, widely dispersed communities in the coastal zone.

5.8 ANAEROBIC DIGESTION

Anaerobic digestion (A/D) is the use of microbial organisms in the absence of oxygen for the stabilization of organic materials by conversion to methane and inorganic products including carbon dioxide.

$$\text{Organic matter} + H_2O \xrightarrow{\textit{Anaerobes}} CH_4 + CO_2 + \text{New biomass} + NH_3 + H_2S + \text{heat}$$

The objective is to use A/D to reduce the high organic loads to magnitudes of COD that can be accommodated in conventional aerobic processes, most typically activated sludge. As such A/D is not a complete processor of wastewaters on its own. It is an addendum to existing conventional aerobic processes.

5.8.1 Process Stages

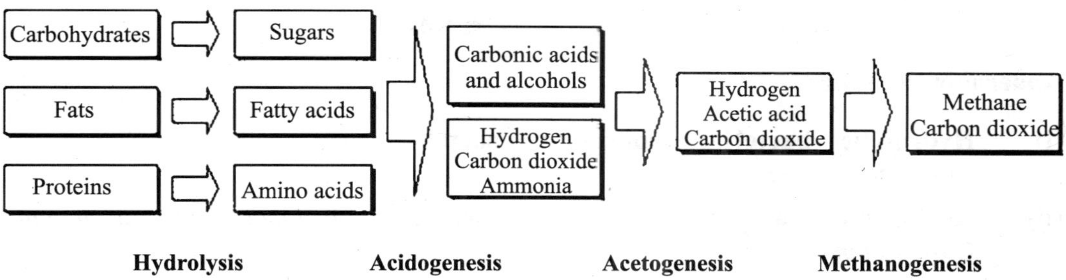

Fig. 5.24 Key Process Stages of Anaerobic Digestion

There are four key biological and chemical stages of anaerobic digestion (Fig. 5.24):

(i) Hydrolysis

(ii) Acidogenesis

(iii) Acetogenesis

(iv) Methanogenesis

In most cases, biomass is made up of large organic polymers. For the bacteria in anaerobic digesters to access the energy potential of the material, these chains must first be broken down into their smaller constituent parts. These constituent parts, or monomers, such as sugars, are readily available to other bacteria. The process of breaking these chains and dissolving the smaller molecules into solution is called **hydrolysis**. Therefore, hydrolysis of these high-molecular-weight polymeric components is the necessary first step in anaerobic digestion. Through hydrolysis the complex organic molecules are broken down into simple sugars, amino acids, and fatty acids.

Acetate and hydrogen produced in **the first stages can** be used directly by methanogens. Other molecules, such as volatile fatty acids (VFAs) with a chain length greater than that of acetate must first be catabolised into compounds that can be directly used by methanogens.

The biological process of **acidogenesis** results in further breakdown of the remaining components by acidogenic (fermentative) bacteria. Here, VFAs are created, along with ammonia, carbon dioxide, and hydrogen sulfide, as well as other by-products. The process of acidogenesis is similar to the way milk sours.

The **third stage** of anaerobic digestion is **acetogenesis**. Here, simple molecules created through the acidogenesis phase are further digested by acetogens to produce largely acetic acid, as well as carbon dioxide and hydrogen.

The **terminal stage** of anaerobic digestion is the biological process of **methanogenesis**. Here, methanogens use the intermediate products of the preceding stages and convert them into methane, carbon dioxide and water. These components make up the majority of the biogas emitted from the system. Methanogenesis is sensitive to both high and low pHs and occurs between pH 6.5 and pH 8. The remaining, indigestible material the microbes cannot use and any dead bacterial remains constitute the digestate. A simplified generic chemical equation for the overall processes outlined above is as follows:

$$C_6H_{12}O_6 \rightarrow 3CO_2 + 3CH_4$$

5.8.2 Anaerobic Microbes

(i) **Hydrolytic bacteria:** Obligatory anaerobic and facultative *Butry vibrio, Clostridium, Lactobacillus.*

(ii) **Acidogenic bacteria:** Obligatory anaerobic, *Acetobacterium acetogenium, Clostridium.*

(iii) **Proton Reducing acetogenes:** *Methanobacillus,* synthrophic bacteria, *Syntrophomonas* etc.

(iv) **Methanogenesis bacteria:** Obligatory anaerobic, *Methanobacterium, Methanococcus, Methanomicrobium, Methanogenium.*

5.8.3 Application of Anaerobic Digestion

Using anaerobic digestion technologies can help to reduce the emission of greenhouse gases in a number of key ways (Table 5.3):

(i) Replacement of fossil fuels.
(ii) Reducing or eliminating the energy footprint of waste treatment plants.
(iii) Reducing methane emission from landfills.
(iv) Displacing industrially produced chemical fertilizers.
(v) Reducing vehicle movements.
(vi) Reducing electrical grid transportation losses.
(vii) Reducing usage of LP Gas for cooking.

5.8.3.1 Waste treatment

Anaerobic digestion is particularly suited to organic material, and is commonly used for effluent and sewage treatment. Anaerobic digestion, a simple process, can greatly reduce the amount of organic matter which might otherwise be destined to be dumped at sea, dumped in landfills, or burnt in incinerators.

5.8.3.2 *Power generation*

In developing countries, simple home and farm-based anaerobic digestion systems offer the potential for low-cost energy for cooking and lighting.

5.8.3.3 *Vehicle fuel*

The biogas (transformed into biomethane) can be used as vehicle fuel in adapted vehicles. This use is very extensive in Sweden, where over 38,600 gas vehicles exist, and 60% of the vehicle gas is biomethane generated in anaerobic digestion plants.

5.8.3.4 *Fertilizer and soil conditioner*

The solid, fibrous component of the digested material can be used as a soil conditioner to increase the organic content of soils. Digester liquor can be used as a fertiliser to supply vital nutrients to soils instead of chemical fertilisers that require large amounts of energy to produce and transport. The use of manufactured fertilisers is, therefore, more carbon-intensive than the use of anaerobic digester liquor fertiliser. In countries such as Spain, where many soils are organically depleted, the markets for the digested solids can be equally as important as the biogas.

5.8.3.4 *Cooking gas*

By using a bio-digester, which produces the bacteria required for decomposing, cooking gas is generated. The organic garbage like fallen leaves, kitchen waste, food waste etc. is fed into a crusher unit, where the mixture is conflated with a small amount of water. The mixture is then fed into the bio-digester, where the bacteria decompose it to produce cooking gas. This gas is piped to kitchen stove. A 2 cubic meter bio-digester can produce 2 cubic meter of cooking gas. This is equivalent to 1 kg of LPG. The notable advantage of using a bio-digester is the sludge which is rich organic manure.

Table 5.3 Comparison between Aerobic & Anaerobic Processes

Parameter	Anaerobic	Aerobic
Energy requirements	Low	High
Degree of treatment	Moderate (60 to 90%)	High (95 %)
Sludge production	Low	High
Process stability(to toxic compounds & load changes)	Low to moderate	Moderate to high
Start-up time	2 to 4 months	2 to 4 weeks
Nutrient requirements	Low	High for certain industrial wastes
Odor	Potential odor problems	Less opportunity for odors
Alkalinity requirements	High for certain industrial wastes	Low
Biogas production	Yes (net benefit is contingent on the need for reactor heating)	No

Wastewater treatment processes at a glance

Process	Application	Quality improvement
Coagulation and sedimentation	Primary treatment (Physical method). Removal of phosphorus. Lime-soda application	Reduction in turbidity. Reduces nutrients phosphorus. Reduces hardness (removes Ca^{2+} and Mg^{2+}).
Gravity filtration	Primary treatment	Reduction in suspended particles; low turbidity.

Filtration and chlorination	Municipal water supplies treatment	Reduction in turbiditty and bacteria
Fine screening	Removal of raw sewage	Reduction in BOD
Microstraining	Clarification prior to filtering	Reduction in microscopic particles
Carbon filters	Adsorption	Removal of taste and odor
Ion exchange	Demineralization	Removal of ions

5.9 REACTOR CONFIGURATION

The basic reactor configuration is a complete mix reactor with no media (Fig. 5.25).

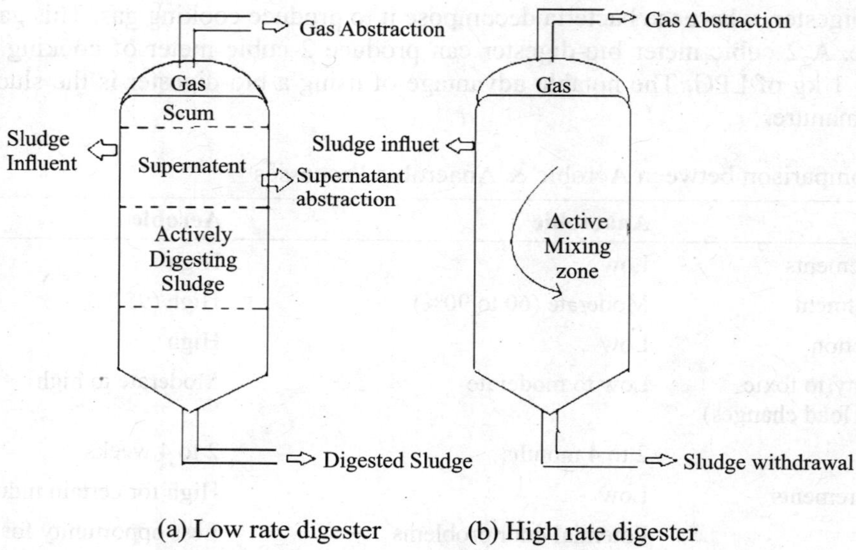

(a) Low rate digester (b) High rate digester

Fig. 5.25 Basic Anaerobic Digestion Process

5.9.1 The Low – Rate Conventional System

It is made up of several layers. The influent sludge enters the tank close to the top at the location of the supernatant layer. Below this is a layer of actively digesting sludge and at the bottom of the tank sits the stabilized sludge, ready for abstraction (withdrawal). They are characterized by intermittent mixing, intermittent sludge feeding and intermittent sludge withdrawal.

5.9.2 High –Rate Digester

They are characterized by continuous mixing, except at times of sludge withdrawal. They have HRTs about one –half those of low rate and gas production is as much as twice. Higher cost of operation because power required for mixing operations and heating requirement.

5.9.3 Classification of Reactors

(i) *First generation reactor type* (Fig. 5. 26)

(a) **Batch digester:** It is characterized by one time influent flow and one time effluent withdrawal in a simple batch. The contents are intimately mixed for some time (HRT).

(b) **Continuously stirred tank reactor (CSTR):** This is characterized by continuous inflow and outflow. The reaction contents are thoroughly mixed. A portion of the effluent may be recycled.

(c) **Plug flow reactor (PFR) with recycle:** The influent flows inside the reactor longitudinally like plugs with no lateral mixing. As a result there is high conversion in the beginning, rate falls with the distance and bacteria decay phase occurs at the end. There may be recycling options.

(d) **Anaerobic contact reactor with CSTR (aerobic):** All the first generation reactor types are characterized by no recycling. Hence hydraulic retention time (HRT) Ø is almost equal to mean cell residence time *i.e.,* $Ø = Ø_c$

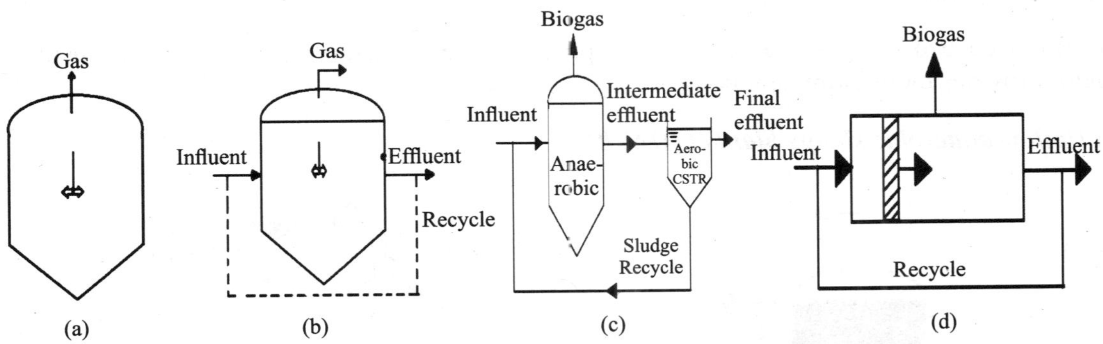

Fig.5.26 (a to d) First Generation Reactors type a, b, c, d

(ii) *Second generation reactor type* (Fig. 2. 27 and 2.28)

This is characterized by system and $Ø_c > Ø$

(a) Upflow or downflow anaerobic filter.

(b) The downflow stationary fixed film reactor, here the filter bed consists of fixed media which provides more time to sludge solid.

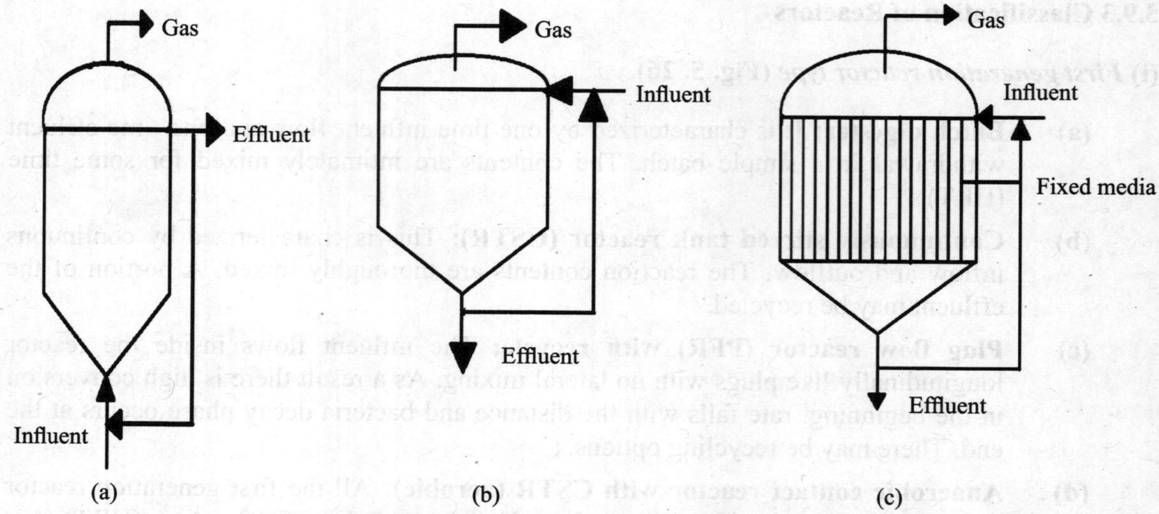

Fig. 5.27 (a to c) Second Generation Reactor type a, b, c

(c) *The fluidized bed reactor*

The fluidized bed designs are becoming popular, as the fluidization of the sludges inside the reactor provides more intimate mixing.

(d) *Upflow anaerobic sludge blanket (UASB)*

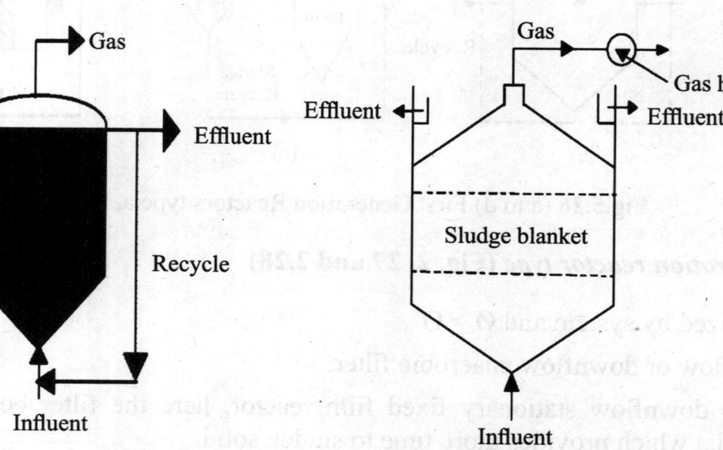

Fig. 5.28 Second Generation Reactor type d

5.9.4 Anaerobic Reactor Design The process design involves several parameters including:

(i) Solids retention time (SRT), days.

$$SRT = \frac{\text{mass of solid in tank, Kg}}{\text{rate of solids removed, Kg/day}}$$

The working range is typically 15 – 30 days.

(ii) Hydraulic retention time (HRT), days-

$$HRT = \frac{\text{working volume, L}}{\text{rate of sludge removed, L/d}}$$

The working range is typically 5 – 10 days.

(iii) Volatile solids (VS) loading rate, Kg VS / m³ / day-

$$VS \text{ loading rate} = \frac{\text{volatile solids added daily, kg VS/d}}{\text{working volume of the digester, m}^3}$$

Typical design values are 2 – 3 kg VS / m^3/d

(iv) Solids production rate, Kg SS / m^3 / day- Design values may peak at 3.2, because above this ammonia toxicity or metal toxicity may limit operation.

(v) Gas production, m^3 of CH_4 / day- Gas production rates ranges from 0.5 to 1.5 m^3 of gas per kg of VS destroyed.

(v) Tank configuration – The tank configurations are now primarily cylindrical with diameters of 5 to 50 m and heights of 3 to 25 m. Tank construction is now almost always of steel.

Mixing and heating systems

The mixing and heating systems may have interdependency. All reactor configurations have thermally insulated containers. Heating mechanisms include external water baths, external jacketed pipes, internal exchangers, steam injection or direct flame. Heat is supplied to achieve mesophilic range of temperature, 38°C. The mixing of digester contents is essential to maintain uniform sludge quality and prevent blockages of inlet and outlet ports for sludge and gas. The simplest mixing mechanism is reintroducing the gas to the tank.

5.10 METHANE PRODUCTION

The amount of methane produced will depend on several parameters, but most especially on the waste type, reactor type, temperature and solids content. Bushwell and Mueller (1952) indicated that the stochiometric equation for methane production was:

$$C_nH_aO_b + \left(n - \frac{a}{4} - \frac{b}{2}\right)H_2O \rightarrow \left(\frac{n}{2} - \frac{a}{8} + \frac{b}{6}\right)CO_2 + \left(\frac{n}{2} + \frac{a}{8} - \frac{b}{4}\right)CH_4$$

For mixed reactors, methane production is calculated by using an empirical formula:

$$M_{CH_4} = 0.35\left(nQCi - 1.4\, r_g v\right)$$

Where n = fraction of biodegradable COD converted (~ 0.85)

Q = flow rate, m^3/s

C_i = COD loading Kg/l

r_g = growth rate, g/m^3/s

V = volume, m^3

M_{CH4} = methane production, m^3/s

As a preliminary estimate of gas production, if growth rate is ignored, then

$$M_{CH4} \approx 0.3Qci$$

Other efficient methods of wastewater treatment

- **Bioremediation**: It involves removal and disposal of soil (hazardous waste) and treatment of the water using a process known as vapor extarction. However, if microrganisms are used that consume the oil, the process becomes inexpensive. The involvement of microrganisms in treatment of wastes is known as bioremediation.
- **Naturalization**: An emerging technology that can treat urban run-off before it is discharged into the water bodies. In this, a closed loop local landscape is created that does not allow run-off to leave particular areas.
- **Nanotechnology**: Nanoparticles (10^{-9} m size) can capture heavy metals such as lead, mercury and arsenic from water because they have a large surface area to volume.

Example 5.9 Find the volume of bio-gas produced in an anaerobic digester with a flow of 360 m^3/h. COD in waste sludge is 125,000 mg/l. Neglect growth rate. Take biodegradable fraction of organic waste (n) as 0.85.

Solution. $M_{CH_4} = 0.35 \times n \times Q \times C_i$

$$Q = 360\ m^3/h = \frac{360}{3600} = 0.1\ m^3/s$$

$$n = 0.85,\ C_i = 125000\ \text{mg/l} = 0.125\ \text{kg/l}$$

$$M_{CH_4} = 0.35 \times 0.85 \times 0.1 \times 0.125 = 3.72\ \text{l/s} = 13.4\ m^3/h$$

Assuming methane in biogas around 60%

$$\text{Biogas generation} = \frac{3.71875}{0.6} = 6.2\ \text{l/s} = 22.3\ m^3/\text{h}$$

Example 5.10 An anaerobic digester is designed to treat 38000 m^3/day wastewater. Find the volumetric loading, amount of biogas generated.

Given COD removal 0.15 kg/m^3 sludge contains 95% water, sp gravity of sludge 1.02 HRT 10 days digester efficiency (η) = 0.7, VSS loading (Y) =0.08 *kg VSS/kg*, COD k_d = 0.03 day^{-1}.

Biogas contains 65% methane.

Solution. Sludge volume $= \dfrac{0.15 \times 38000}{1.02 \times 10^3 \times 0.05} = 111.8\ m^3$

Biodegradable COD loading

$$0.15 \times 0.85 \times 38000 = 4845\ \textit{kg/day}$$

Digester volume = $(V) = T \times Q = 111.8\ m^3 \times 10 = 1118\ m^3$

Volumetric loading $\text{kg/COD/m}^3 d = \dfrac{4845}{1115} = 4.76\ \text{kg/m}^3\text{d}$

Neglecting growth rate $M_{CH_4} = 0.35 \times 0.85 \times \dfrac{38000}{24 \times 3600} \times \dfrac{4845}{38000} = 0.0168\ \text{l/s} = 1.45\ \text{m}^3/\text{day}$

Assuming methane in biogas as 65%

Biogas production 1.45/0.65 = 2.23 m^3/day

5.11 TRICKLING FILTERS

The secondly used common biological waste treatment process is the trickling filter method. The trickling filter is also called **percolating filter**. It has good adaptability to handle peak shock loads and the ability to function satisfactorily after a short period of time. Milk processing, paper mill and pharmaceutical wastes are among those treated by trickling filters. **Conventional trickling filters** consist of a rock bed 1 to 3 meters depth, with enough opening between rocks to allow air to circulate easily. The influent is sprinkled over the packing, which is coated with a biological slime. When the liquid trickles over the packing, oxygen and the dissolved organic matter diffuse into the film to be metabolized by the micro-organisms in this slime layers. End products such as CO_2, NO_3^- etc diffuse back out of the film and appear in the filter effluent. As the micro-organisms utilize the organic matter, the thickness of the slime film increases to a point where it can be no longer be supported on the solid media and gets detached from the surface. This process is known as **sloughing**. A tank following the trickling filter removes the detached bacteria film and suspended matter. Handling and disposal of sludge from biological waste water treatment plants is important problem and represents about half of the cost of most sewage treatment plants. The concentration of solids in the primary sewage sludge is about 5%; the activated sludge contains less than 1% solids and sludge from trickling filters has about 2% solids. The common unit operations of sludge treatment and disposal involve concentration or thickening, digestion, conditioning, dewatering, oxidation and safe disposal.

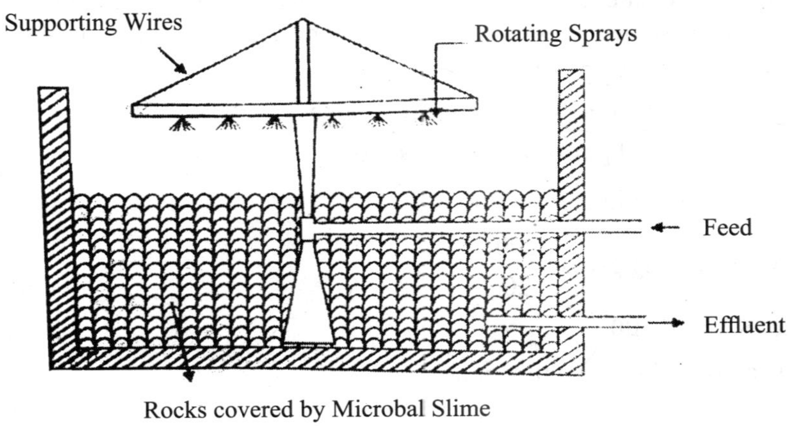

Fig. 5.29 Trickling Filters

(i) Advantages of trickling filters

(a) Give good effluent under wide variations of influent quality.
(b) Simple in operation.
(c) Operating cost is less.
(d) Good capability to function under extreme weather condition.
(e) About 60 to 90% efficiency in removal of BOD and the suspended solids.

(ii) Disadvantages of trickling filters

(a) Foul odour.
(b) Nuisance due to psychoda fly during warm weather.
(c) Require large area.
(d) Not suitable to build-up area.
(e) Head loss through the filter is high.
(f) Primary treatment of sewage is necessary for the sewage input.

Example 5.11 Calculate the runoff from a contributing catchment area of 100 Hectare. Assume run off coefficient is 0.62. Intensity of rainfall (l) as 1400 mm/year.

Solution. Using rational formula:

$$Q = 0.278\ CIA$$

$$C = \text{run off coefficient} = 0.62$$

$$A = \text{Catchment area} = 100\ \text{Ha} = 100 \times 10{,}000\ \text{m}^2 = 10^6\ \text{m}^2 = 1\ \text{km}^2$$

$$I = 1400\ \text{mm/year} = \frac{1400}{365 \times 24} = 0.16\ \text{mm/hour}$$

$$Q = 0.278 \times 0.62 \times 0.16 \times 1 = 0.027\ \text{m}^2/\text{s} = 99.3\ \text{m}^3/\text{hour}$$

Example 5.12 Calculate the sludge volume index for mixed liquor with 2000 mg/lit suspended solids having settled volume of 200 ml with a litres sample. Is this sludge volume index good or bad?

Solution. $\text{S.V.I.} = \dfrac{\text{ml. of settled volume of sludge}}{\text{1 gm. of dry wt. of solid}}$

2000 mg settling gives 200 ml.

Or 1 gm settling gives $\dfrac{200}{20}$

Hence S.V.I. = 100

This is a good sludge volume index.

Review questions

1. **What is BOD and how it is calculated?**

Ans. BOD is Biochemical Oxygen Demand and defined as the amount of oxygen required by aerobic microorganisms to stabilize the organic matter in water or waste water.

BOD can be calculated by formula: BOD= p (DOi - DO_f)

Where, p= Dilution factor, DOi & DO_f = initial and final DO concentration.

2. **Differentiate BOD and COD and give the relation between them.**

Ans. BOD stands for biochemical oxygen demand and can be defined as the amount of oxygen required by aerobic microorganisms to stabilize the organic matter.

COD can be defined as the amount of oxygen required to fully oxidize the organic matter to carbon dioxide with a strong oxidizing agent under acidic conditions. The amount of oxygen required to oxidize an organic compound to carbon dioxide, ammonia and water is given by:

$$C_nH_aO_bN_c + \left(n + \frac{a}{4} - \frac{b}{2} - \frac{3}{4}c\right)O_2 \rightarrow nCO_2 + \left(\frac{a}{2} - \frac{3}{2}c\right)H_2O + cNH_3$$

$$\text{BOD} \cong 1.6\ \text{COD}$$

3. **What are the different types of waste water treatment?**

Ans. Pre-treatment, primary, secondary and advanced treatment.

4. **In waste water pre-treatment, what is a grit channel and what are its types?**

Ans. Grit is defined as sand, gravel or other mineral matter that has a nominal diameter of 0.15-0.20 mm or larger. Grit channel is an equipment to remove the grit after screening of the waste water.

Types

(i) The helical flow aerated grit chamber

(ii) The horizontal flow grit chamber.

5. **What is SRT and HRT?**

Ans. SRT stands for solid retention time and HRT stands for hydraulic retention time. These two are important design parameters of activated sludge system.

6. **What is SVI?**

Ans. SVI is "sludge volume index" and can be defined as volume occupied by one gram of settleable sludge. It is settle ability of the activated sludge.

$$\text{SVI} = \frac{\text{Vs} \times 1000}{\text{MLSS}}$$

Where,

SVI = sludge volume index,

Vs = volume of settled sludge,

MLSS= Mixed liquor Suspended Solids

7. Differentiate between low rate and high rate digesters.

Ans. In the standard rate digestion process, the contents of the digester are usually unheated and unmixed. In this acidification, methane fermentation and sludge thickening takes place in single tank. Standard or low rate digesters have intermittent mixing, intermittent sludge feeding and intermittent sludge withdrawal Detention time for this process vary from 30 to 60 days.

In high rate digestion process, the contents of the digester are heated and completely mixed. Continuous or intermittent sludge feeding and withdrawal is practiced in these digesters. The required detention time is 15 days or less.

8. Define anaerobic digestion and mention its application.

Ans. Anaerobic digestion is the use of microbes in absence of oxygen for stabilization of organic materials in the form of methane and other products.

$$\text{Organic matter} + H_20 \text{ -------}\rightarrow CH_4 + CO_2 + NH_3 + H_2S + \text{new biomass} + \text{heat}$$

Methane is a highly flammable gas which can be used for generating electricity, burner flame etc.

9. With respect to anaerobic treatment, answer the following question.

(a) Which one is the important step and why?

(b) Why pH is a very important process parameter?

(c) What is the principal advantage and suitability?

(d) What is anaerobic denitrification?

Ans.

(a) Methanogenesis is the most important step as it aims at the maximum production of by-product methane which is the fuel component of biogas. It is the slowest and hence the rate determining step.

(b) pH is likely to be lowered because of acid formation in hydrolysis. If there is no sufficient natural alkalinity, then liming is to be done. pH less than 6.2 is detrimental to anaerobic microbes.

(c) Methane is by product fuel gas. It is suitable for high organic wastewater like effluent of sugar mill, paper will, breweries and other agrochemical units.

(d) It is the conversion of nitrite – N and nitrate –N to nitrogen gas. Also it is called anoxic process

$$NO_2^- / NO_3^- \xrightarrow{\text{Anaerobic}} N_2 \text{ gas}$$

10. Prove that, BODt = Lo $\left(1-e^{-k_1 t}\right)$

Where, BODt = oxygen consumed by a waste water in first t days

Lo= ultimate carbonaceous oxygen demand

K_1=BOD deoxygenation rate co-efficient/ day.

■■■■

Chapter 6

Air Pollution

6.1 INTRODUCTION

Air is one of the essential factors making life on the earth possible. Protection of air from pollution is a matter of great concern. Air pollution may be defined as any atmospheric condition in which substances are present at concentrations high enough above their normal ambient levels to produce a measurable effect on man, animals, vegetation, or materials. Substances mean any natural or anthropogenic (man-made) chemical compounds capable of being airborne. They may exist in the atmosphere as gases, liquid drops, or solid particles. **Air pollution** is an undesirable change in the physical, chemical or biological characteristics of air.

Air pollutants are the substances which pollute the air. Some of the common pollutants are dust, soot, ash, carbon monoxide, excess of carbon dioxide, sulphur dioxide, oxides of nitrogen, hydrocarbons, chlorofluorocarbons (CFC), lead compounds, asbestos dust, cement dust, pollens and radioactive rays.

6.1.1 Sources of Air Pollution

The sources of air pollution are classified into two groups such as natural and man-made sources as discussed below:

(i) **Natural sources of air pollution:** They are dust storms, forest fires, ash from smoking volcanoes, decay of organic matters and pollen grains floating in air.

(ii) **Manmade sources of air pollution:** They are population explosion, deforestation, urbanisation and industrialisation, whose effects can be explained as follows:

(a) Burning of fuels like wood, cow dung cakes, coal and kerosene in homes pollute the air.

(b) Exhaust gases emitted by motor vehicles which pollute the air are the major source of air pollution in big cities.

(c) Industries pollute air by releasing various types of pollutants such as sulphur dioxide, oxides of carbon, nitrogen oxide, chlorine, asbestos dust and cement dust.

(d) Thermal power plants pollute air by emitting sulphur dioxide and fly-ash.

(e) Nuclear power plants pollute air by releasing radioactive rays.

(f) Use of fertilisers and pesticides in agriculture pollute the air.

(g) Mining activities releases particulate matter into the air and pollutes it.

(h) Indiscriminate cutting of trees and clearing of forests increases the amount of carbon dioxide in the atmosphere and thereby pollutes it.

(i) Use of chlorofluorocarbons in refrigeration, fire extinguishers and aerosol sprayers pollutes air by depleting the ozone layer.

(j) Smoking pollutes air by emitting carbon monoxide and nicotine.

6.1.2 Harmful Effects of Air Pollution

(i) Air pollution affects respiratory system causing breathing difficulties and diseases such as bronchitis, asthma, lung cancer, tuberculosis and pneumonia.

(ii) Air pollution affects the central nervous system causing carbon monoxide poisoning. CO has more affinity for haemoglobin than oxygen and thus forms a stable compound carboxy haemoglobin (COHb), which is poisonous and causes suffocation and death.

(iii) Air pollution causes depletion of ozone layer due to which ultraviolet radiations can reach the earth and cause skin cancer, damage to eyes and immune system.

(iv) It causes acid rain which damages crop plants, trees, buildings, monuments, statues, metal structures and also makes the soil acidic.

(v) It causes greenhouse effect or global warming which leads to excessive heating of earth's atmosphere, further leading to weather variability and rise in sea level. The increased temperature may cause melting of ice caps and glaciers, resulting in floods.

(vi) Air pollution from certain metals, pesticides and fungicides causes serious ailments.

(vii) Lead pollution causes anaemia, brain damage, convulsions and death.

(viii) Certain metals cause problem in kidney, liver, circulatory system and nervous system.

(ix) Fungicides cause nerve damage and death.

(x) Pesticides like DDT (Dichloro diphenyl trichloroethane) which are toxic enter into our food chain and get accumulated in the body causing kidney disorders and problems of brain and circulatory system.

6.1.3 Units of Concentration

The different ways of expressing concentrations of air pollutants are: ppm (V/V), ppb (V/V), mg/m^3, mg/Nm^3 (Nm^3 = normal dry m^3, i.e. at STP, Temp =0°C, Pressure = 1013mb).

6.2 AIR POLLUTION CLASSIFICATION

(i) According to chemical composition

(a) Sulfur-containing compounds

(b) Nitrogen-containing compounds

(c) Carbon-containing compounds

(d) Halogen-containing compounds

(e) Toxic substances

(f) Radiative compounds

(ii) According to physical state

(a) Gaseous

(b) Liquid (aqueous)

(c) Solid

(iii) According to the manner in which they reach the atmosphere

(a) Primary pollutants (those emitted directly from the sources).

(b) Secondary pollutants (those formed in the atmosphere by chemical interactions among primary pollutants and normal atmospheric conditions).

(iv) According to the space scales of their effects

(a) Criteria air pollutants are six major pollutants defined by Environmental Protection Agency (EPA) for which ambient air standards have been set to protect human health and welfare. Criteria pollutants (defined by EPA): 1. Ozone, O_3. 2. Carbon monoxide, CO. 3. Sulfur dioxide, SO_2. 4. Nitrogen oxides, NOx. 5. Lead (Pb). 6. Particulates, PM_{10}.

(b) Non Criteria air pollutants

6.2.1 Primary Pollutants

Those air pollutants are directly emitted by an identifiable source, e.g sulphur dioxide, oxides of nitrogen, carbon monoxide etc.

6.2.2 Secondary Pollutants

Those pollutants formed in the atmosphere by chemical reactions occurring among the primary pollutants and the atmospheric air constituents. The reaction mechanisms and various steps involved in the process are influenced by many factors such as concentration of reactants, the amount of moisture present in the atmosphere, degree of photo activation, meteorological parameters like pressure, temperature and local topography. They include:

(i) Peroxyacetyl nitrate (PAN)

$$HC + NO_x \xrightarrow{\text{UV Radiation}} PAN$$

(ii) O_3

(iii) Oxidized hydrocarbons

6.2.3 Criteria Pollutants

These pollutants are likely to exist in all urban areas. Their concentrations will vary depending on the level of industrial and traffic activity and the degree of sophistication of control. Six criteria pollutants are given in Table 6.1 below

Table 6.1 Properties and Pollution significance of Criteria Pollutants (adapted from WHO, 1987)

Pollutant	Properties	Pollution Significance
Carbon monoxide	Colourless, odourless gas	Formed during incomplete combustion of hydrocarbons, causes greenhouse effects and climatic changes
Nitrogen dioxide	Brown – orange gas	Significant component of photochemical smog and acid deposition
Ozone	Highly reactive	A secondary pollutant, produced during formation of photochemical smog. Damages flora and materials
Sulphur dioxide	Colourless, choking gas, Soluble in H_2O to produce Sulphurous acid, H_2SO_3	Principal component of acid deposition. Damages humans, flora, fauna and Materials
PM_{10}	Particulate matter < 1 μm in diameter – blacksmoke	Coal burning power station, traffic, domestic coals, quarrying, incineration: can cause respiratory problems
Lead	Heavy metal, bioaccumulative	Principal source leaded petrol. Also from lead pipes, quarrying, incineration. Damages humans and fauna when in excess

6.2.4 Non – criteria Pollutants

Those pollutants are discussed below for which emission limits from industry are set:

(i) Many of these are carcinogenic, mutagenic and damaging to the central nervous system (CNS) as well as having a lot of other negative health effects.

(ii) Some of them are man – made synthetic chemicals such as 1, 2 –dichloroethane however, others are the products of industry and more specifically the chemical, pharmaceutical and petroleum product industries.

(iii) The heavy metals of Cd, Cr and Hg are ubiquitous in the manufacturing and chemical industries, particularly electroplating, plastics, paints, chlorine and chemicals.

Air quality standards prescribed by Central Pollution Control Board (India)

Pollutant	Industrial and multiple use zone	Residential and rural zone	Wildlife and ecologically sensitive zone
SPM	500	200	100
SO_2	120	80	30
CO	5000	2000	1000
NO_X	120	80	30

6.3 ACID DEPOSITION

Rain fall by nature is slightly acidic due to the tendency to chemically occurring reactions with atmospheric CO_2. It forms a weak solution of carbonic acid with pH 5.6; by definition any precipitate measuring less than 5.6 on the pH scale is considered acid rain (Fig. 6.1). Carbonic acid created by CO_2 in air from:

(i) Volcanic emission

(ii) Biological decomposition

(iii) Chlorine & sulfates from ocean spray can drop the pH of precipitate below 5.6 while, alkaline dust can raise the pH above 7. Acid rain is only one form in which acid deposition occurs. Fog, snow, mist, and dew also trap and deposit atmospheric contaminants. Furthermore, fall out of dry sulfate, nitrate, and chloride particles can account for as much as half of the acidic deposition in some areas.Emissions of acidic oxides like CO/CO_2, SO_2/SO_3, NO_x (NO and NO_2) and hydrocarbons from industry, transport, households and power production interact with water vapour and sunlight and are chemically converted to acidic compounds like H_2SO_4/ H_2SO_3, HNO_3 / HNO_2, H_2CO_3 etc. These acids in turn return to the earth as dew, drizzle, fog, sleet, snow or rain. The deposition of acids is either as dry deposition, which may be in particulate or gaseous form or a wet deposition in rain or snow.

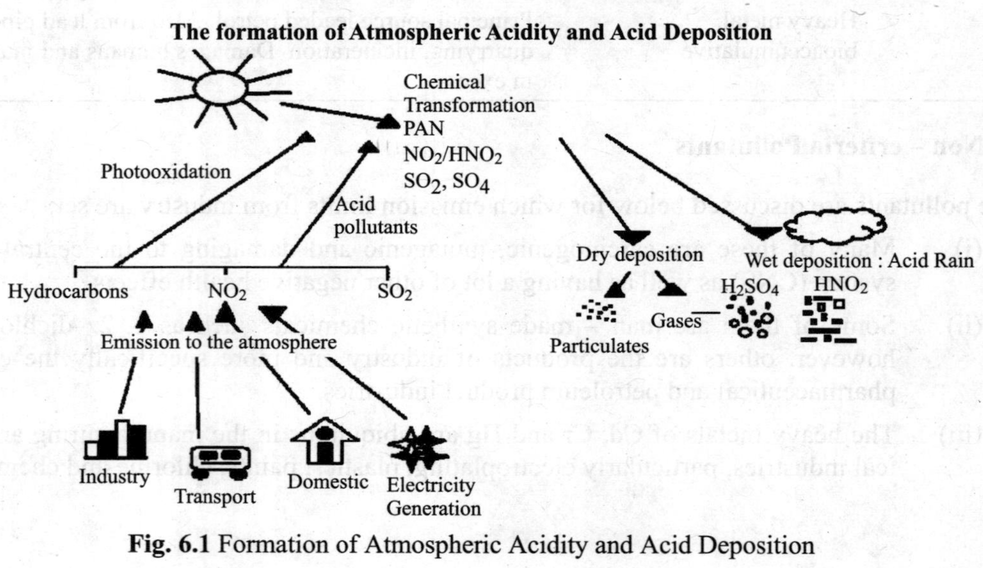

Fig. 6.1 Formation of Atmospheric Acidity and Acid Deposition

Formation of acid rain

SO_2------H_2SO_4 (62%)

NO_2-------HNO_3 (32%)

Cl---------HCl (6%)

In urban areas, where transportation is the major sources of pollution, nitric acid is equal to or slightly greater than sulphuric acid in the air.

Formation of H_2SO_4

(i) Oxidation of sulfur when coal burns

$$S + O_2 \rightarrow SO_2$$

(ii) Formation of SO_3

$$SO_2 + O \rightarrow SO_3$$

$$2SO_2 + O_2 \rightarrow 2SO_3$$

(iii) Dissolved in water droplets

$$SO_3 + H_2O \rightarrow H_2SO_4$$

Formation of HNO_3

NO_2 dissolved in water droplets:

$$O_2 + 4NO_2 + 2H_2O \rightarrow 4HNO_3$$

6.3.1 Environmental effect of Acid Rain

(i) Damage to aquatic life.

(ii) Reproduction is the most sensitive stage in the life cycle. Eggs of many fish species are killed at pH 5.0. Disrupt the food chain by killing:

(iii) **Plants**

(iv) **Aquatic insects**

(v) **Invertebrate on which fish depend for food**

(a) At pH level less than 5

- Adult fish Trout, Salmon, Game fish etc die

(b) Acidity

- Alters body chemistry
- Destroy gills and prevent oxygen up take
- Causes bone decalcification
- Disrupt muscle contraction

(c) Phytoplankton population are reduced, and many common water, dwelling invertebrates such as may flies and stone flies cannot survive when the pH falls below 5.5. Acid dead lakes have pH below about 3.5

(vi) Deterioration of buildings and monuments

(a) Most glorious buildings and works of art are being destroyed by air pollution:

- Smoke and soot coat buildings
- Paintings
- Textiles
- Lime stone and marble are destroyed by atmospheric acid at an alarming rate

$$CaCO_3 + H_2SO_4 \rightarrow CaSO_4 + CO_2 + H_2O$$

(vii) **Mobilization of toxic metal**

Acid rain can cause lightly bound toxic metals such as aluminum which can kill fish by damaging their gills and causing asphyxiation. It also cause leaching of heavy metals: Hg, Cd in to drinking water and results in Bioaccumulation

(viii) **Damage to forest productivity**

It causes:

(a) Root necrosis

(b) Lack of seeding growth

(c) Premature tree death

(d) Growth reduction

(e) Defoliation

(f) High altitude forests are subjected to especially intense doses of these acids because clouds saturated with pollutants tend to hang on, mountain tops, bathing forests in a toxic soap for days even weeks at a time.

(ix) Visibility reduction

Particulates in the atmosphere reduce the visibility due to scattering and absorption of light. The dust particles of $2000/cm^3$ can obscure a mountain at 75 km; while a concentration of NO_2 of 20 ppm would probably reduce the visibility to 1 km. NO_2 causes the sky to appear brownish in color in addition to reducing visibility.

6.3.2 Control Measures

(i) Source control measure includes fuel change (low sulphur), fuel desulphurization, fluidised bed combustion, flue gas desulphurization , CO_2 sequestration.

(ii) Short – term control of acid rain may be achieved by use of lime.

6.4 GLOBAL CLIMATE CHANGE: GREEN HOUSE GASES

Global warming is the greatest challenge faced by our planet. It is, in fact, the increase in the temperature of the earth's neon- surface air. It is one of the most current and widely discussed factors.

In a greenhouse, visible light passes through the glass and heat up the soil warming the plants (Fig. 6.2). The warm soil emits radiation in longer wavelength, particularly infra-red. Glass partly reflects and partly absorbs infrared radiation, keeping the greenhouse warmer than the outer atmosphere.

In a similar way the earth's atmosphere acts like green house (thermal blanket).

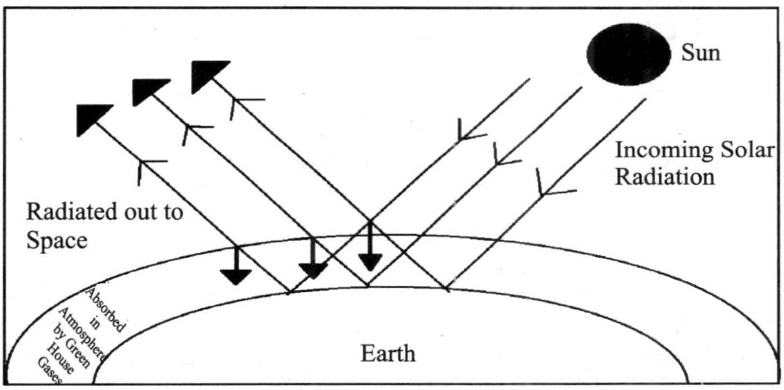

Fig. 6.2 Green House Effect

6.4.1 Green house gases

The major greenhouse gases in order of greenhouse contribution are:

(i) CO_2 --- 55%

(ii) CFCs ---- 22%

(iii) CH_4 ---- 15%

(iv) N_2O ----5%

(v) O_3 ---- 3%

These gases have the effect of absorbing incoming short-wave solar energy (at wavelength < 4 μm). They also have the ability of absorbing some of the outgoing earth's radiated energy of long wavelength (>4 μm). Each greenhouse gas has its own ideal wavelength bands at which it absorbs solar or earth's radiation energy best. As such several of the greenhouse gases absorb radiation at the same wavelengths. The term **greenhouse gases** is reserved for those gases that absorb radiation at wavelengths >4 μm, i.e. gases that absorb the long wavelength earth's radiation. Thus they trap much of the outgoing earth's radiative energy, thereby heating the atmosphere. These greenhouse gases have the effect of acting like a thermal blanket around the globe, rising its temperature --- Global warming. The consequences of the rise in temperature are being felt all over the globe. The finding of scientific research done in this field reveals that the temperature of the earth is likely to rise from 1.4°C to 5.8°C within a period of 100 years.

6.4.2 Sources of Greenhouse gases in Atmosphere

(i) *Carbon dioxide (CO_2)*

Carbon dioxide (CO_2) is not a pollutant but an important constituent of environment (0.033%) which is responsible for causing a warming effect on the earth's surface. The CO_2 level is expected to rise in future due to ongoing burning of fossil fuels and land use change. Large scale deforestation, burning of chemical and fossil fuel have created this increase. Trees play a significant role in the global carbon cycle. They are the largest land-based mechanism for removing carbon dioxide from the air. Deforestation is checking these positive processes. It is the second principle cause of atmospheric carbon dioxide.

(ii) *Chlorofluorocarbons (CFCs)*

Use of air conditioners, refrigerators, fire extinguishers produces CFCs. Chlorine is freed from CFCs by the process of photolysis by short-wave radiation and drifts upwards to damage the ozone layer. Due to their long residence time in atmosphere, it is assumed that the level of CFC will increase.

(iii) *Methane(CH_4)*

Methane is a naturally occurring gas that is produced under anaerobic conditions. This occurs in swamps, rice fields, cattle stocks and in the production and consumption of fossil fuels. It has long residence time of about ten years, after which it may be oxidized with hydroxyl radicals.

(iv) *Nitrous oxide (N_2O)*

It is produced in the nitrogen cycle via nitrification i.e. $NH_4 \Rightarrow N_2 \Rightarrow N_2O$

Also produced from emissions of diesel engines, lightening, use of fertilizers in agricultural lands, waste water treatment plants. It has a residence time of about 150 years and is 200 times as potent a greenhouse gas as CO_2. The quantities of N_2O produced are insignificant by comparison with CO_2.

(v) *Tropospheric ozone (O_3)*

They are also responsible for global warming. Increase in ozone is due to explosive population growth, increased industrial expansion, technological advancement, deforestation, growing urbanisation, etc.

(vi) *Water vapour (H_2O)*

Water vapour is often ignored as a greenhouse gas. The power plant emissions show that the amounts of H_2O produced are similar to the amounts of CO_2. Also the long wavelength energy radiated by the earth is absorbed not only by CO_2, N_2O and CH_4 but also by H_2O.

6.4.3 Environmental Effects of global warming

(i) *Direct effects*

Thermal extreme effect: (heat waves) (a) the skin, CNS and circulatory system are the most affected: thermoregulation disturbance; heat stroke and exhaustion, decreased male fertility, cerebro-vascular stroke, etc. (b) Effect on the respiratory organs: persons with chronic diseases like asthma, bronchitis, cardiovascular are the most affected. (c) Weather instability and natural calamities: cyclones, landslides, draught, flooding, etc.

(ii) *Indirect effects*

(a) Vector borne diseases: mosquitoes (malaria, yellow fever); Tsetse fly (African sleeping sickness); Fly (faeco-oral related diseases, onchocerciasis, leishmaniasis and schistosomiasis).

(b) Waterborne diseases: due to water shortage & contamination

(c) Sea-level rise: damages fresh water, affects soil salinity

(d) Agricultural productivity:

- Reduced rainfall; increased pests; decreased soil fertility; decreased farming land, etc.
- Affects both staple foods & horticulture harvesting; and animal husbandry.

(e) Food shortages and malnutrition

(f) Air pollution impact: increased pollen; concentration of pollutants; impacts to socio-economic development

6.4.4 Control Measures

Forestation can be of great help in this regard. Planting more trees and reducing timber cuts worldwide will help restore the imbalance. Secondly, we must follow an environmental policy of 'reduce, reuse, recycle', i.e. promoting the reuse of anything. Thirdly, the use of fuel-efficient vehicles should be promoted as these vehicles have lower emissions of harmful gases. Fourthly, every individual should be aware of the importance of the protecting environment. Besides, eco-friendly technologies must be promoted, and must be substituted with the technologies which cause great emission of global warming gases. Public awareness campaign can be of great help in this regard because unless each and every individual is aware, only governments' effect cannot bring desired difference.

Example 6.1 A stack emits SO_2 at the rate 16 gm/sec. Determine ambient SO_2 concentration at 2 km downwind direction. Take wind speed $\mu_{10} = 6$ m/s. Assume stability class as D take σ_y = 150m, σ_z = 50m.

Solution: $$C(x,0,0) = \frac{Q}{\pi \sigma_y \sigma_x U}$$

$$x = 2\ \text{km},\ \sigma_y = 150\text{m},\ \sigma_z = 50\text{m}$$

$$C(x,0,0) = \frac{16\times10^6}{3.14\times150\times50\times5} = 135.88\ \mu g/m^3$$

$$C(x=2, y=0, z=40) = \frac{Q}{\pi\sigma_y\sigma_z U}\exp\left[\frac{1}{2}\times\left(\frac{40}{50}\right)^2\right]$$

$$= \frac{16\times10^6}{3.14\times150\times50\times6}\exp\left[-\frac{1}{2}\times\left(\frac{40}{50}\right)^2\right] = 82\ \mu g/m^3$$

Example 6.2 Find out SO_2 emission rate from a stack of coal based power plant consuming 34 tons per hour coal with sulfur content 0.5%. Find out the stack height using empirical formula

$$H = 14 \times Q^{0.3}$$

H = stack height in m.

Q = SO_2 emission in kg/hours.

Solution: Coal consumption 34 TPH

Sulfur content: $34\times\frac{0.5}{100}\times1000 = 34\times5$ kg/h

SO_2 emission from stack = $34\times5\times\frac{64}{32} = 340$ kg/hr

Stack height = $14\times(340)\,0.3 \approx 80$ metre

Stack height of 85-90 m should be provided for effective dispersion of SO_2.

Example 6.3. Find out the stack height from the given set of data.

Ground level concentration of SO_2 at a location (2 km) is found to be 36 µg/m^3.

Velocity of flow U= 6 ms^{-1}, SO_2 emission from stack 15 *g/s*.

Assume stability class with $\sigma_y = 120$ m $\sigma_z = 40$ m

Here x = 2 km, y = 2 km C (x, y, 0) = 36×10^{-6} g/m

To find stack height H in m.

Solution. $C(x, y, 0)\pi = \frac{Q}{\pi\sigma_y\sigma_z U}\exp\left[-\frac{1}{2}\times\left(\frac{H}{\sigma_z}\right)^2\right]$

$$\Rightarrow 36\times10^{-6}\frac{\text{gm}}{\text{m}^3} = \frac{15\ \text{gm/s}}{3.14\times120\times40\times6}\exp\left[-\frac{1}{2}\times\left(\frac{H}{40}\right)^2\right]$$

$$\Rightarrow \exp\left[-\frac{1}{2}\times\left(\frac{H}{40}\right)^2\right] = \frac{36\times10^{-6}\times3.14\times120\times40\times6}{15}$$

$$\Rightarrow -\frac{1}{2}\times\left(\frac{H}{40}\right)^2 = \text{Ln}\,(0.217) = -1.53$$

$$\Rightarrow \frac{H}{40} = (1.53\times2)^{1/2} = 1.75$$

$\Rightarrow H = 40\times1.75 \approx 70\text{ m}$ (Ans.)

6.5 OZONE DEPLETION

Ozone is found in stratosphere; its detection in troposphere (ground level) is an indicator of pollution.

6.5.1 Role of Ozone Layer

(i) Solar radiation: 55% IR; 40% visible light; 5% UV.

(ii) Acts as a UV protective blanket (layer).

(iii) UV radiation includes three bands: UVA (400-320nm); UVB (320-280nm); UVC (280-200nm), far UV (< 50nm).

(iv) It blocks nearly all UVC, half UVB, and small part UVA.

6.5.2 Mechanism for UV blocking

Equilibrium in nature between destruction and production of O_3

(i) $O_2 \rightarrow O+O$ (energized by UVC)

(ii) $O+O_2 \rightarrow O_3 + IR$ (production of ozone)

(iii) $O_3 + UV \rightarrow O_2+O$: (destruction of ozone as energized by UVB)

6.5.3 Causes of Ozone depletion

(i) Human activity: release of halogenated hydrocarbons: CFCs; CCL_4;

(ii) Use of refrigerants, propellants, solvents, foam production, etc.

(iii) Are stable under normal conditions in the troposphere. The volatile chemicals become active with the presence of UV to react with O_3;

The reaction:

(i) $CCL_2F_2 + UV \rightarrow CCLF_2 + CL^-$

(ii) $CL^- + O_3 \rightarrow CLO + O_2$

(iii) $CLO + O \rightarrow CL^- + O_2$

(iv) Net: $O_3+O\rightarrow2O_2$ (Chlorine atom acts as a catalyst).

6.5.4 Potential adverse effects

(i) UVC is absorbed by O_3, (does not reach the earth);

(ii) UVB and UVA reach the earth.

(i) *Direct effect*

(a) **Skin damage and cancer**

- Acute exposure: thermal extremes due to UV: sunburn; skin lesions;
- Cumulative (chronic) exposure: skin cancer in fair skinned humans;
- UNEP estimates: 5% skin cancer increase during 2070's (extra 100 cases of skin cancer per million population per year) in Europeans living at around latitude 45 degrees N).

(b) **Effects on eye:** by UVB; cornea, lens, and retina: cataract (snow blindness): UNEP estimates: 1.75 million additional cataract cases worldwide each year due to 10% loss of ozone.

(c) **Effects on immune system**

- Skin sensitivity

(ii) *Indirect effects*

(a) Effects on plants: impairs photosynthesis & UVB; farm productivity declining

(b) Effects on aquatic system: Phytoplankton & UVB; aquatic farm productivity declining. CO_2 take up will be diminished.

EI Nino

EI Nino is an abnormal warm condition that develops in the eastern Pacific Ocean, mostly in the southern coasts. It occurs mainly near the coasts of Peru and Equador due to ocean-atmospheric interaction. The impact of EI Nino has disastrous effects on the economics of many countries. The effect of EI Nino is likely to cause severe floods, droughts, cyclones, monsoon etc. Though centered in the South Pacific, EI Nino phenomenon can affect the weather and climate conditions in much of the tropics. A new pattern is expected to emerge for the Asian summer monsoon. This can affect large areas of temperate and tropical Asia. These conditions prevailed in 1982-83, in the Pacific region and Indonesia. It also led to failure of winter monsoon in the Indian Ocean nations of Sri Lanka, South India and parts of south-eastern Asia. In addition, the warm ocean water spread over eastern Pacific and cause decline in finishing. EI Nino also results in global climate change upsetting the normal climate pattern in many regions.

6.6 WIND

The air in motion is called **wind,** air which is rushing from an area of high pressure towards an area of low pressure.

6.6.1 Wind Measurement –Wind rose diagram

Wind speed and direction can be measured continuously with the help of an instrument called **anemometer**. Wind rose is a pictorial or graphical representation of data of wind speed and direction. It can be in 16 possible directions starting from N. The different nodes represent the percentage of wind blowing from that direction (Fig. 6.3).

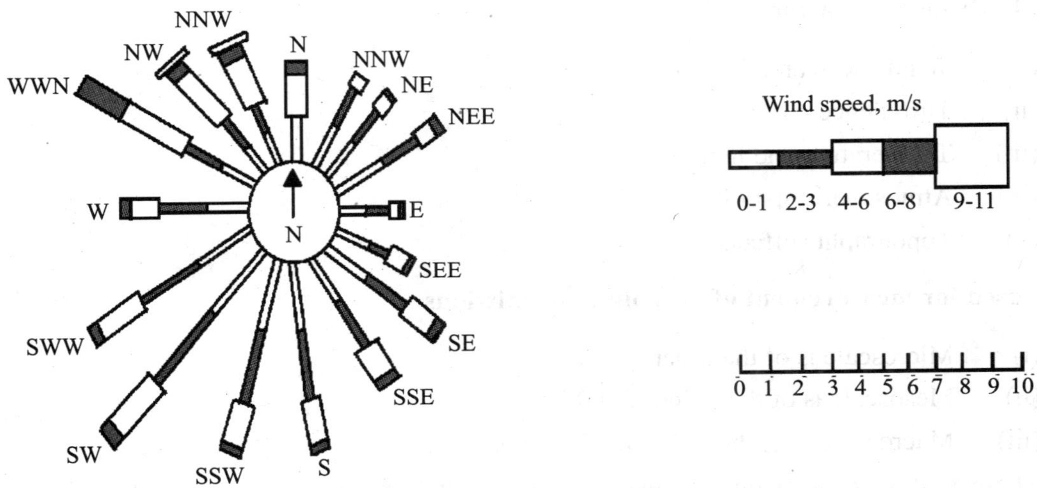

Fig. 6.3 Wind Rose

6.6.2 Variation of wind direction with altitude

Fluctuating lateral wind direction is termed as standard deviation of lateral wind direction. It also varies with atmospheric stability class and to some extent with height. More unstable condition means greater is the standard deviation σ_θ in lateral wind direction. Similarly vertical standard deviation σ_ϕ is defined along the wind direction. σ_ϕ varies with height (Fig. 6.4).

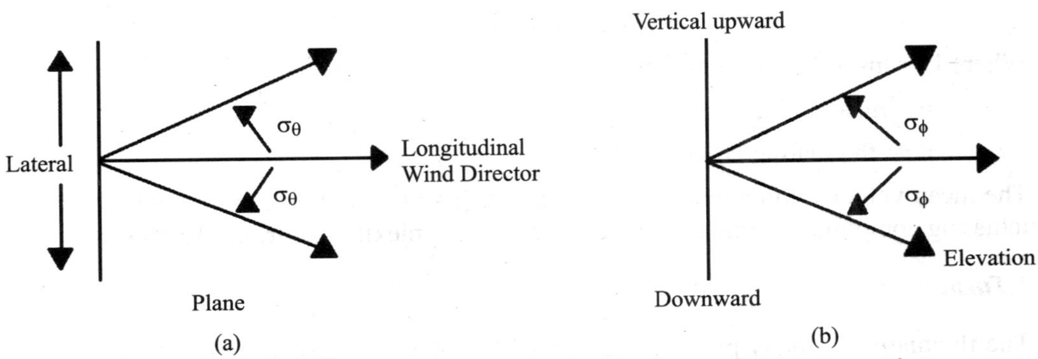

Fig. 6.4 Definition of σ_θ and σ_ϕ Standard Deviations of Wind Direction

6.7 AIR POLLUTION METEOROLOGY

When a gaseous or particulate emission, be it from a vehicle exhaust, an industrial stack or other source, is released into the atmosphere its fate is almost impossible to predict. This is because of the complex factors such as meteorology, sources and process influence its subsequent pathways. The influencing factors are primarily discussed below:

6.7.1 Meteorological factors

(i) Wind speed and direction
(ii) Turbulence
(iii) Temperature and humidity
(iv) Atmospheric stability
(v) Topographic effects

Scales used for measurement of air pollution emissions

(i) Microscale is of the order of 1 km.
(ii) Mesoscale is of the order of 100 kms.
(iii) Macroscale is of the order of 1000 kms.

These scales are also time related and since wind speeds are ~5.0m/s, the micro scale meteorological effects occur at durations of minutes to hours, the mesoscale from hours to days and the macroscale at days to weeks.

6.7.1.1 *Wind speed and Direction*

Wind speed at the earth's surface is zero due to friction of the surface roughness. The wind speed increases with distance due to the earth's motion. In the near surface Atmospheric boundary layer (ABL), random fluctuations of wind speed (and direction) are set up due to the surface roughness. In this zone, an instantaneous velocity measurement will have a mean and a fluctuating component. This is represented by:

$$U = u + u'$$

Where U = instantaneous velocity

u = mean velocity component

u′ = fluctuating component

The mean velocity component is due to the earth's motion, while the magnitude and sign of the fluctuating component is due to the presence and complexity of surface roughness.

6.7.1.2 *Turbulence*

The fluctuating velocity phenomena with eddies is called turbulence and because of surface roughness is called **mechanical turbulence**. Mechanical turbulence is partly responsible for air pollution dispersion. Turbulence is also produced by the vertical temperature gradient. The heated surface of the earth causes the hot air or thermals to rise and cool air falling, producing currents of air. This turbulent motion is called **thermal turbulence** or **buoyancy**.

Atmospheric turbulence is characterized by different sizes of eddies. These eddies are primarily responsible for diluting and transporting the pollutants injected in to the atmosphere. If the size of the eddies is larger than the size of the plume or a puff then the plume or the puff will be transported downwind by the eddy with little dilution. Molecular diffusion will ultimately dissipate the plume or the puff. If the eddy is smaller than the plume or the puff, the plume or the puff will be disperse uniformly as the eddy entrains fresh air at its boundary.

6.7.1.3 *Temperature*

In the lower troposphere, the temperature of the ambient air usually decreases with altitude. The rate of temperature change or gradient is known as the **lapse rate**. Ambient lapse rate: A balloon equipped with a thermometer, when released, will move upwards through the atmosphere and record the temperature of the ambient air. This temperature gradient is known as the **ambient lapse rate** and it varies from day to day, day to night and season to season. Most of the time the ambient lapse rate decreases with altitude, but there can be occasions when the reverse occurs. The reference lapse rate by which the ambient lapse rate is compared is the dry adiabatic lapse rate (DALR).

6.7.1.4 *Atmospheric Stability*

The ability of the atmosphere to disperse the pollutants emitted in to it depends to a large extent on the degree of stability. A comparison of the adiabatic lapse rate with dry adiabatic lapse rate gives an idea of stability of the atmosphere.

A neutrally stable atmosphere occurs when the ambient lapse rate (ALR) is equal to the DALR i.e., the rate of cooling with altitude is ~ 1°C/100 m. Here if a parcel of air is moved upwards or downwards, its temperature adjusts to that of its surroundings.

An unstable atmosphere occurs when the ALR exceeds DALR, i.e. the rate of cooling with altitude is > 1°C/100 m. Here, if a parcel of air is moved upwards, within itself it cools by ~ 1°C/100 m, so it is warmer than its new surroundings. If a parcel of air is moved downwards, it is cooler and denser than its surroundings and will thus continue to sink. This ambient condition is said to be unstable with a superadiabatic lapse rate.

A stable atmosphere occurs when ALR is less than DALR, i.e. the rate of cooling is <1°C/100 m. If a parcel of air is moved upwards, it will within itself cool at ~ 1°C/100 m. It will find itself cooler than the surroundings and therefore be forced to sink. Similarly, if a parcel of air is moved downward it will heat at ~ 1°C/100 m. It will find itself warmer than its surroundings and due to buoyancy will be forced back upwards. This ambient condition is said to be stable with a subadiabatic lapse rate (Fig. 6.5).

A stable inversion condition is a variant of the stable atmosphere. Here the temperature increases with altitude. If a parcel of air is moved upwards it will within itself cool at ~ 1°C/100 m. It will find itself much cooler than its surroundings, and so will be forced to sink. If a parcel of air is moved downwards, it will heat at 1°C/100 m. It will find itself warmer than its surroundings and due to buoyancy will be forced back upwards. An inversion temperature condition is a very stable condition, forcing air pollutants to remain trapped in the atmosphere for long periods. The occurrence of the different stability classes depends on meteorological conditions.

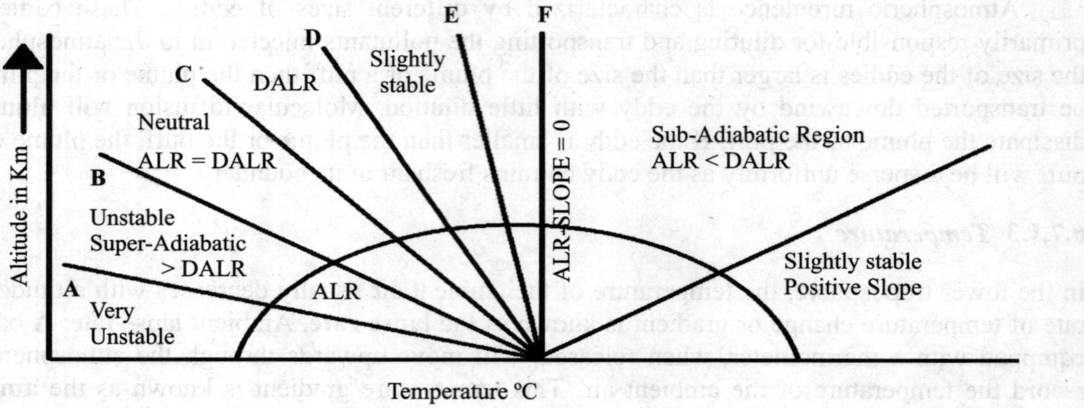

Fig. 6.5 Pasquill-Gifford Stability Classes- Vertical Temperature Profile

6.7.2 Lapse Rate and Dispersion

If the ambient temperature profile [ambient / environmental lapse rate (ELR)] is known and is compared to the dry adiabatic lapse rate, the fate of an air pollutant emission can be estimated. Stack emissions produce plumes.

6.7.2.1 Plume dispersion

Dispersion is the process by which contaminants move through the air and a plume spreads over a large area, thus reducing the concentration of pollutants it contains. The plume spreads both horizontally and vertically. If it is gaseous, the motion of the molecules follows the flow of gaseous diffusion. The most commonly used model for the dispersion of gaseous air pollutants is the Gaussian, developed by Pasquill, in which gases dispersed in the atmosphere are assumed to exhibit ideal gas behavior

6.7.2.2. The Gaussian plume model

The present tendency is to interpret dispersion data in terms of the Gaussian model. The standard deviations are related to the eddy diffusivities.

6.7.2.3 Plume rise

Generally, effluent plumes from the chimney stacks are released in to the atmosphere at elevated temperatures. The rise of the plume after release to the atmosphere is caused by buoyancy and the vertical momentum of the effluent. Under windless conditions, the plume rises vertically but more often it is bent as a result of the wind that is usually present. This rises of the plume adds to the stack an additional height ΔH, such that the height H of the virtual origin is obtained by adding the term ΔH, the plume rise, the actual height of the stack, Hs. The plume center line height H = Hs +ΔH is known as the effective stack height and it is this height that is used in the Gaussian plume calculations.

6.7.2.4 Plume behaviour

The behavior of a plume emitted from an elevated source such as a tall stack depends on the degree of instability of the atmosphere and the prevailing wind turbulence.

6.7.3 Classification of plume behaviour (Fig. 6.6)

(i) **Looping:** It occurs under super adiabatic or unstable conditions with light to moderate wind speeds on a hot summer afternoon when large scale thermal eddies are present. The eddies carry portion of a plume to the ground level for short time periods, causing momentary high surface concentration of pollutants near the stack. Thus the plume moves about vertically in a spastic fashion and the exhaust gases disperse rapidly. Looping favours the mixing of pollutants both upside and down.

(ii) **Conning:** It occurs under cloudy skies both during day and night, when the lapse rate is essentially neutral. The plume shape is vertically symmetrical about the plume line and the major part of the pollutant concentration is carried down -wind fairly far before reaching the ground level.

(iii) **Fanning:** It occurs when the plume is dispersed in the presence of very light winds as a result of strong atmospheric inversions. The stable lapse rate suppresses the vertical mixing, but not the horizontal mixing entirely. Plume width is very small indicating minimum dispersion. For high stacks, fanning is considered a favourable meteorological condition because the plume does not contribute to ground pollution.

(iv) **Fumigation:** Here a stable layer of air lies a short distance above the release point of the plume and the unstable air layer lies below the plume. This unstable layer of air causes the pollutant to mix down-wind toward the ground in large lumps, but fortunately this condition is usually of short duration lasting for about 30 minutes. Fumigation is favoured by clear skies and light winds, and it is more common in the summer seasons.

(v) **Lofting :** The condition for lofting plume are the inverse of those for fumigation , when the pollutants are emitted above the inverse layer, they are dispersed vigorously on the upward direction since the top of the inversion layer acts as a barrier to the movement of the pollutants towards the ground. On clear winter nights, air close to ground is cooler than the air above. So the stable condition prevail below and unstable conditions above. Stable below, neutral/ unstable above – lofting. There is mild mixing in the vertical, above the discontinuity. Below the discontinuity, there is no mixing and the plume remains aloft.

(vi) **Trapping:** It occurs when the plume effluent is caught between two inversion layers. Unstable condition between two stable / neutral conditions. Stable inversion aloft inhibits any mixing above the stack height, while the near neutral condition below the stack tip encourages mixing in the vertical downwards direction. As such, the plume impacts the ground at some distance downward.

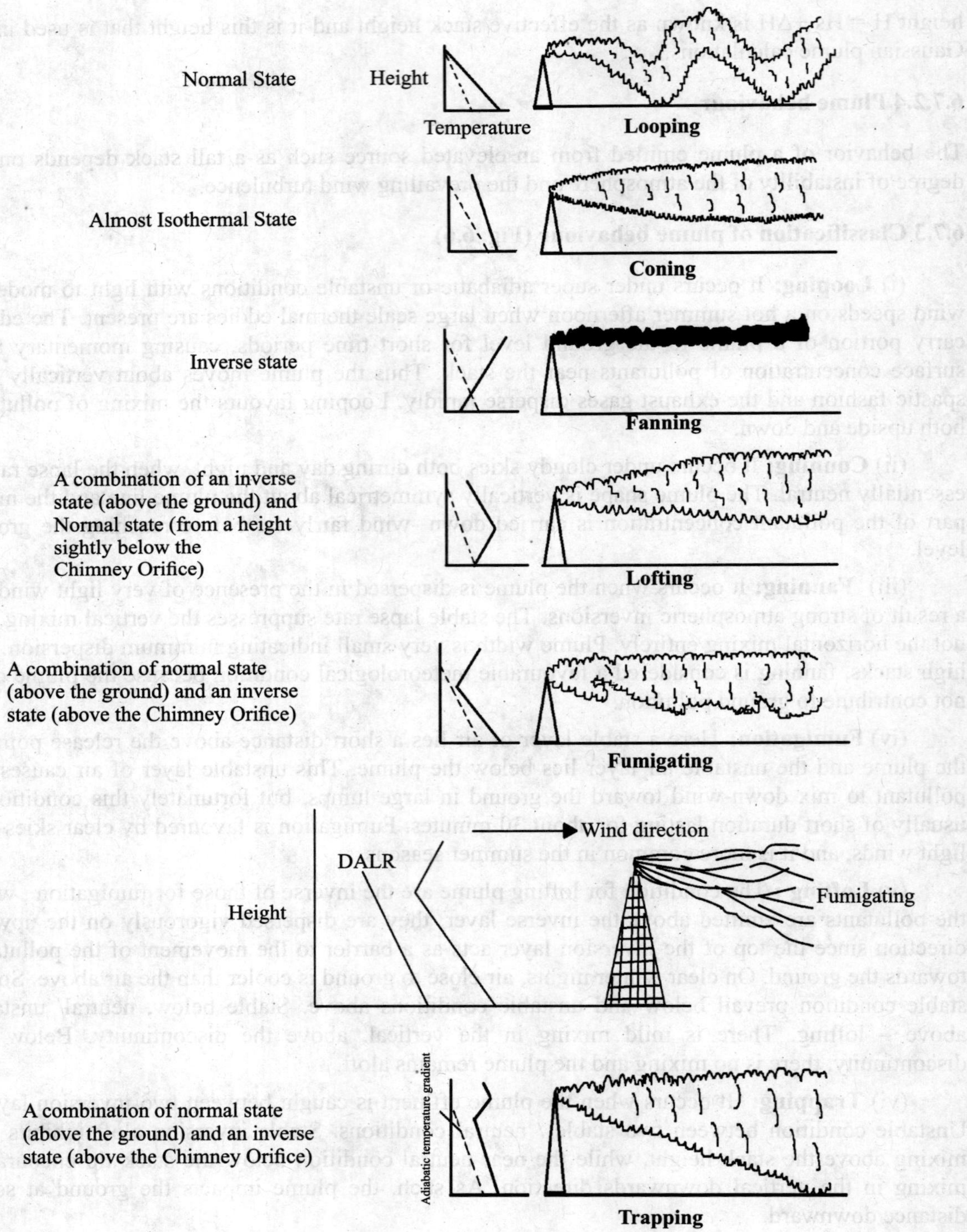

Fig. 6.7 Types of Plumes

6.7.4 *Types of Plumes developed*

(i) Unstable or super adiabatic condition (ALR > DALR)

Plume behavior – looping favors the mixing of pollutants both upside and down. Plume is forced upward and downward by thermal turbulence.

6.8 ATMOSPHERIC DISPERSION

Plume dispersion from a chimney stack is a convenient phenomenon to discuss atmospheric dispersion, although many other area are also of interest, e.g. fugitive emissions from pipes and vents, vehicle exhaust emissions, plumes from fires or explosions and emissions from dump sites. In an effort to predict concentrations of pollutants remote from the source, many techniques of air quality modeling have evolved (Fig. 6.8). These include:

(i) Physical
(ii) Statistical
(iii) Empirical
(iv) Numerical
(v) Gaussian

Traditionally, **physical modelling** utilizing wind tunnels was used. However, only a few centres in the world had adequately sized wind tunnels to investigate urban pollutant plumes. To overcome scaling effects, models of urban areas required to be of adequate size so as to properly represent the variation in vertical velocity and temperature profile. Physical modeling is desirable in multi source emissions on a complex terrain.

Statistical or **empirical** techniques are used if inadequate information on the physico-chemical processes exists to satisfy input requirements of numerical or Gaussian modeling. Such modeling will tend to produce information to assist with understanding of the fundamentals of the air quality problem. Such an evaluation may require the data monitoring to assist with understanding of the problem.

Numerical modelling requires the solution of the equations for conservation of mass, energy and momentum in three dimensions. Such modeling capability is most desirable but is as yet not readily available. This is due to the complexity of fluid and mass transport and particularly the turbulent terms are yet not fully amenable to solution.

Gaussian modeling is the more widely used technique for estimating the impact of non-reactive pollutants (USEPA, 1986). Gaussian modeling is far from being exact, as some of the model assumptions compromise accuracy. These assumptions include the followings:

(a) There is no variation in wind speed and direction between the source and the receptor.

(b) All effluent remains in the atmosphere and no provision is made for wet or dry deposition or chemical conversion. Any plume impacting on the ground is totally reflected.

(c) Dispersion does not occur in the downwind direction. It only occurs in the vertical and cross wind direction. The dispersion is stochastic and is described exactly by the Gaussian distribution.

(d) Emission rates are assumed constant and continuous.

6.8.1 Gaussian Distribution

Figure 6.9 shows the normal (Gaussian) curve, used to describe an occurrence governed by probability. The height of the central ordinate is called the 'mean' and the width of the curve is described in 'standard deviation'.

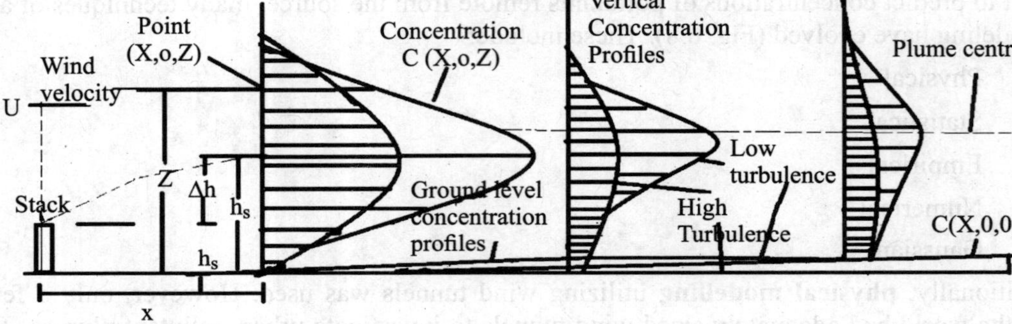

Fig.6.8 Schematic Elevation of Plume Concentration Profile

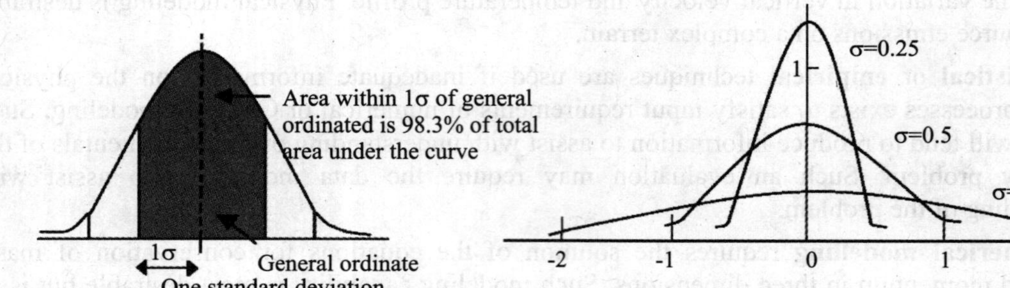

Fig. 6.9 Normal (Gaussian) Distribution Curve

For instant, 68.3 percent of the total area under the curve resides within 'one' standard deviation of the centre, 95.5 percent corresponds to 'two' standard deviations and 99.75 per cent corresponds to 'three' standard deviations. The Gaussian distribution in the lateral y direction is described by

$$G_y = \frac{1}{\sqrt{2\pi\sigma_y}} \exp\left[-\frac{1}{2}\left(\frac{y}{\sigma_y}\right)^2\right] \tag{6.1}$$

Similarly in the *z* direction

$$G_z = \frac{1}{\sqrt{2\pi\sigma_z}} \exp\left[-\frac{1}{2}\left(\frac{z}{\sigma_z}\right)^2\right] \tag{6.2}$$

As the distance downwind increases, the peak centre-line concentration decreases as the plume spreads out wider in the *y* and *z* directions. Figure 6.9(*b*) indicates that t' increases as the distance downwind increases.

Diffusion Equation

The concentration of a pollutant plume at any *x, y, z* location can be computed from the diffusion equation below (Fig. 6.10).

$$C(x, y, z) = \frac{Q}{2\pi\sigma_y\sigma_z U} \exp\left[-\frac{1}{2}\left(\frac{y}{\sigma_y}\right)^2\right] \exp\left[-\frac{1}{2}\left(\frac{z}{\sigma_z}\right)^2\right] \tag{6.3}$$

For the coordinate system : $x = 0$ at stack

$y = 0$ at plume centre-line

$z = 0$ at ground level

If the effective stack height is *H*, then the *z* coordinate of the plume centre-line is *(z-H)*, where is measured upwards from ground level:

$$C(x, y, z) = \frac{Q}{2\pi\sigma_y\sigma_z U} \exp\left[-\frac{1}{2}\left(\frac{y}{\sigma_y}\right)^2\right] \exp\left[-\frac{1}{2}\left(\frac{z}{\sigma_z}\right)^2\right] \tag{6.4}$$

In the conservation of mass assumption, that all plume contact with the ground is totally reflected, a second term must be added to account for this. This is shown schematically in Fig. 6.9. then Eq. (6.4) becomes:

$$C(x, y, z) = \frac{Q}{2\pi\sigma_y\sigma_z U} \exp\left[-\frac{1}{2}\left(\frac{y}{\sigma_y}\right)^2\right] \left\{\exp\left[-\frac{1}{2}\left(\frac{z - H}{\sigma_z}\right)^2\right] + \exp\left[-\frac{1}{2}\left(\frac{z + H}{\sigma_z}\right)^2\right]\right\} \tag{6.5}$$

The appropriate units are:

Q = any property per unit time, e.g. kg/s, m^3/s

C = that property per volume, e.g. kg/m^3, m^3/m^3 or ppb

t'_y, t'_z = diffusion coefficients, in m, as functions of downwind distance *x*

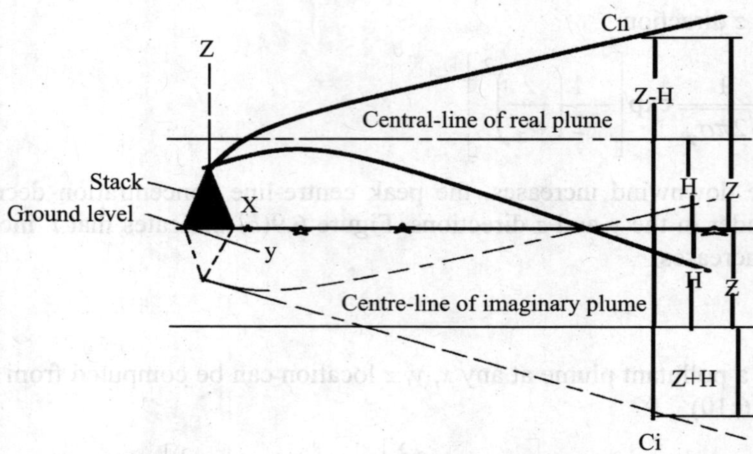

Fig. 6.10 Coordinate System and Ground Reflection of Plume Development

Knowing the downwind distance *x* and the stability criteria, σ_y and σ_z can be read from the figures.

$$C(x, y, o) = \frac{Q}{\pi \sigma_y \sigma_z U} \exp\left[-\frac{1}{2}\left(\frac{y}{\sigma_y}\right)^2\right] \exp\left[-\frac{1}{2}\left(\frac{H}{\sigma_z}\right)^2\right] \tag{6.6}$$

The maximum ground level concentration occurs at the plume centre-line, at $y = 0$, so

$$C(x, o, o) = \frac{Q}{\pi \sigma_y \sigma_z U} \exp\left[-\frac{1}{2}\left(\frac{H}{\sigma_z}\right)^2\right] \tag{6.7}$$

6.9 INDUSTRIAL AIR EMISSIONS CONTROL

Industrial air pollution occurs as the result of the release of pollutants (called emissions) into the atmosphere. The pollutants mix in air and are diluted but may travel long distances on slow, steady winds if an industrial chimney is tall enough to propel them high into the atmosphere. It is a fundamental problem of air pollution science that pollutant concentrations are exceedingly difficult to measure accurately.

Industrial air emissions are regulated, so as to keep the air quality acceptable to national and international standards. The design of air emissions abatement equipment is a difficult and complex task. It requires not only knowledge of the physical and chemical properties of the stream but also an appreciation of the vast array of equipment available and how this equipment operates.

There are three general options in air emissions.

- Waste minimization (of raw materials, a product or by abatement)
- Recovery and recycling
- Destruction or disposal

6.9.1 Characterizing the Air Stream

Purification of a gaseous stream involves manipulating the physical, chemical and sometimes the biological properties of the stream. Further information required

(i) The variability in the composition, flow rate, temperature and pressure of the stream (i.e. due to startup, shutdown).

(ii) The explosivity of the stream. This is especially important where VOCs are concerned e.g. petrol. Properties such as the flash point, auto – ignition temperature and the concentration at which it forms an explosive mixture with air, known as the lower explosive limit (LEL) should be known.

(iii) The corrosiveness of the steam in both liquids and gases; e.g. if SO_2 combines with water vapour and condenses, it forms corrosive sulphuric acid (H_2SO_4).

6.9.2 Equipment Selection

Equipment selection first considers the type of compounds which are required to be removed. These may be considered under three broad categories:

(i) VOCs, defined as 'any organic substance or mixture which can release vapour to the atmosphere and with the potential to cause environmental effects at low atmospheric levels.

(ii) Inorganic compounds

(iii) Particulate matter

The stream contents will determine the type of equipment which may be used in any application. Major types of equipment:

(i) Incinerators

(ii) Adsorbers

(iii) Condensers

(iv) Filters

(v) Scrubbers

(vi) Absorbers

(vii) Various particle collection devices

6.9.3 Control at source by Equipment

The most effective means of dealing with the problem of air pollution is to prevent emission at the source itself. Different techniques are used for controlling air pollution caused by 'gaseous pollutants' and that caused by 'particulate pollutants'.

6.9.3.1 *Methods of controlling gaseous pollutants*

The air pollution caused by gaseous pollutants like hydrocarbons, sulphur dioxide, ammonia, carbon monoxide etc. can be controlled by using three different methods- combustion, absorption and adsorption.

(i) **Combustion:** This technique is applied when the pollutants are organic gases or vapours. The organic air pollutants are subjected to 'flame combustion or catalytic combustion' when they are converted to less harmful product carbon dioxide and a harmless product water.

(ii) **Absorption:** In this method, the polluted air containing gaseous pollutants is passed through a scrubber containing a suitable liquid absorbent. The liquid absorbs the harmful gaseous pollutants present in air.

(iii) **Adsorption:** In this method, the polluted air is passed through porous solid adsorbents kept in suitable containers. The gaseous pollutants are adsorbed at the surface of the porous solid and clean air passes through.

6.9.3.2 *Methods of controlling particulate emissions*

The air pollution caused by particulate matter like dust, soot, ash etc. can be controlled by using fabric filters, wet scrubbers, electrostatic precipitators and certain mechanical devices.

6.9.3.3 *Some other methods of controlling air pollution*

(i) Tall chimneys should be installed in factories.

(ii) Better designed equipment and smokeless fuels should be used in homes and industries.

(iii) Renewable and non- polluting sources of energy like solar energy, wind energy, etc., should be used.

(iv) Automobiles should be properly maintained and adhere to emission control standards.

(v) More trees should be planted along roadsides and houses.

6.9.4 Classification of pollution control Equipments

Pollution control equipments are generally classified into two types.

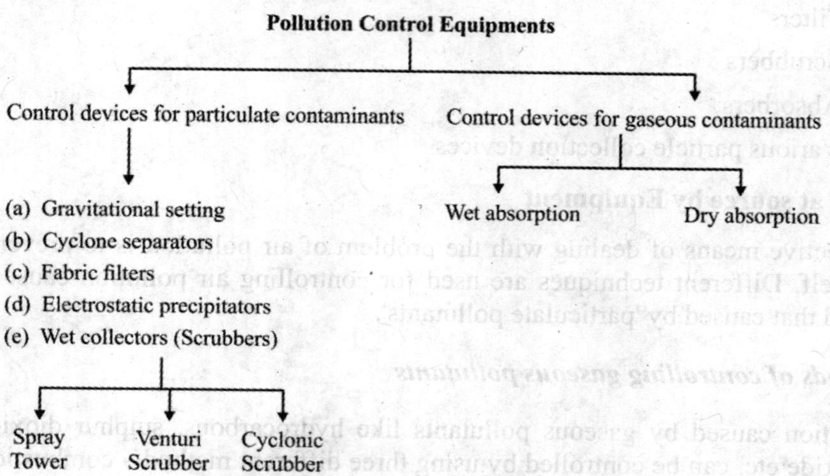

To select the required equipment certain basic data must be available. The required data are:

(i) Quantity of gas to be treated and its variation with time

(ii) Nature and concentration of matter to be removed

(iii) Temperature and pressure of gas stream

(iv) Nature of the gas phase

Mechanical Devices: It works on the basis of following:

6.9.4.1 Gravity Settling Chamber

In this device, the particulate settle down by the action of gravitational force and get removed.

It is the simplest and crudest of all pollution control devices (Fig. 6.11). It often serves as a preliminary screening device. The particulates having higher density obey Stokes's law and settle at the bottom of the chamber from where they are removed. The presence of several horizontal shelves or trays improves the collection efficiency by shortening the settling path of the particles. Gravitation force may be employed to remove particulates in settling chambers when the settling velocity is greater than about 0.12m/s.

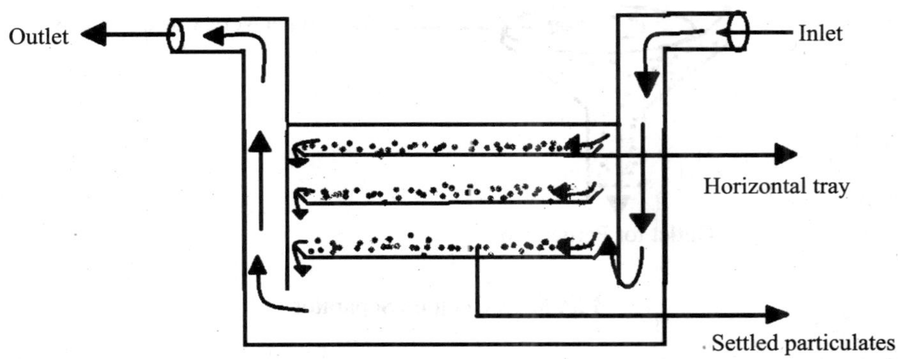

Fig. 6.11 Gravitational Settling Chamber

Advantages

(i) Low initial cost

(ii) Simple construction (in brick or stone masonary concrete)

(iii) Low maintenance cost

(iv) Dry and continuous disposal of solid particulates

Disadvantages

(i) Large space requirement

(ii) Comparatively large size (> 10 μm) particles can be collected.

6.9.4.2 Cyclone Separators

Centrifugal force is utilized to separate the particulate matter from the polluted gas. Centrifugal force several times greater than gravitational force can be generated by a spinning gas stream and this quality makes cyclone separators more effective in removing much smaller particulates than cannot be removed by gravitational settling chambers (Fig. 6.12). The dust laden gas enters tangentially, receives a rotating motion and generates a centrifugal force due to which the particulates are thrown to the cyclone walls as the gas spirals upwards inside the cone (i.e. flow reverses to form an inner vortex which leaves through the outlet). The greater the centrifugal force, the greater the separating efficiency.

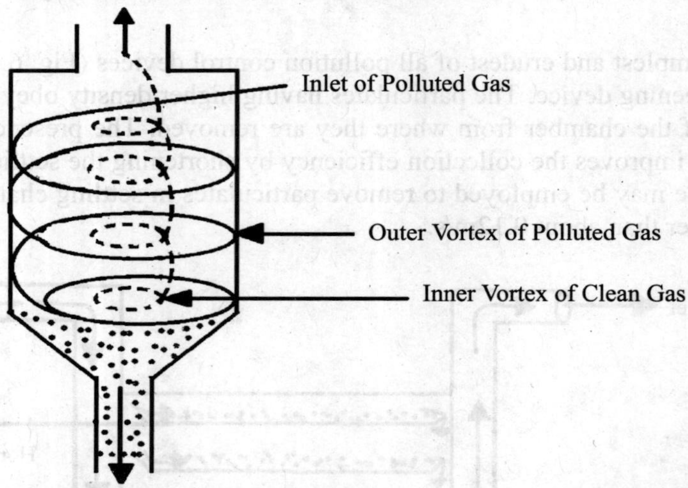

Fig. 6. 12 Cyclone Separator

Advantages

(i) Low initial cost
(ii) Simple construction and operation
(iii) Low pressure drop
(iv) Low maintenance requirements
(v) No moving parts
(vi) Can be constructed of any material, which will meet the temperature and pressure requirements, and the corrosion potential of the flue gas.

Disadvantages

(i) Low efficiency for particles below 5 to 10 μm in size.
(ii) Equipment is subjected to severe abrasive deterioration.

6.9.4.3 Fabric Filters

A filter is a porous structure composed of a granular or fibrous material, which retains the particulates and allows the gas to pass through the voids of the filter (Fig. 6.13). Fabric filters are made in the form of tubular bags or cloth envelops and is suitable for a dust loading of the order of 1gm/m^3. They are capable of removing dust particles as small as 0.5 microns and will remove substantial quantities of particles as small as 0.1 microns also. The process of controlling air pollution by using fabric filters is called 'bag filtration'.

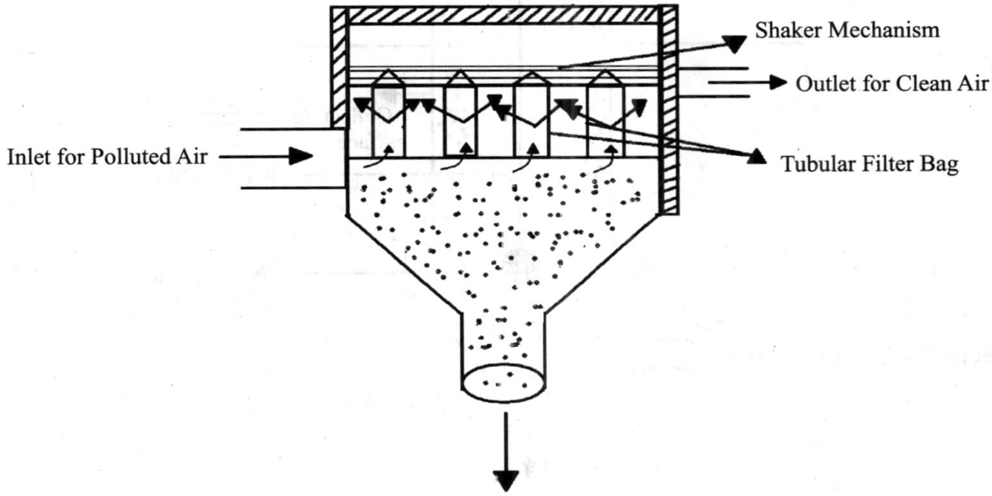

Fig. 6.13 Fabric Filter

Advantages

(i) High collection efficiency for particles less than 10 μm in diameter

(ii) Simple construction and operation

(iii) Nominal power consumption.

Disadvantages

(i) High maintenance and fabric replacement cost

(ii) Large size of equipment

(iii) Operating limits are imposed by high temperature of carrier gas and high humidity.

6.9.4.4 Electrostatic Precipitators (ESP)

When the polluted air containing particulate pollutants is passed through an electrostatic precipitator, it induces electric charge on the particles and then the aerosol particles get precipitated on the electrodes. It works on the principle of electrostatic precipitation i.e., electrical charged particulates present in the polluted gas are separated from the gas stream under the influence of the electrical field (Fig. 6. 14)

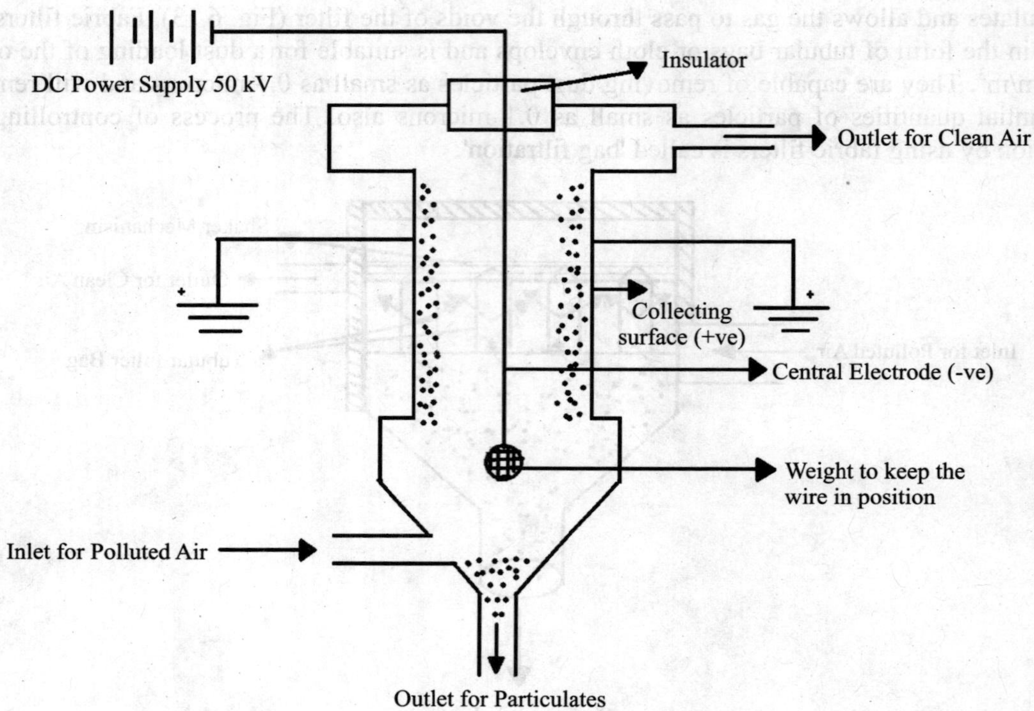

Fig. 6.14 ESP (Pipe – type)

A typical wire and pipe precipitator consists of:

(i) A positively charged collecting surface (grounded)

(ii) A high voltage (50 kV) discharge electrode wire

(iii) Insulator to suspend the electrode wire from the top

(iv) A weight at the bottom of the electrode wire to keep the wire in position

Efficiency of an ESP with plate collectors is given by:

$$\text{Efficiency} = 1 - \exp(-AW/Q)$$

(Anderson – Deutsch equation)

Where, A = area of the plates, m^2

W = particle sedimentation velocity in an electric field, m/s

Q = gas flow rate, m^3/s

When gas containing aerosol is passed between two electrodes that are electrically insulated from each other, and between which there is a considerable difference in electric potential that attracts the oppositely charged ions or particulates, which are subsequently removed.

Classification (Basing on electrode arrangement)

(i) *Pipe – type Precipitator*

In this type, most of the pipes act as collecting electrodes. The pipes may be round, square or octagonal. The populated gas enters from the bottom, flows upwards (i.e. between the high voltage wire and the grounded collecting surface).The high voltage in the wire ionizes the gas. The negative ions migrate towards the grounded surface and pass on their negative charge to the dust particles also. Then these negatively charged dust particles are electrostatically drawn towards the positively charged collector surface, where they finally get deposited. The collecting surface is rapped or vibrated to periodically remove the collected dust – particles so that the thickness of the dust layer deposited does not exceed 6mm, otherwise the electrical attraction becomes weak and efficiency of the electrostatic precipitator gets reduced.

(ii) *Plate – type Precipitator*

Here the collecting electrodes consist of parallel plates of size 1 – 2 m wide and 3 – 6 m high.

These plates are spaced at 20 – 30 cm. The no. of plates in the precipitator depends upon the inlet flow, so that the inlet gas velocities are 1 -2 m/s in each channel. The collection of the aerosols takes place on the inner sides of the parallel plates. The dust material collected can be removed by rapping and vibrating periodically. As the electrostatic precipitator has 99% efficiency and can be operated at high temperatures (600°C) and pressure and less power requirement, it is economical and simple to operate as compared to other devices.

Advantages

(i) High collection efficiency

(ii) Particles as small as 0.1 μm can be removed

(iii) Low maintenance and operating cost

(iv) Treatment time is negligible (about 0.1 to 10 sec.)

(v) Satisfactory handling of large volume of flue gases

Disadvantages

(i) High initial cost

(ii) Space requirement is more

(iii) Poisonous ozone gas is produced by high voltage discharge electrodes during gas ionization

6.9.4.5 Wet Collectors (Scrubbers)

They are used to trap SO_2, NH_3 and metal fumes by passing the fumes through water. Common wet scrubbers such as spray tower, venturi scrubber, cyclone scrubber are discussed below:

(i) *Spray Tower*

Water is introduced into a spray tower by means of a spray nozzle (i.e. there is downward flow of water). As the polluted gas flows upwards, the particulates (size exceeding 10 µm) present collide with the water droplets being sprayed downward from the spray nozzles. Under the influence of gravitational force, the liquid droplets containing the particulates settle to the bottom of the spray tower.

(ii) *Venturi Scrubber*

It removes submicron particulates (size 0.5 to 5µm) associated with smoke and fumes. The particles get entrapped in the droplets and fall down to be removed later on. They also remove soluble gaseous contaminants. Due to the atomization of water, there is proper contact between the liquid and the gas, increasing the efficiency of the venturi scrubber (their power cost is high because of the high inlet gas velocity i.e. 60 to 180 m/s).

(iii) *Cyclone Scrubber*

The way of introducing the polluted gas stream tangentially into the scrubber makes the stream to revolve like a cyclone and the centrifugal force drives the particles to fall apart towards the wall while the clean gas leaves through the top. The dry cyclone chamber can be converted into a wet cyclone scrubber by inserting high pressure spray nozzles at various places within the dry chamber.

6.9.4.6 Control of Gaseous Pollutants

Absorption

The principle is a gas-liquid contacting process for gas separation that utilizes the preferential solubility or chemical reactivity of the pollutant gas in the liquid phase. Efficient absorption depends upon:

(i) The amount of surface contact between the gas and the liquid – the fume gas is allowed to remain in contact with the liquid

(ii) The concentration of the absorbing media

(iii) The speed of the reaction between the absorbent and the gas.

Removal of HCs by absorption is employed in many industries, notably asphalt plants, coffee roasters, varnish etc. It is also used for recovery of valuable by products like acetic acid, formic acid, amines and ketones.

Advantages

(i) Low initial cost

(ii) High collection efficiency for small particles

(iii) Simultaneous removal of particulates as well as gases

(iv) Applicable for high temperature installations

Disadvantages

(i) High power consumptions for higher efficiency

(ii) Moderate to high maintenance cost due to corrosion and abrasion

(iii) Wet disposal of the collected material

Example 6.4 Recommend the number and type of bags required in a bag house to be attached with a furnace stack with flue gas temperature of 200°C. Assume air to cloth ratio as 2.0 m/min. gas flow to bag house $12\,m^3/s$.

Solution. Gas flow (Q) $= 12\,m^3/s = 720\,m^3/min$.

Filtration rate = Air to cloth ratio = 2.0 m/min

$$\text{Filtration rate} = \frac{720\,m^3/min}{2.0\,m/min} = 360\,m^2$$

Assume bag dimension as 0.3 *m* diameter and 7 *m* long.

$$\text{Area of 1 bag } \pi DL = 3.14 \times 0.3 \times 7 = 6.594\,m^2$$

$$\text{Number of bags required} = \frac{360}{6.594} = 54.6 \approx 55 \text{ numbers.}$$

Provide 55 numbers of silicon coated glass fibre bags for this high temperature application.

Example 6.5 A quality of 50 m^3/s of air flows from a cement manufacturing facility. It contains cement particles whose velocity is 0.12 m/s. If 99% removal efficiency is required, calculate the surface area of the ESP.

Solution. The efficiency 'E' of the ESP is given by $E = 1 - e^{-A\omega/Q}$, given that E = 99%

Flow rate of gas Q = 50 m^3/s, The particle settling velocity ω = 0.12 m/s, out plate area A=?

$$E = 1 - e^{-A\omega/Q}, \qquad 0.99 = 1 - E = 1 - e^{-Ax0.12/50}$$

$$\Rightarrow -A \times 0.12/50 = \text{Ln}\,(1 - 0.99)$$

$$\Rightarrow -A = \frac{50}{0.12} \times \text{Ln}\,(1 - 0.99)$$

$$\Rightarrow A = -416.67\,\text{Ln}\,(0.01) = 1919\ m^2$$

Example 6.6 Calculate the settling velocity of a particle, given that its diameter = 1μm, viscosity of air = 2.0 x10^{-5} kg/m.s and the difference between the densities of the particle and the air = 1.0 x 10^3kg/m^3.

Solution. Using stoke's theorem the settling velocity 'V' of a particle is given by,

$$V = \frac{g d_p^2\,(\rho p - \rho a)}{18\,\mu}$$

d_p = diameter of the particle

ρ_p = density of the particle

ρ_a = density of the air

μ = viscosity of the air

$g = 9.8\ m/sec^2$, $d_p = 1\ \mu m = 1 \times 10^{-6}\ m$

$(\rho_p - \rho_a) = 1.0 \times 10^3\ kg/m^3$ and $\mu = 2.0 \times 10^{-5}\ kg/m.s.$

Substituting these values in formula, we get:

$$V = \frac{9.8 \times (1 \times 10^{-6}) \times 1.0 \times 10^3}{18 \times 2.0 \times 10^{-5}} = 2.73 \times 10^{-5}\ m/s$$

Example 6.7 Determine the stack height required when SO_2 emissions are 100 x 10^6 g/s and SO_2 concentration 30 $\mu g/m^3$ are permitted at a distance of 500m from the stack. Assume wind velocity of 5 m/s and stability coefficients as 35 and 19.

Solution. Height of the stack is obtained from using the formula;

$$30 = \frac{100 \times 10^6}{2 \times \frac{22}{7} \times 5 \times 35 \times 19} \text{ x exp}\left[-\frac{1}{2}\left(H/_{19}\right)^2\right]$$

$$\Rightarrow H = 45.6\ m$$

Example 6.8 Determine the stack height when the particulate emissions are at the rate of 3×10^{-3} tonne/h.

Solution. Using the formula, $H = 74\ (Q_p)^{0.27}$, we have

$$\text{Stack height} = 74 \text{ x } \left(3 \times 10^{-3}\right)^{0.27} = 15.4m$$

(But the minimum stack height should be 30m)

Example 6.9 Find the diameter of the particles that will be removed from a settling chamber, given that V (Settling Velocity) = 0.33 m/s, μ = 2.0 x 10^{-5} kg/m.s, Sp. Gr. Of the particle = 2.0, L (Settling Chamber Length) = 8m, H (Height) =2m, ρ_p = 2000 kg/m³.

Solution. Minimum size of particles that are removed is obtained using the formula;

$$d_p = \left(\frac{18\,\mu VH}{gL\rho_p}\right)^{1/2} = \left[\frac{18 \times 2.0 \times 10^{-5} \times 0.33 \times 2}{9.81 \times 8 \times 2000}\right]^{1/2} = 40\ \mu m$$

Example 6.10 Determine the diameter of the particle that is removed in a cyclone with 50% efficiency, given that V (inlet velocity) = 16 m/s, d × b (inlet dimension) = (1×0.5m), ρ_p =1500 kg/m³, μ = Viscosity of air = 2.0 x 10^{-5} kg/m.s, Assume N (Number of turns) = 5 or 6.

Solution. $d = \left(\dfrac{9\,\mu(b)}{2\pi N V \rho_p}\right)^{1/2} = \left(\dfrac{9\times 2\times 10^{-5}\times 0.5}{2\times 3.14\times 5\times 16\times 1500}\right)^{1/2} = 12\,\mu m$

Example 6.11 Determine the size and number of bag filters needed, given that Air flow rate = 10 m^3/s, Velocity through fabric filter = 0.025 m/s.

Solution. Total filter cloth are required = $\dfrac{10}{0.025} = 400\ m^2$

For a filter of 0.25m diameter and 5m length, filtering surface area = $\pi \times 0.25 \times 5 = 4m^2$

Total number of bags required = $\dfrac{400}{4} = 100$

Example 6.12 Calculate the ESP (99% efficiency) plate area required when the drift current velocity of charged particles is 0.1 m/s and the flow rate of air is 10 m^3/s.

Solution. Using the formula, $1 - 0.99 = e^{\left(-\frac{0.1}{10}A\right)}$

Plate area required (A) = 460.5 m^2

6.10 FLUE GAS DESULPHURIZATION (FGD)

It is divided into two types:

Non – regenerative System

It generates a residue which must be disposed. Approx. 95 to 97 percent of the world's FGD systems are of this type.

Regenerative System

It converts sulphur oxides (SO_2, SO_3) to a marketable product. It is not likely that regenerable and catalytic processes will emerge as the desirable method of desulphurization for 2 reasons:

(i) The quantities of sludge that are required to be land filled are growing with the increasing numbers of FGD plants. Also the cost of land filling is likely to increase substantially with increased monitoring and control requirements.

(ii) Public are unlikely to accept that a non – regenerable waste is the best practicable environmental option when a number of regenerable stream options also exist. Most FGD processes centre around an absorption tower in which the sulphur dioxides are chemically absorbed into an alkaline liquid stream.

Classification

Wet or Dry depending on the phase in which the reaction occurs.

Most FGD systems in use today are wet non – regenerable processes.

6.10.1 Chemistry of FGD

Non – regenerative systems are based around the reaction of sulphur – dioxide with one or more of the following compounds: lime (CaO), caustic soda (NaOH), soda ash (Na_2CO_3), ammonia (NH_3), slaked lime ($Ca(OH)_2$), limestone ($CaCO_3$) and dolomite ($MgCO_3$+ $CaCO_3$, $MgCO_3$ mainly). Some of the reactions are as follow:

$$SO_2 + \text{CaO} \rightarrow \text{CaSO}_3$$
$$2CaSO_3 + \text{O}_2 \rightarrow 2\text{CaSO}_4$$
$$\text{CaCO}_3 + \text{SO}_2 \rightarrow \text{CaSO}_3 + \text{CO}_2$$

(i) Sulphite is a thixotropic gel, difficult to landfill.

(ii) Sulphate is a stable solid; hence settling ponds are often aerated to separate the sulphate from the gel.

(iii) Sulphate (known as gypsum) is the prime ingredient in plasterboard.

6.11 NO_X REMOVAL

NOx implies two major oxides, nitrogen oxide (NO) and nitrogen dioxide (NO_2). Mechanism of NOx production from combustion processes:

(i) From the reaction of N_2 in the fuel air with oxygen at the high temperatures of a burner chamber.

(ii) From nitrogen existing in the fuel.

(iii) From reactions of fuel – derived radicals with N_2 ultimately leading to NO.

To control NOx emissions effectively, the dominant formation mechanism must be known.

6.11.1 *Chemistry of NO_x formation*

The simplest, and most widely used, model of NOx formation is the Zeldovich mechanism, which may be expressed chemically as follows:

$$N_2 + \text{O} \rightarrow \text{NO} + \text{N}$$
$$\text{O}_2 + \text{N} \rightarrow \text{NO} + 0$$
$$\text{OH} + \text{N} \rightarrow \text{NO} + \text{H}$$

(i) These reactions may be considered to be independent of the main fuel burning reaction which takes place in boilers and furnaces.

(ii) These reactions occur because of the high temperature that exists in the flame itself.

(iii) NO_x formation does not occur to a significant extent below 1800 K and that the Nox present is dependent on temperature (Lefebvre, 1983).

6.11.2 *Minimizing NO_x Formation*

Burner design needs to balance the combustion process requirements with NOx formation. This is done in a number of ways:

(i) By minimizing the quantity of oxygen present to minimize NO_x formation (Table 6.2).

(a) Recirculation of gas is also common. This acts by reducing the peak flame temperature and the quantity of oxygen present.

(b) By injecting water into the combustion chamber, the temperature is reduced due to the energy being taken up by the latent heat of water.

(c) Finally, staged combustion can be used to reduce the peak temperature. Upto three stages can be used, but this requires very tight control over both the fuel and air flow rates to each stage.

(d) Ammonia is used in conjunction with a catalyst in a technique known as selective catalytic reduction (SCR). Ammonia is injected into the flue gas upstream of the catalyst bed. The NOx combine with ammonia on the catalyst surface where they decompose to harmless nitrogen and water. Such systems give good NOx removal rates but are expensive.

(e) Selective non-catalytic reduction (SNCR) is a post – combustion method of eliminating NO_x emissions. It involves the injection of ammonia or urea into the combustion chamber or an area of high temperature downstream. It requires a high temperature to achieve the reaction activation energy and hence is not applicable to certain types of equipment.

Table 6.2 Methods of NOx Reduction

Method	**Applicabity**	**NO_x Reduction (%)**
Flue Gas recirculation	Thermal NOx Reduction	70 – 80
Low NOx burners	Fuel and thermal NOx Reduction	10 - 25
Staged burners	Fuel and thermal NOx Reduction	40 - 70
Selective Catalytic reduction	Fuel and thermal NOx Reduction	80 - 90
Selective Non-catalytic reduction	Not applicable to small boilers or gas turbines	60 - 80

6.12 FUGITIVE EMISSIONS

Fugitive emissions are industrial emissions from both point and non – point sources. These sources may be considered the equipment and methods associated with the transferring, conveying, loading, unloading, storage, packaging and processing of materials.

Control Fugitive Emissions

It depends on:

(i) The compounds involved

(ii) The quantities involved

(iii) The equipment being used

(iv) Equipment modification

(v) Local ventilation / extraction to an emission control unit from where the cleaned gas is emitted as a point source

6.12.1 *Fugitive Dust sources and their control*

(i) ***Sources:*** Open mining / quarrying, unpaved roads, construction activity and the open or semi – open storage or movement of solid materials.

(ii) ***Control:*** Repeated application of water to chemical stabilization, surface cleaning, windbreaks or simply good management.

6.12.2 *Fugitive VOC Sources and their Control*

(i) **Sources:** Releases of organic and halogenated organic compounds to the atmosphere are defined as fugitive. Industrial valves, pumps and flanges have been identified as the main contributors to fugitive emissions from industrial facilities.

(ii) **Control**

(a) A rule is now in place where industries whose fugitive emissions increase must inspect more frequently and replace troublesome equipment instead of simply paying a fine.

(b) Quality Improvement Programme (QIP) requires a company to identify the problem, investigate alternative solutions and apply the most appropriate one. This must all be completed within a given time scale. Also it provides a compelling incentive for companies to do better.

(c) VOC abatement technologies are being adapted for control of petroleum fume emissions at marine loading / unloading stations and filling stations.

Review questions

1. Differentiate between primary and secondary air pollutants with examples.

Ans. Primary air pollutants are those which are emitted by identified, manmade source, whereas secondary air pollutants are formed in the atmosphere by chemical or photochemical reactions of primary pollutants. e.g. of primary pollutants are: SO_2, CO, NO_X, SO_X, particulates, hydrocarbons etc. e.g. of secondary pollutants are: O_3, PAN, Smog, acid rain etc.

2. What is acid deposition?

Ans. Acid deposition can be defined as the deposition or settling of the acid (H_2SO_4, HNO_3) on the earth formed in the environment due to SO_X & NO_X. Acid deposition mainly occurs as wet deposition (e.g. Rain, snow etc) and dry deposition (with particulate deposition).

3. Name some of the Green House Gasses (GHG)

Ans. Carbon dioxide (most important)

Methane

Water vapour

Nitrous oxide

4. **Name some of ozone depleting substances.**

Ans. Nitrous oxide and chlorofluorocarbon (CFC) group of chemicals

5. **List some of the global air pollution problem.**

Ans. Acid rain, Green house effect, Ozone Layer Depletion, etc.

6. **What are the factors that affect the dispersion of air pollutants in the atmosphere?**

Ans: These factors are:

(i) Wind speed and direction
(ii) Temperature and humidity
(iii) Turbulence
(iv) Atmospheric stability
(v) Terrain effect on dispersion includes
   (a) Heat island
   (b) Land/ sea interface
   (c) Valley and hill slopes

7. **Show the Pasquil-Gilford diagram for the various classes of stability condition of the atmosphere.**

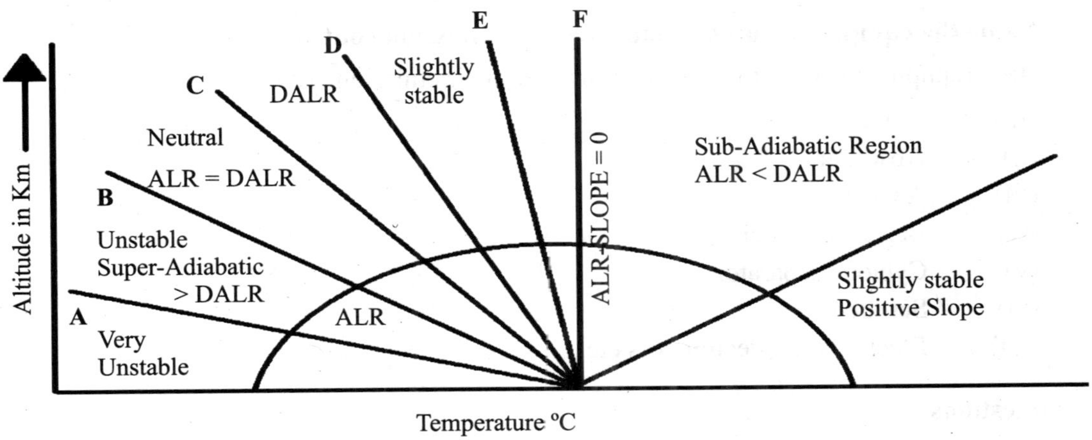

8. **What do you mean by super adiabatic condition of atmosphere?**

Ans. An unstable atmosphere occurs when the ambient lapse rate exceeds Dry adiabatic lapse rate. This ambient condition is said to be unstable with a superadiabatic lapse rate.

9. **Does the air temperature lapse rate have anything to do with air quality?**

Ans. Yes, because it is related to amount of vertical mixing of emitted air pollutants.

10. What is flue gas desulphurization?

Ans. Flue gas is bad or highly polluted gases and flue gas desulphurization (FGD) stands for removal of sulphur oxides from emissions.

FGD process may be of two types:

1. Generative process
2. Non-generative process

Generative process mainly based on reactions which results in some marketable products such as calcium sulphate (Gypsum), H_2SO_4.

$SO_2 + CaO \rightarrow CaSO_3$

$2CaSO_3 + O_2 \rightarrow 2CaSO_4$ (Calcium sulphate/Gypsum)

$2SO_2 + \text{Excess } O_2 \rightarrow 2SO_3$

$2SO_3 + 2H_2O \rightarrow 2H_2SO_4$

Non-generative process generates residue which must be disposed and the reaction based on the reaction of SO_2 with compounds like:

(a) Lime (CaO) (b) Lime stone ($CaCO_3$)

(c) Soda ash (Na_2CO_3) (d) Caustic soda ($NaOH$)

Some of the reactions of non-regenerative process are

$SO_2 + CaO \rightarrow CaSO_3$ (calcium sulphite)

$CaCO_3 + SO_2 \rightarrow CaSO_3 + CO_2$

11 Name the equipments used in industrial air emission control

Ans. Equipments used to control the industrial air emission are:

(i) Condenser
(ii) Absorber
(iii) Adsorber
(iv) Bag house filter
(v) Cyclone separator
(vi) Scrubbers
(vii) Electrostatic precipitators etc.

Long questions

1. What is global warming?
2. What is acid rain and its effect?
3. Describe the direct and indirect health effects of global warming?
4. List the cause of ozone depletion? And its potential health effects?

Chapter 7

Solid Waste Management

7.1 INTRODUCTION

Management of solid wastes is a major challenge these days for the administrators, engineers and planners. Solid wastes are those wastes arising from human & animal activities that are normally discarded as useless or unwanted. They are the heterogeneous mass of throwaways from residences and commercial activities as well as more homogeneous accumulation of agricultural, industrial, and mineral wastes. It may be called as refuse, rubbish, garbage, trash etc. Huge volumes of solid wastes are generated and need to be collected, transported and finally disposed off. However, the system for managing wastes is primitive and cannot cope with the huge volumes of wastes being generated.

7.1.1 Types of Solid Wastes

Solid wastes are materials that are considered cheaper to throw away than to store or use and are of different types as discussed below:

(i) Municipal solid waste: These are wastes produced by operations and maintenance of municipal facilities such as food wastes, ashes and residues, demolition & construction wastes, special wastes (street sweeping, road-side litter, catch basin-debris, dead animals, abandoned vehicles), treatment plant waste.

(ii) Industrial solid waste: These wastes are generated from various processes in small and large scale industries. These are heterogenous in nature such as: rubbish, ashes, demolition & construction wastes, special waste, hazardous waste.

(iii) **Hazardous solid waste:** Wastes such as radioactive substances, chemicals, corrosives, reactive, toxic, biological wastes from hospitals, flammables, explosives etc.

7.1.2 Trend of Urbanization

Table 7.1. Trends of urbanization

S.No.	Urbanization trends	1951	1991	2001	2021
1	Number of Urban (Agglomerations/Town)	2795	3768	5161	-
2	Urban Population (In million)	62.0	217.0	285.0	550.0
3	Total Population (in %)	17.3%	25.72%	27.8%	41%

7.1.3 Magnitude of Problem

Urbanization is global concern but more pronounced in developing countries. Increasing population and shifting from rural areas to high paying urban culture is the leading cause of urbanization (Table 7.1) which directly contributes to waste generation. Therefore, it is evident that change in lifestyle, increase in consumerism leads to higher waste generation in India. The per capita waste generation is increasing by 1.3% per annum with urban population increasing between 3-3.5% per annum. The yearly increase in waste generation is around 5% annually. India produces 42.0 million tons of municipal solid waste annually at present.

The per capita generation of waste varies from 200 gm to 600 gm per capita/day. Average generation rate at 0.4 kg per capita per day in 0.1 million plus towns. Collection efficiency ranges between 50% to 90% of the solid waste generated.

7.2 SOURCE, CLASSIFICATION & COMPOSITION OF MSW

7.2.1 Source of Solid Wastes

Table 7.2 Sources and types of solid waste (according to WHO, 1991)

SOURCES	FACILITY	TYPE OF WASTE
Domestic	Single family dwelling, multifamily dwelling, low, medium and high rise apartments	Food, paper, packaging, glass, metals, ashes, bulky household waste, hazardous household waste.
Commercial	Shops, restaurants, markets, office buildings, hotels and moles, institutions.	Food, paper, packaging, glass, metals, ashes, bulky household waste, hazardous household waste.
Industrial	Fabrication, light and heavy manufacturing refineries, chemical plants, mining, power generation.	Industrial process waste, metals, lumber, plastics, oils, hazardous waste.
Construction and demolition		Soil, concrete, timber, steel, plastics, glass, vegetation.
Agricultural waste	Activities such as planting and harvesting of trees, animal farms, poultry farms.	Feed lots, animal manure and other agricultural waste.
Municipal waste	Operations and maintenance of municipal facilities such as street sweeping.	Roadside litter, tree trimmings, yard wastes from public parks and play ground, dead animals.

7.2.2 Composition of Solid Wastes

The solid wastes are broadly of 2 types: organic and inorganic (Table 7.3). Industries tend to identify the composition of their waste according to its industry type. For example paper products, metal products. The World Health Organization (WHO)1991 organize industrial waste into 3 broad categories:

(i) Non-hazardous industrial waste

(ii) Hazardous waste

(iii) Hospital waste

Table 7.3 Physical Composition of Solid Waste

General Composition	Typical Composition	Detailed Composition
Organic	Food putrescible Paper and cardboard	Food items, fruits, vegetables, paper, cardboard.
	Plastics	Polyethylene terephthatate (PETE), High density polyethylene (HDPE), Polyvinyl chloride (PVC), Low density polyethylene (LDPE), Polypropylene (PP), Polystyrene (PS), other multilayer plastics.
	Clothing/Fabric Yard Waste, wood	Textiles, carpets, rubber, leather, Garden trimmings wood
	Miscellaneous organics	Bone
Inorganic	Metals	Tin cans, Ferrous metals, Aluminum, Non-ferrous metals.
	Glass, Dirt, Ash etc.	Colorless, colored dirt, screenings, ashes, stones, bricks.
	Unclassified	Bulky items

7.2.3 Classification of Solid Wastes

The wastes generated within a municipality (excluding industrial and agricultural) will vary widely, depending on the community and its level of commercialism and institutionalism. The data on waste water will also depend on level of sophistication of the waste management operation.

Analysis of MSW

Physical, geotechnical, chemical and biological analysis of municipal solid wastes are important for the design of an integrated solid waste management system. these are described below:

(i) *Physical analysis*

a) Composition: Paper, plastics, textile, leather, metals, glass, ash, fine earth and others, Specific weight, particle size, moisture content, field capacity, and permeability etc.

The specific weight can be calculated as:

$$\text{Specific weight}=(W_{WC} - Wc)/Vc$$

Where, W_{wc} is weight of container filled with waste (kg)

Wc is weight of empty container (kg); and

Vc is volume of container (m^3)

(ii) Chemical Analysis

This analysis is required in the design of various processes such as energy recovery or composting.

(a) ***Proximate analysis***: *It is carried out to determine the* moisture content, ash content, volatile matter and fixed carbon.

(b) ***Ultimate analysis***: The ultimate analysis of waste consists of determination of percentage of its various chemical constituents such as C, H, N, O and S

(c) ***Energy content***: Energy content of municipal solid waste is the net calorific value of wastes. It is the heat produced by a unit quantity of waste, at constant volume and at constant pressure of one atmosphere.

Modified Dulong formula can also be used for determination of energy content of municipal solid wastes e.g.,

$$\text{Energy content (Btu/lb)}= 145C +610(H_2\text{-}O_2/8)+40S+10N$$

The energy components of MSW components:

- Food wastes: 4600 KJ/Kg
- Paper and Cardboard: 16000 KJ/Kg
- Plastics: 32000 KJ/Kg
- Textiles: 17000 KJ/Kg
- Wood: 18500 KJ/Kg
- Glass: 140 KJ/Kg
- Leather: 17000 KJ/Kg

(3) Biological Analysis: This include water soluble constituents such as sugars, starches, amino acids, and other organic acids, proteins, hemicelluloses, cellulose, lignin. Biodegradability, production of odour and breeding of flies.

Characteristics of MSW

(i) Compostable / Bio-degradable matter = 30% - 55% (can be converted into manure)

(ii) Dry and Inert material = 40% - 45% (goes to combustion & landfill)

(iii) Recyclable materials = 5% - 10% (Recycling)

These percentages may vary from city to city depending on food habits (Table 7.4)

Table 7.4. Sample Industrial Waste Type (Standards Industrial Classification Manual 1972)

SIC Classification	Waste Generating Process	Wastes
Food & food products	Processing, packaging, shipping.	Metals, fats, oils, bones, vegetables, fruits etc.
Paper & applied products	Paper manufacture, cardboard boxes etc.	Paper, cardboard, inks, glues, chemicals etc.
Electrical products	Manufacture of electrical appliances.	Scrap metal, plastics, rubber, textiles etc.
Chemical products	Manufacturing of drugs etc.	Organic and inorganic chemicals, metals, plastics, solvent paints etc.

7.3. SEPARATION, STORAGE & TRANSPORTATION OF MSW

Solid Waste Management System (India)

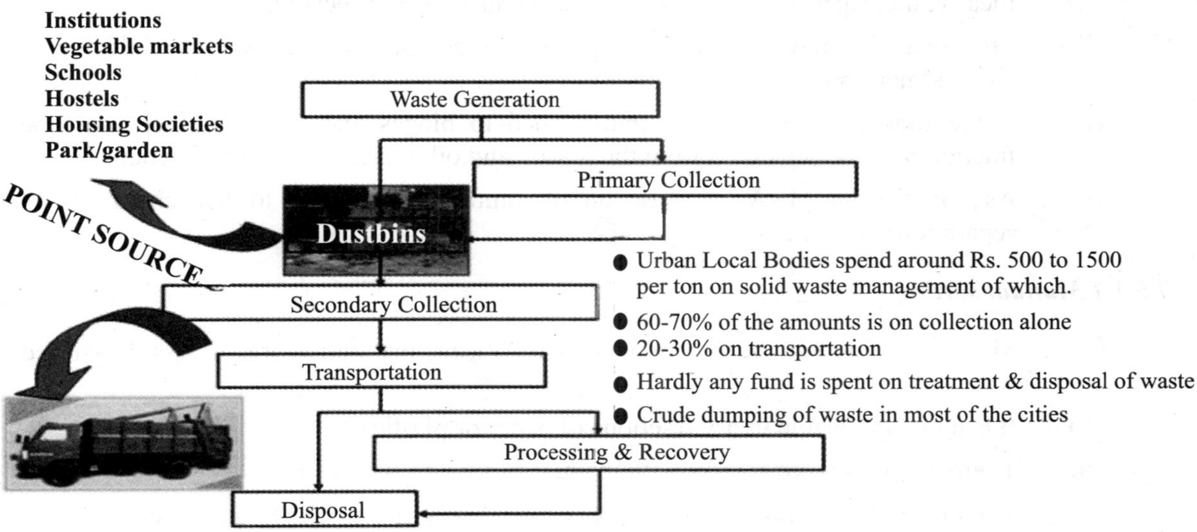

Fig. 7.1 Solid Waste Management

7.3.1 Separation of MSW

Separation can be done either at source in the household (or industry) or at the transfer station or at final destination where mechanical separation / sorting is possible (Fig. 7.1). Overall MSW can be divided into:

(i) Wastes that are desirably separated at source (for regular collection by public/ private lanlier)

(ii) All other household waste.

The wastes that are desirably separated at source are:

(i) **Food wastes** – Household source separation.

(ii) **Paper and cardboard** - Household source separation.

(iii) **Plastic** - Household source separation.

(iv) **Metal ferrous** - Household source separation.

(v) **Metals non-ferrous** – Community recycling at drop-off centers.

(vi) **Glass** -Community recycling at drop-off centers, household source separation.

All household waste that can be delivered to drop-off centers include:

(i) Bulky waste (furniture, tyres etc.)

(ii) Yard waste

(iii) Hazardous household waste (in some countries collected at sources).

Requirements for source separation

(i) The level and type of source separation will depend on people's attitudes and the end use.

(ii) Ideally, the paper, glass and non-ferrous metals could e-recycled.

(iii) The remainder may only need to separate into combustibles and non-combustibles if the treatment process is incineration.

(iv) If the food fraction is to be transformed to biogas and compost, then the food fraction must be separated from the plastic and other non-degradable fractions.

(v) As commercial MSW is made up of similar components to household MSW, separation at source is also desirable.

7.3.1.1 *Manual sorting*

(i) Manual sorting can be recommended for clean, dry and more or less pre-sorted waste.

(ii) It is more applicable in dry fractions of paper or plastics.

(iii) There are occupational risks with manual sorting on wet waste.

(iv) The methodology of manual sorting is "negative sorting" i.e. the unwanted materials are removed.

7.3.1.2 *Mechanical Sorting*

(i) It is most commonly used.

(ii) Magnetic separators are used to separate the ferrous and non-ferrous components.

(iii) Vibrating screens or rotating screens are used to separate fractions by particle size.

(iv) Air screening and ballistic screens are used to separate light from heavy material.

(v) Volume reduction by compaction is carried out on different waste fractions, particularly paper, cardboard, cans, and plastics.

7.3.2 Storage of MSW

The type of storage used depends on the collection facilities, which may be as discussed below (Fig. 7.2):

(i) Doorstep collection.

(ii) Regular kerb collection.

(iii) Civic amenity drop-off.

(iv) Haulier for skip collection of bulky items.

(v) Community recycles bins.

(vi) Vacuum trucks.

(vii) Skips or open top containers are regularly used for builders and demolition rubble.

(viii) They are also used for large garden waste and community clean ups.

(ix) Containers are used for the largest of bulk transport, generally over large distances.

(x) Containers are used in commercial application, department stores, warehouses etc. where in many cases the wastes are compacted with either static or dynamic compactors to further increase their capacity.

Fig. 7.2 Storage of MSW

Collection of Commingled Waste

Curb services

The home owner is responsible for placing the containers to be emptied at the curb on collection day and for returning the empty containers to their storage location until the next collection.

Alley services

Alley storage of containers used for solid waste is common. The collection is similar to curb services.

Set out-setback services

Containers are set out from the homeowner's property and set back after being emptied by additional crews that work in conjunction with the collection crew responsible for loading the collection vehicle.

Setout services

Setout service is essentially the same as setout-setback service, expect that the homeowner is responsible for returning the containers to their storage location.

7.3.3 Transportation of MSW

Transport vehicles used for transport of MSW are discussed below:

(i) Traditional open top trucks used in low income countries.

(ii) Tradition refused trucks taking primary bagged waste.

(iii) Modern single compartment trucks taking wheeled bins from single unit dwellings.

(iv) Single compartment truck taking larger wheeled bins from multiunit apartment buildings and commercial outlets.

(v) Multi-compartment trucks taking source separated waste. A multi-compartment truck capable of also taking several different recyclables is the most modern an efficient (one operator) system for collection

(vi) Skip trucks taking skip loads.

(vii) Trucks taking container loads, either closed or open looped.

(viii) Vacuum trucks used in areas with limited accessibility, with tube lengths up to 100m.

7.3.1.1 Types of collection systems

7.3.1.2 *Haul Container System (HCS)*

These are collection systems in which the containers used for the storage of wastes are hauled to a materials recovery facility (MRF), transfer station, or disposal site, emptied, and returned to either their original location or some other location.

Hauled container systems are ideally suited for the removal of wastes from sources where the rate of generation is high because relatively large containers are used. The use of large containers eliminates handling time as well as the unsightly accumulations and unsanitary conditions associated with the use of numerous smaller containers.

Another advantage of hauled container systems is their flexibility: Containers of many different sizes and shapes are available for the collection of all types of wastes. It required only one truck and driver to accomplish the collection cycle.

7.3.1.3 *Collection truck*

There are three main types of vehicles used in hauled container systems as shown below (Fig. 7.3):

(1) Hoist truck

(2) Tilt- frame container

(3) Truck tractor trash-trailer

Hoist truck | Tilt-frame | Truck tractor trash-trailer

Fig. 7.3 Collection Trucks

7.3.1.4 Stationary Container System (SCS)

In the stationary container system, the containers used for the storage of wastes remain at the point of generation, except when they are moved to the curb or other location to be emptied. The systems vary according to the type and quantity of wastes to be handled as well as the number of generation points. There are two main types:

(1) Systems in which manually loaded collection vehicles are used (Fig. 7.4).

(2) Systems in which mechanically loaded collection vehicles are used (Fig. 7.5).

The major application of manual loading collection vehicles is in the collection of residential source-separated and commingled wastes and litter. Manual loading is used in residential areas where the quantity picked up at each location is small and the loading time is short.

Manually loaded Collection trucks

Fig. 7.4 Manually Loaded Collection Trucks

(a) Side-loaded right-hand (b) Side-loading vehicle used for (c) Rear-loaded collection vehicle for standup drive collection of source-separated vehicle for commingled collection solid waste commingled solid wastematerials

Mechanically loaded Collection trucks

Fig. 7.5 Mechanically Loaded Collection Trucks

7.3.2.1 *Transfer Stations*

7.3.2.2 *Objectives*

(i) Reduction in transport cost.

(ii) Reduction in trafficking of smaller vehicles at the treatment (disposal) sites.

(iii) Reduction in refuse crew cost due to waiting time for vehicles in transit.

7.3.2.3 *Elements of a Transfer Station*

(i) Level 1 is a concrete paved area where refuse is tipped from reuse vehicles. This area may have the facility for two or more vehicles to unhold simultaneously.

(ii) The waste falls into a hopper where a control breaker (gate) permits the waste to further fall in controlled volumes into a compacting unit. This unit is on Level 2, about 5 m below Level 1.

(iii) The compacting unit pushes the compacted waste horizontally into a waiting container (in the docking area).

(iv) The containers which can be 30 to 40 m^3 capacity are loaded on to waiting multiaxle trucks. The containers are on a 'rail-line' which permits easy movement and storage capacity for a number of containers.

7.3.2.4 *Types of Transfer Stations*

(i) Direct-Load Transfer System: At direct-load transfer stations, the wastes in the collection vehicles are emptied directly into the vehicle to be used to transport them to a place of final disposition or into facilities used to compact the wastes into transport vehicles that are transported to the disposal site.

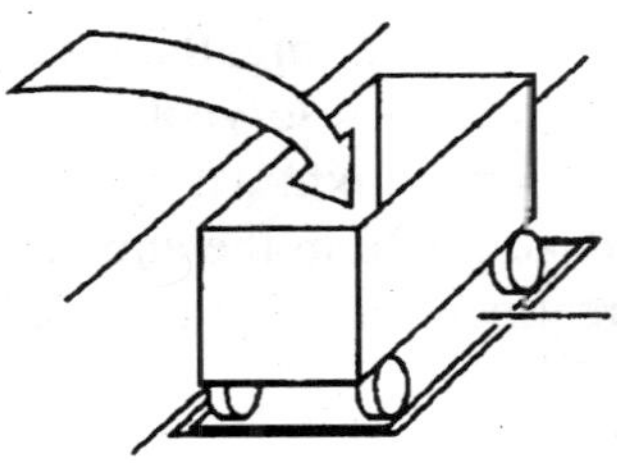

Fig. 7.6 Direct-Load Transfer System

(ii) Storage-Load Transfer System: In the storage-load transfer station, wastes are emptied directly into a storage pit from which they are loaded into transport vehicles by various types of auxiliary equipment

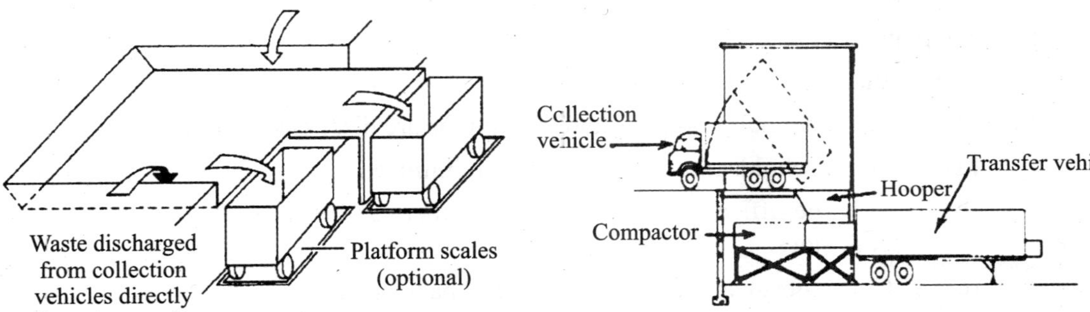

Fig.7.7 Storage-Load Transfer System

7.4 MSW MANAGEMENT

Various technologies of treatments and ultimate disposal of MSW include as discussed below:

Imbalance in solid waste management kills vulture

Since long birds have played an important role in the earth's detritus –based ecological cycles. In India, the issue of MSW disposal and vultures first came to light in the late 1970s when a number of aircrafts suffered bird-hits.

Studies revealed that vultures were frequent victims because of the close proximity of MSW disposal sites to airport where slaughterhouse wastes were being feely dumped. Vultures soaring over these dumps proved to be easy victims, mainly because of their larger size and slow flight.

Government of India along with the Airports Authority in support with Bombay Natural History Society addressed the issue in a scientific manner and issued formal notification to municipalities and corporation banning the disposal of MSW within a radius of 10 km surrounding airport and air force runways. However the same zeal was not extended in regulating the disposal of dead cattle in India. Hundred of dead cattle and buffaloes are dumped in open fields and MSW disposal sites.

It has been estimated that in South Asia, during the past two decades, the vulture population thus declined by over 90%. Vultures used to visit open disposal sites to enjoy the large supply of meat till they succumbed to some unknown contaminant. It was analyzed that vultures were killed by residues if dichlofenac found in the animal carcasses. Dichlofenac is a widely used pain-killer, especially in the treatment of sick cattle and buffaloes.

7.4.1 Waste Minimization Techniques

It deals with reduction of the amount of MSW that is operated at source. Some of these sources are discussed below:

(i) Production units for food and households products

(ii) Production units for commercial products

(iii) Shopping outlets

(iv) Households

(v) Offices, commercial properties and institutions are discussed below:

 (a) In households, food wastes can be minimized.

 (b) In home-office or other institution, Paper can be printed (or photocopied) on both sides.

 (c) Consumers can buy goods with least packaging etc.

7.4.2 Reuse and Recycling of MSW

7.4.2.1 Aluminum Cans

(i) It constitutes about 1% of MSW.

(ii) Recycling of the aluminium cans is about 85%

(iii) It is most successful of the recyclables and is so because of economics.

(iv) It is most expensive to import raw aluminium (to the US) than it is to recycle it.

7.4.2.2 Paper and Cardboard

(i) It constitutes about 35% of MSW.

(ii) Recycling has been slow.

(iii) All corrugated cardboard is now made from recycled paper.

(iv) It is cheaper (in many countries) to use raw wood for paper production than it is to use recycled paper.

(v) Some paper/cardboard is recycled and used as fibre board.

7.4.2.3 Glass

(i) Recycling glass is getting attention.

(ii) There are many advantages to glass recycling as it accounts for 5-10% of the waste streams.

(iii) Recycled glass has many uses.

(iv) If not, it can be crushed and adopted for producing new glasses, with energy savings.

7.4.2.4 Plastics

(i) It constitutes 5-15% of the MSW streams

(ii) They are non-biodegradable and most undesirable in landfills.

(iii) They have a high heating value and are desirable for incineration plants.

(iv) They are potentially reusable as other forms

(v) The potential is for reforming the plastics into other shape.

(vi) To date, 5% of plastics are recycled and these consists of PET and HDPE drinks containers.

7.4.2.5 Yard Waste

(i) They can be shredded and composted and subsequently reused as soil amendments or other agricultural uses.

(ii) Composting produces problems with odours and heavy metals.

(iii) If not composted, the wood fraction especially has a relatively good heating value for incineration plants.

7.4.2.6 Organic food fraction

(i) This fraction is about 50% or more in developing countries.

(ii) Four option for this stream are:

   (a) Composting
   (b) A/D (Anaerobic/Digestion)
   (c) Reuse as animal feedstock
   (d) Incineration

7.4.2.7 Miscellaneous Items of MSW Stream

Items like waste oil, car batteries, computers, wood, TVs, refrigerators, clothes, electronics appliances, etc. are the recyclables and in the case of waste oil and batteries, the trade is successful.

Solid waste management in Chandigarh- A case control study

- The Central Pollution Control Board (CPCB) executed one demonstration project with the city corporation of Chandigarh to deal municipal solid waste such as sweeping roads and collection, storage and disposal of wastes.
- The hazardous household waste such as batteries and empty chemical containers are disposed of at the existing land fill site along with the other solid waste which is hazardous.
- The municipal corporation has set apart an area about half an acre land for developing the requisite facilities for the disposal of such waste.
- A high importance is given to collection and transportation of waste form slaughter houses and meat and fish markets as they can be decomposed readily.
- A biogas plant has been installed or scientific disposal of wastes.

7.4.3 Biological MSW Treatment

At least three options exist for biological MSW treatment as discussed below:

(i) Aerobic or composting

(ii) Anaerobic or biogas

(iii) Combined aerobic or anaerobic

7.4.3.1 *Composting*

It is an anaerobic process where micro-organisms in an oxygen environment, decomposes the organic food waste as follows:

$$\text{Organic matter} + O_2 \rightarrow \text{new cells} + CO_2 + H_2O + NH_3 + SO_4$$

The key inorganic nutrients required are nitrogen, phosphorous, sulphur, potassium, magnesium, calcium and sodium.

7.4.3.2 *Process Requirements*

(i) Temperature

(ii) Moisture content

(iii) Oxygen

(iv) C/N ratio

(v) pH

(vi) Biochemical composition and texture

7.4.3.2.1 *Temperature*

(i) Best results are achieved if the thermophilic stage can be reached within the first few days and the compost then maintained at this temperature.

(ii) Temperatures above the optimum thermophilic range inhibit biological activity but also improve the hygienic condition of compost.

7.4.3.2.2 *Moisture content*

(i) Optimum moisture content is 50% to 60%, below which metabolic activities slows .

(ii) Above 50% to 60%, water fills the voids between particles, inhibiting oxygen access and causing temperature reduction and anaerobic conditions and malodours to occur.

7.4.3.2.3 *Oxygen*

(i) Optimum oxygen levels are 15% to 20%.

(ii) Turning and ventilating compost are meant to keep the oxygen content at sufficient level.

7.4.3.2.4 *C/N ratio*

(i) It is the measure of optimum biochemical conditions and occur at a C/N ratio of 30.

(ii) This ensures adequate nitrogen for cell synthesis and carbon as the energy source.

7.4.3.2.5 *pH*

(i) The optimum pH range is 6 to 8.

(ii) In the initial days the ph reduces to about 5 as organic acids are formed. Then the pH rises as these acids are consumed in the thermophilic stage.

(iii) Anerobiotics sets in at pH below 4.5.

7.4.3.2.6 *Biochemical Composition and Texture*

(i) The composition of water influences the process rate.

(ii) Some materials are easily degradable, like plants manure, primary wastewater sludge, and food wastes.

(iii) Straw, wood, leaves; yard wastes and paper with high contents of lignin are slowly biodegradable.

(iv) The texture influences the process in its variable surface area as habitats for the micro-organisms and its ability to retain moisture or oxygen.

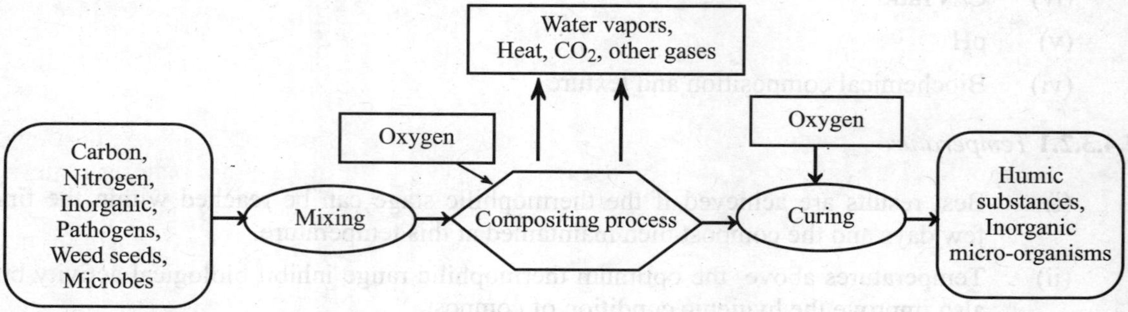

(a)

(b) Anaerobic digestion (Biomethanation)

Fig. 7.8 Biological Treatment of MSW (a) Aerobic composting (b) Anaerobic digestion

Two stage Aerobic and Anaerobic Methods

The two stage process using high solids anaerobic digestions followed by in-vessel aerobic composition is currently an active area of research.

Advantages includes

(i) No external water is needed for producing a slurry feedstock to the anaerobic digester.

(ii) No effluent water is produced.

(iii) There are high rates of bio-gas production of 5 to 10 m^3 of reactor.

(iv) A stabilized humus product is product is produced.

(v) A refuse derived fuel (RFD) is produced.

7.4.4 Thermal Treatment

Combustion/Incineration: Thermal degradation of organic material can be carried out with or without oxygen.

(i) If thermal degradation takes place with a deficit of oxygen (partial combustion), some of the energy is stored as chemical energy from the organic material will be released as burnable gases. It is also called **gasification**.

(ii) Gasification in dry distillation (heating without access for air, oxygen or steam), the process is called **pyrolysis**.

(iii) Gasification has the advantage that the gases are more economical to purify than the flue gases in pyrolysis, due to the large volume of the flue gases generated from combustion.

(iv) Gases are often be utilized directly in gas engines.

(v) Problem is caused by high content of water or a low density of biomass.

(vi) Difficulties encounter in crushing and feeding are reasons that the technology has not yet been developed to a commercial level.

(vii) Since municipal solid waste is more non-homogeneous than the other biomass fuel (straw and wood), it is also more difficult to treat by gasification.

Thermal degradation with excess oxygen is called combustion.

When the fuel is waste, this is called as incineration.

(i) Incineration is chemical oxidation at high temperatures, where organic material is converted into energy (heat), flue gas and slag.

(ii) Waste differs from other solid fuels in being non-homogeneous and a higher content of water than coal or wood.

(iii) This has been solved by turning waste into pellets, named RDF (refused derived fuel).

(iv) RDF pellets can be used in furnances which burn traditional solid fuels, often in a mixture with coal.

(v) They can also be stored just as coal, or other solid fuels.

Operation of the thermal treatment systems involves not only higher cost, but also a relatively higher degree of expertise.

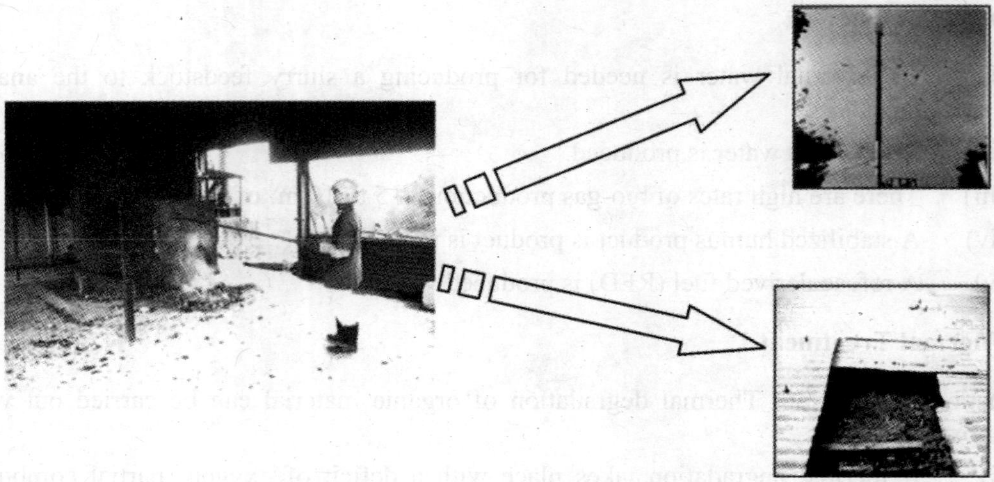

Fig. 7.9 Incineration

7.4.5 Disposal/ Landfill of MSW

Modern landfills practices include monitoring programs for incoming waste for gases, for leachate etc. To control pollution of the surrounding environments, particularly ground water, surface water and air.

(i) **Area landfill (Above ground level):** The area landfill is used when the terrain is unsuitable for the excavation of trenches in which to place the solid waste. High-groundwater conditions necessitate the use of area-type landfills.

(ii) **Excavated cell/trench landfill (Below ground level):** The trench method of landfilling is ideally suited to areas where an adequate depth of cover material is available at the site and where the water table is not near the surface.

(iii) **Canyon/valley landfill:** Depressions, low-lying areas, valleys, canyons, ravines, dry borrow pits etc. have been used for landfills.

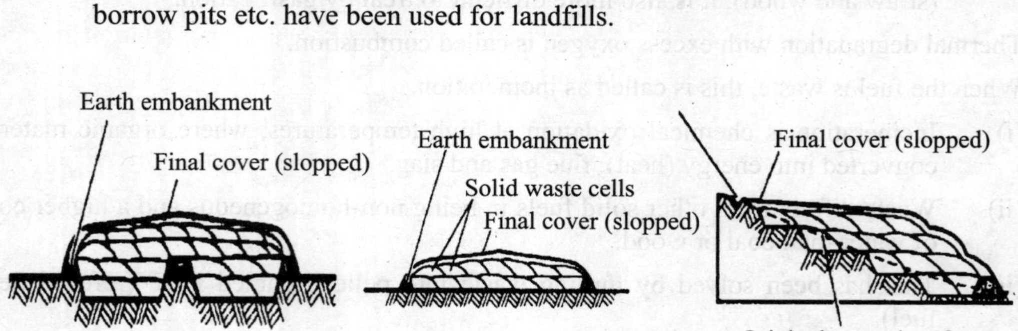

(a) (b) (c)

Fig. 7.10 Types of MSW landfill (a) Area landfill (b) Trench landfill (c) Valley landfill

(i) Leachate is the water type liquid that seeps out of landfill.

(ii) It is from rain infiltration through the landfill and also from the moisture fraction of MSW.

(iii) Since it has a long residence time in a landfill, in a pre-dominantly anaerobic environment, it is contaminated with organics and heavy metals and as such is highly toxic.

(iv) It is essential in landfill practice that leachate is not allowed to enter the surface water or groundwater, but that should be collected and treated in a conventional municipal waste water treatment plant.

(v) Landfill gas is produced in landfill sites due to the anaerobic degradation of biodegradable organic wastes.

(vi) The gas is produced is typically about 60% methane and 40% CO_2.

(vii) Landfill gas, with its high content of methane is potentially explosive and as such needs to be controlled.

7.5 MSW MANAGEMENT RULES IN INDIA

(i) ***Waste Management & Handling Rules (1987)*** by Ministry of Environment and Forest (MoEF) modified in year 2000 by **Honorable Supreme Court** for scientific MSW management, ensuring proper collection, segregation, transportation, processing and disposal of MSW and upgrade of the existing facilities to arrest contamination of soil and ground water.

(ii) **Center for Pollution Control Board (CPCB)**has been assigned to monitor the implementation of these rules, and the municipalities will be required to submit annual reports regarding the status of MSW in their areas to the CPCB.

(iii) In addition, there are Municipal Corporation Acts by different states such as the **Delhi Municipal Corporation Act (1957, Uttar Pradesh Municipal Corporation Act (1959)** and **Karnataka Municipal Corporation Act (1976)**.

(iv) The Delhi Plastic Bag (Manufacture, Sales and Usage) and non-biodegradable garbage (control) Act, 2000, was enacted to prevent contamination of foodstuff carried in recycled plastic bags, reduce the use of plastic bags, throwing or depositing non-biodegradable garbage in public drains, roads and places open to public view.

The Delhi Municipal Corporation Act, 1957 contains the following sections relevant to solid waste management in Delhi:
• Section 42(C): Obligatory function of the corporation-the scavenging removal and disposal of filth, rubbish and other obnoxious or polluted matter
• Section 350 : Provisions for daily cleansing of streets and removal of rubbish/filth
• Section 352 : Provisions or appointments of receptacles, depots and places of rubbish etc.

• Section 353 : Duty of owners and occupiers to collect and deposit rubbish etc
• Section 358 : Commissioner's power to get premises scavenged and cleaned

Criminal laws and the Waste Management

There are two major criminal laws dealing with solid waste management are discussed below:

(i) The Indian Penal Code, 1860

(ii) The Criminal Procedure Code, 1973

The Indian Penal Code of 1860 has dealt with solid waste management under Chapter XIV 'of offences affecting the public health, safety, convenience, decency and morals'.

Section 133 of the Criminal Procedure Code, 1973 deals with 'removal of nuisance' and empowers the Sub-Divisional Magistrate or any executive Magistrate, on receiving report/ information, to make order to remove the public nuisance and desist from carrying any trade, business which is causing public nuisance **(Section 188, I.P.C.)**.

Article 47 of the Indian Constitute makes it a paramount principle of governance that 'steps are taken for the improvement of public health as amongst its primary duties.'

INTEGRATED SOLID WASTE MANAGEMENT (ISWM)

Integrated solid waste management (ISWM) can be defined as the selection and application of suitable techniques, technologies and management program to achieve specific waste management objectives and goals. The specific and economical approaches for management of solid waste is integrated and sustainable (solid) waste management (ISWM) which provides the interfaces and connections between three physical elements and three governance features.

Three physical elements	Three governance features
(1) Public health	(1) Inclusive
(2) Environment	(2) Financially sustainable
(3) Resource management	(3) Sound institutions and pro-active policies

The three physical elements that all needed to be addressed for an ISWM system to work well and to work sustainably over the long period are:

(i) **Public health (Collection):** Maintaining healthy conditions in urban conglomerates, particularly through good collection services.

(ii) **Environment (Waste treatment and disposal):** Protection of the environment through the waste chain, especially during treatment and disposal.

(iii) **Resource management (Recycling and organic material, "4Rs"):** Closing the loop by returning both materials and nutrients to beneficial use, through preventing waste and striving for high rates of organic recovery, reuse and recycling.

The governance strategies to deliver a well functioning system of ISWM until the 1990s would probably have been framed primarily around technology; but there is consensus today on the need for a much broader approach. Three interrelated requirements for delivering ISWM are distinguished under the framework of **good waste governance**. There is a need for the system to:

(i) Be **inclusive**, providing transparent spaces for stakeholders to contribute as users, providers and enablers.

(ii) Be **financial sustainable**, which means cost effective and affordable.

(iii) Rest on a base of **sound institutions and pro-active pol*icies***, without such a strong and transparent institutional framework the system will not work well over a long term.

Hierarchy of ISWM

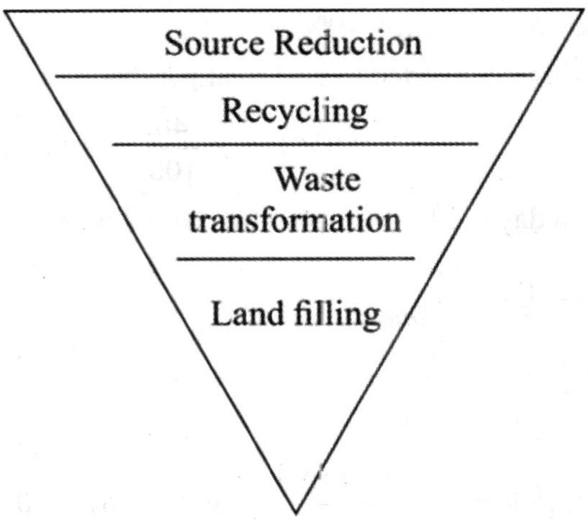

Example 7.1 Calculate the maximum working area of a landfill if the average annual waste input is 30000 tonnes and the average annual rainfall is 1.1m. Assume the absorption capacity of the waste to be 0.9 m^3/tonne.

Solution. $$A_{max} = \frac{CW}{R}$$

Where C = absorption capacity of the waste,

W = average annual waste input and

R = average annual rainfall

In this case, C = 0.9 m^3/tone

W = 30000 tonnes/year and R = 1.1m

Therefore, $$A_{max} = \frac{0.9 \times 30000}{1.1} = 24545 \text{ m}^2$$

Maximum working area of the landfill = 24545 m^2

Example 7.2 The solid waste from the city of a 3 million people is used to fuel a 40% efficient steam-electric power plant. Assuming that half of the waste is combustible with a heating value of 15×10^6 J/kg, and assuming that each person's contribution to the solid waste is 0.4 kg/day, calculate the size of the power plant.

Solution. Number of persons = 3 million = 3×10^6

Each person's contribution to solid waste = 0.4 kg/day

Therefore, total waste per day = $3 \times 10^6 \times 0.4 = 1.2 \times 10^6$ kg/day

Therefore, Combustible waste = $1.2 \times 10^6 \times \frac{1}{2} = 0.6 \times 10^6$ kg/day

Heating value of waste = 15×10^6 J/kg

Therefore, total heat energy available per day = $0.6 \times 10^6 \times 15 \times 10^6 = 9.0 \times 10^{12}$ J/day

Efficiency of the plant = 40%

Therefore, Thermal energy converted to electrical per day

$$= 9.0 \times 10^{12} \times \frac{40}{100} = 3.6 \times 10^{12} \text{ J/day}$$

Number of seconds in a day = $24 \times 60 \times 60 = 86400 = 8.64 \times 10^4$ s

$$\text{Now Power P} = \frac{\textit{Energy E}}{\textit{Time t}}$$

We have E = 3.6×10^{12} J/day

t = one day = 8.64×10^4 s

$$\text{Therefore, Power P} = \frac{3.6\times10^{12}}{9.64\times10^4} \text{ watt} = 4.167 \times 10^7 \text{ watt}$$

$$\text{Converting to megawatts, we have P} = \frac{4.167\times 10^7}{10^6} = 41.67 \text{ MW}$$

The size of the plant = 41.67 Megawatts.

Example 7.3 A suspended solid analysis is run on a sample of water. The total mass of the crucible and filter as 55.3520 gm. A sample of 260 ml. is less then filtered and the residue dried to constant mass at 103°C. If the constant mass of the crucible, filter and the residue is 55.389 gm., what is the suspended solid content in the sample?

Solution. The total mass of the crucible and filter = 55.352 gm.

Mass of the crucible + Filter + residue = 55.389 gm.

Since the residue is obtained after filtration, therefore, it contains only Suspended Solids (SS).

Therefore, Mass of suspended solid = (55.389 – 55.352) = 0.037 gm. =37 mg,

37 mg of "SS" was present in 260 ml of water sample.

$$\text{Therefore, In 1000 ml of water sample "SS"} = \frac{1000\times 37}{260} = 142.3 \text{ mg.}$$

Therefore, The Suspended Solid content is 142.3 mg/L.

Example 7.4 Calculate the volumetric flow rate of Leachate through a compacted clay Liner if the area of the landfill is 15ha. And the Linear thickness is 1m. The hydraulic conductivity is 7.5×10^{-10} m/sec. Assume that the head of water is 0.6m.

Solution. The Darcy velocity for the Leachate,

$$V = k\left(\frac{\Delta h}{\Delta L}\right) = \frac{7.5 \times 10^{-10}\ ms^{-1} \times 0.6\ m}{1\ m} = 4.5 \times 10^{-10}\ m/sec$$

Now to determine flow we have to apply continuity equation.

$$Q = vA$$

$$= (4.5 \times 10^{-10}) \times 15ha \times 10^{-1}\ m^2\ ha^{-1}$$

$$= 6.75 \times 10^{-5}\ m^3\ /sec.$$

Example 7.5. Determine the Landfill area required to dispose of 50 m^3/day, 5 days/week at 25% DS using wide trenches. Assume the life required is 20 years and the trenches are 10 m wide, 3m deep and 100m long. Assume 3m solid ground spacing between trenches and a 20m buffer to the property boundary.

Solution. Required trench volume = 50 × 5 × 52 × 20 = 2,60,000 m^3

$$\text{Number of trenches required} = \frac{2,60,000}{10 \times 3 \times 100} = 86.6 \approx 87 \text{ numbers}$$

Actually the usable area required = 87 × (10 + 3) (100 + 3) = 11649 m^2 = 11.6 ha

1164990 m^2 = 341m × 341m

Therefore, the minimum area required = (341 + 40) (341 + 40) = 145161 m^2= 14.5 ha

Review questions

1. **What is waste management?**

   Ans. Waste management is the collection, transportation and disposal of waste materials.

2. **What are the rules and regulations guiding waste management in India**

   Ans. **Municipal Solid Waste (Management and Handling) Rules 2000** regulate the management and handling of the municipal solid wastes and are applicable to every municipal authority responsible for collection, segregation, storage, transportation, processing and disposal of municipal solid wastes

   Bio-Medical Waste (Management and Handling) Rules, 1998 regulate the management and handling of bio-medical waste and are applicable to all persons who generate, collect, receive, store, transport, treat, dispose, or handle bio medical waste in any form.

   E-Waste (Management and Handling) Rules, 2010 regulate the management and handling of electrical and electronic waste and is applicable to every producer, consumer involved in manufacture, sale, purchase and processing of these equipments or its components.

3. What are the common methods of waste disposal?

Ans. The commonly practiced technologies for SWM can be grouped under three major categories, i.e., bio-processing, thermal processing and sanitary landfill. The bio-processing method includes aerobic and anaerobic composting. Thermal methods are incineration and pyrolysis. Sanitary landfill is generally used to dispose off the final rejects coming out of the biological and thermal waste processing units.

4. What is aerobic composting?

Ans. Aerobic composting is the creation of fertilizing compost using bacteria that thrive in an oxygen-rich environment. Aerobic composting is considered the fastest method of composting, but involves more work interms of rotating the organic material periodically.

5. What is anaerobic composting?

Ans. Anaerobic composting is the creation of fertilising compost using bacteria that cannot thrive in the presence of oxygen. Anaerobic composting is known to work slowly, but also requires lesser work.

6. What is incineration?

Ans. Incineration is a waste treatment process that involves the combustion of organic substances contained in waste materials. Incineration of waste materials converts the waste into ash, flue gas, and heat. In some cases, the heat generated by incineration can be used to generate electric power.

7. What is a sanitary landfill?

Ans. A sanitary landfill is a low-lying area that is filled with waste rejects. It has a liner at the bottom to prevent the groundwater from contaminating with the mix of the liquid that oozes from the waste that is buried called the leachate. Waste is buried in-between layers of soil and is compacted nicely to make it a hard surface. When the landfill is completed, it is capped with a layer of clay or a synthetic liner in order to prevent water from entering. A final topsoil cover is placed, compacted and graded, and various forms of vegetation may be planted in order to reclaim the otherwise useless land.

8. How to practice waste management at home?

Ans.

- Keep separate containers for dry and wet waste in the kitchen.
- Keep two bags for dry waste collection- paper and plastic, for the rest of the household waste.
- Keep plastic from the kitchen clean and dry and drop into the dry waste bin. Keep glass/ plastic containers rinsed of food matter.
- Keep a paper bag for throwing sanitary waste.

9. What are the first few steps to initiate a waste management programme in your apartment complex?

Ans.

- Form a group with like-minded people.
- Explain waste segregation to your family / neighbours in your apartment building.
- Get the staff in the apartment building to also understand its importance.
- Get separate storage drums for storing dry and wet waste.
- Have the dry waste picked up by the dry waste collection centre or your local scrap dealer.

10. What are the different types of waste?

Ans.

(i) Wet waste: Wet waste consists of kitchen waste - including vegetable and fruit peels and pieces, tea leaves, coffee grounds, eggshells, bones and entrails, fish scales, as well as cooked food (both veg and non-veg).

(ii) Dry Waste: Paper, plastics, metal, glass, rubber, thermocol, styrofoam, fabric, leather, rexine, wood – anything that can be kept for an extended period without decomposing is classified as dry waste.

(iii) Hazardous waste: Household hazardous waste or HHW include three sub-categories – E-waste; toxic substances such as paints, cleaning agents, solvents, insecticides and their containers, other chemicals; and biomedical waste.

(iv) E-waste: E-waste or electronic waste consists of batteries, computer parts, wires, electrical equipment of any kind, electrical and electronic toys, remotes, watches, cell phones, bulbs, tube lights and CFLs.

(v) Biomedical waste: This includes used menstrual cloth, sanitary napkins, disposable diapers, bandages and any material that is contaminated with blood or other body fluids.

11. What are ways of storing the waste at homes?

Ans.

(i) Dry waste : Store it in a bag in the utility area after cleaning and drying till it is picked up. No food residue must be left in the bottles and packets. Clean them as you would to reuse them. If clothes are totally unusable, or very damaged, they are categorized as dry waste. If clothes are soiled with body fluids, they become sanitary waste. If they are soiled with paint, or any chemicals, they are HHW (household hazardous waste).

(ii) E-waste: Store those in separate container which is kept closed, away from moisture and in which nothing else is put.

12. How will you dispose your home waste?

Ans.

(i) Compost your wet waste at home. Home composting can easily be done in any aerated container.

(ii) Compost your wet waste at the community level

If you live in a large apartment building, a community composting system like tank composting could be set up for all the wet waste from the residents. If not, the wet waste can be given out every day to your Municipality collection system.

(iii) Biomedical waste has to be wrapped separately in a newspaper and handed over to the municipality waste collection system. Expired medicines and injections, used syringes also have to be disposed in the same manner.

Paint and other hazardous waste like cosmetics, mosquito repellents, tube lights etc have to be stored separately and handed over to the Municipal collection system.

Chapter 8

Hazardous Waste Management

8.1 INTRODUCTION

A waste that cause significant increase in mortality or an increase in serious irreversible illness, or pose a potential hazard to human health or the environment when it is improperly treated, stored , transported, disposed or otherwise managed.

8.1.1 What Kinds of Wastes are Hazardous?

Hazardous wastes include many kinds of discarded chemicals and other wastes generated from commercial, industrial and institutional activities. Industries that are major producers of hazardous waste include petrochemicals, pharmaceuticals, pesticides, paints and dyes, petroleum, fertilizers, asbestos, caustic soda, inorganic chemicals and general engineering (Table 8.1). Units manufacturing pesticides, drugs, pharmaceuticals, textiles, dyes, fertilizers, paint etc. have a major potential for generating hazardous waste such as heavy metals, cyanides, pesticides, complex aromatic compounds (such as PCBs), and other chemicals, which are toxic, flammable, reactive, corrosive or have explosive properties.

Table 8.1 Major Sources of Hazardous Waste Generation

Industry Type	Hazardous substance
Iron & steel	Acid tar, oily sludge, complex organic sludge. metal waste, flue dust, zinc waste etc.
Aluminum smelter	Spent pot lime (fluent & cyanide)
Fertilizer	Gypsum, spent chemical catalysts
Sulphuric acid	Sulphur sludge, spent catalysts
Sodium dichromate	Leached solid (Cr^{+6})
Tannery	Chromium sludge
Solvent industry	Spent solvent
Chloro alkali	Mercury contaminated sludge
Electroplating	Cyanide, heavy metals
Pharmaceuticals	Organic complex
Pesticides	Toxic organic complex
Paint industry	Complex chemical sludge & Solvents

- These wastes contain one or more of 39 toxic, carcinogenic, mutagenic or teratogenic compounds at levels exceeding the established limited.
- They catch fire easily and are reactive or unstable enough to explode or release toxic fumes.
- They are also capable of corroding metal containers.

8.2. CHARACTERISTICS OF HAZARDOUS WASTE

Hazardous wastes may be divided into two basic groups: characteristically hazardous wastes and listed hazardous wastes:

8.2.1. Characteristics of Hazardous Wastes

These hazardous waste exhibits a certain physical property or characteristic. There are four characteristics that can make a waste hazardous : ignitability, corrosivity, reactivity, and toxicity. A brief of these four characteristics is provided below

(i) **Ignitability** - Ignitable wastes, such as wastes oils and solvents, can create fires under certain conditions, are spontaneously combustible, or have a flash point less than 60°C (140°F).

(ii) **Corrosivity** - Corrosive wastes, such as battery acid, are acids or bases (pH less than or equal to 2, or greater than or equal to 12.5) that are capable of corroding metal containers, such as storage tanks, drums, and barrels.

(iii) **Reactivity** - Reactive wastes, such as lithium-sulfur batteries and explosives, are unstable under normal conditions. They can cause explosions, toxic fumes, gases, or vapors when heated, compressed, or mixed with water.

(iv) **Toxicity** - Toxic wastes are harmful or fatal when ingested or absorbed (e.g., containing mercury, lead, etc.). When toxic wastes are land disposed, contaminated liquid may leach from the waste and pollute ground water.

8.2.2 Listed Hazardous Wastes

These are wastes that are hazardous because they are listed on one or more of four different particular lists of wastes:

(i) **F-Listed wastes**- This list of wastes includes 28 different wastes, including certain spent solvents, metal finishing wastes, dioxin-containing wastes, chemical manufacturing wastes, wood preserving wastes, petroleum manufacturing wastes, and hazardous waste landfill leachate.

(ii) **K-Listed wastes**- This list of wastes includes over 100 wastes from specific industrial processes. The specific processes are in the industries of wood preserving, petroleum refining, primary and secondary metals manufacturing, and the manufacturing of industrial chemicals, inks, pigments, pesticides, explosives, and veterinary pharmaceuticals.

(iii) **U-Listed wastes**- This list of wastes includes several hundred different commercial chemical products. Wastes that fall under this listing include only those products that contain the listed constituent as the sole active ingredient. These wastes include old or off-specification virgin materials that are being discarded, as well as container residues and spill residues of these materials.

(iv) **P-Listed wastes**- This list of wastes includes about 200 different commercial chemical products that are defined as **acutely hazardous**. This means that the wastes are especially toxic. Wastes that fall under this listing include only those products that contain the listed constituent as the sole active ingredient. These wastes include old or off-specification virgin materials that are being discarded, as well as container residues and spill residues of these materials.

8.2.3 Universal Wastes

This is a special subset of hazardous wastes that are regulated under a streamlined set of regulations called **universal waste rule**. These wastes include:

(i) **Batteries**. Covered batteries include lead-acid batteries, nickel-cadmium batteries, silver cells and mercury-containing batteries.

(ii) **Cancelled and recalled pesticides**.

(iii) **Mercury thermostats and other mercury-containing equipment (e.g., mercury switches, barometers, sphygmomanometers, etc.).**

(iv) **Mercury-containing lamps**. This includes fluorescent lamps (including compact fluorescent lamps), mercury vapor lamps, and other lamps that contain mercury.

(v) **Used electronics**. This includes desk top and lap top computers, computer peripherals, monitors, copying machines, scanners, printers, radios, televisions, camcorders, video cassette recorders ("VCRs"), compact disc players, digital video disc players, MP3 players, telephones, including cellular and portable telephones, and stereos.

8.2.4 Used Oil

Just as with universal waste, the hazardous waste regulations also have a special set of requirements for used oil. The term **used oil** means any oil refined from crude oil or synthetic oil, that: (a) has been used and as a result of such use is contaminated by physical or chemical impurities; or (b) is no longer suitable for the services for which it was manufactured due to the presence of impurities or a loss of original properties. This includes both used and unused oils that are being discarded. Common examples of used oils include:

(i) Used crankcase (engine) oil

(ii) Brake fluid, transmission fluid, and power steering fluid

(iii) Used gear, chain, and ball bearing lubricants

(iv) Hydraulic and compressor oils

(v) Metalworking fluids (including water-soluble coolants)

(vi) Drawing and stamping oils

(vii) Heat transfer oils (including quenching oils)

(viii) Dielectric fluid (e.g., transformer oil).

HAZARDOUS-WASTE CHARACTERISTICS

(i) **Hazardous wastes** are classified on the basis of their biological, chemical, and physical properties. These properties generate materials that are toxic, reactive, ignitable, corrosive, infectious or radioactive.

(ii) **Toxic wastes** are poisons, even in very small or trace amounts. They may have acute effects, causing death or violent illness, or they may have chronic effects, slowly causing irreparable harm. Some are carcinogenic, causing cancer after many years of exposure. Others are mutagenic, causing major biological changes in the offspring of exposed humans and wildlife.

(iii) **Reactive wastes** are chemically unstable and react violently with air or water. They cause explosions or form toxic vapours.

(iv) **Ignitable wastes** burn at relatively low temperatures and may cause an immediate fire hazard.

(v) **Corrosive wastes** include strong acidic or alkaline substances. They destroy solid material and living tissue upon contact, by chemical reaction.

(vi) **Infectious wastes** include used bandages, hypodermic needles, and other materials from hospitals or biological research facilities.

(vii) **Radioactive wastes** emit ionizing energy that can harm living organisms. Because some radioactive materials can persist in the environment for many thousands of years before fully decaying, there is much concern over the control of these wastes. However, the handling and disposal of radioactive material is not a responsibility of local municipal government. Because of the scope and complexity of the problem, the management of radioactive waste—particularly nuclear fission waste—is usually considered an engineering task separate from other forms of hazardous-waste management and is discussed.

8.2.5. Impacts of Hazardous Wastes

As per the Waste Management Rules, 1989, and the MoEF's Guidelines (1991), hazardous waste generated by industries has to be disposed of in secured landfills and the toxic organic fraction of the waste needs to be incinerated. Improper storage, handling, transportation, treatment and disposal of hazardous waste results in adverse impacts on ecosystems and the human environment. Heavy metals and certain organic compounds are phytotoxic and at relatively low levels can adversely affect soil productivity for extended periods. For example, uncontrolled release of chromium contaminated waste water and sludge results in the contamination of aquifers. These aquifers can no longer be used as sources of freshwater. Discharge of acidic and alkaline waste affects the natural buffering capacity of surface waters and soils and may result in reduction of a number of species. It is said that one gallon of used oil can contaminate one million gallons of water rendering it unpotable. Marine species can be affected even if exposed to oil levels as low as 1 ppm.

- Of all the hazardous waste, nuclear wastes have caused the greatest concern throughout the world. 1, 000 tons of uranium ore to produce 1 kg of plutonium, the popular radioactive fuel used in commercial and defence nuclear reactors.
- Nuclear wastes are most dangerous because they do not degrade soon. Radioactive materials vary in their half –life, the tome-taken for half the atoms of any element to decay is considered half-life.
- The prairie Island nuclear power plant near Minneapolis (USA) stores radioactive wastes in specially designed cakes. These cakes are made up of 22.5 cm thick skins of carbon steel.

8.3 TRANSPORTATION OF HAZARDOUS WASTE

Hazardous waste generated at a particular site often requires transport to an approved treatment, storage, or disposal facility (TSDF). Because of potential threats to public safety and the environment, transport is given special attention by governmental agencies. In addition to the occasional accidental spill, hazardous waste has, in the past, been intentionally spilled or abandoned at random locations in a practice known as "**midnight dumping**." This practice has been greatly curtailed by the enactment of laws that require proper labelling, transport, and tracking of all hazardous wastes.

The transport of hazardous waste is governed by a number of regulations. The regulations are:

(i) ADR (Accord European relative.......Dangere use Par Route) - transport of dangerous goods by road.

(ii) RID (Regalement International concernant.....Danger use Par chemindefer) – transport of dangerous goods by rail.

(iii) IMDG (International Maritime Dangerous Goods code) - transport of dangerous goods by ship.

(iv) ICAO (technical instruction) – transport of dangerous goods by air.

8.3.1 The ADR/RID Classification System

Both the sets of regulations are operating with a nine-classification system as shown in Table 8.2.

There is a specific warning label: toxic compounds, identified by a white diamond sign with an insert of a skull and cross bones.

(i) Keeping dry identified by an umbrella.

(ii) Handle with care identified by a drinking glass.

(iii) This side up by two upright arrows.

Table 8.2 ADR/RID Classification

Class	Item Description	Open/Restricted
1a	Explosives and explosive items	Restricted class
1b	Items loaded with explosives	Restricted class
1c	Detonators, fireworks and similar type of goods	Restricted class
2	Gases, compressed, condensed or dissolved under pressure	Restricted class
3	Flammable liquids	Open class
4.1	Flammable solids	Open class
4.2	Self-ignitable compounds	Open class
4.3	Compounds which develop flammable gases in contact with water	Open class
5.1	Oxidizing compounds	Open class
5.2	Organic peroxides	Open class
6.1	Toxic compounds	Open class
6.2	Infectious compounds	Open class
7	Radioactive compounds	Restricted class
8.	Corrosive compounds	Open class
9.	Miscellaneous dangerous goods	Open class

8.3.2 Pre-transport and Preparation of Waste

Once the waste has been classified according to the ADR/RID regulations, the waste generator must fulfil a number of other steps before the waste is ready for transport.

In the ADR/RID regulation, specifications in selecting the containers (drums) are given.

A manifest (or declaration) must be filled which includes information on;

(i) The waste generator

(ii) The composition of the waste

(iii) The physical appearance

(iv) The method of packaging

(v) The UN number.

8.3.3 Transport in Bulk

Some waste derived from large-scale waste generators are packed in so-called **IBCs (Intermediate bulk containers)** or tank containers with capacities of 20 to 30 m^3.

The truck carrying a tank container with dangerous goods must be equipped with a set of orange warning signs with a black rim, one in the front of the truck and one in the rear (Fig. 8.1).

In case of normal dangerous goods transport, the sign will show two sets of numbers, one set above a black divider line (a two or three digit number) and one set below (always a four digit number).

The **upper set** of numbers refers to the properties of the goods, e.g. 3 for flammable. The **second digit** (3) indicates that the liquid is very flammable, e.g. 268, which means that the liquid belongs to ADR/RID class 2, compressed gases, it is toxic(6) and finally the compound is corrosive(8).Below the divider line there is a four-digit number giving the identity of the transported compound (Table 8.3).

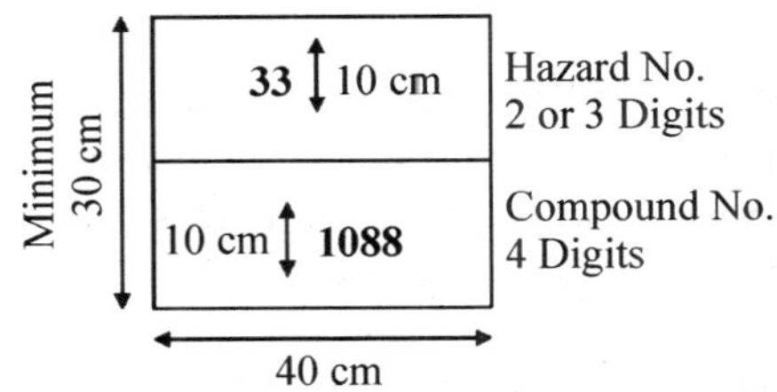

Fig. 8.1 Sample of Warning Number of Bulk Transport of Dangerous Goods

For hazardous waste transport, special numbers are given. Examples of commonly used are:

(i) 1759 corrosive solid compounds

(ii) 1760 corrosive liquids

(iii) 1906 waste acids

(iv) 1992 flammable liquids, toxic

(v) 1993 flammable liquids

(vi) 2920 corrosive liquids, flammable

(vii) 2922 corrosive liquids, toxic

Table 8.3 Sample UN numbers for the transport of hazardous goods

UN number	Hazardous Item
1001	Acetylene (ethylene), dissolved
1002	Compressed air
1003	Liquefied air, refrigerated
1005	Ammonia, dry, liquefied or soluble in water
1008	Boron trifluoride
1009	Trifluobromomethane (Freon)
1010	Butadiene, stabilized
1011	Butane or mixtures with butane
1012	Butylene
1013	Carbon dioxide
1014	Carbon dioxide in mixtures with oxygen
1015	Carbon dioxide in mixtures with dinitrogen oxide

8.3.4 Transport Vehicles

Hazardous waste is generally transported by truck over public highways. Only a very small amount is transported by rail, and almost none is moved by air or inland waterway. Highway shipment is the most common because road vehicles can gain access to most industrial sites and approved TSDFs. Railroad trains require expensive siding facilities and are suitable only for very large waste shipments.

Hazardous wastes can be shipped in tank trucks made of steel or aluminium alloy, with capacities up to about 34,000 litres (9,000 gallons). They also can be containerized and shipped in 200-litres (55 gallon) drums. Specifications and standards for cargo tank trucks and shipping containers are included in governmental regulations.

8.4 TREATMENT, STORAGE, AND DISPOSAL

Several options are available for hazardous-waste management. The most desirable is to reduce the quantity of waste at its source or to recycle the materials for some other productive use. Nevertheless, while reduction and recycling are desirable options, they are not regarded as the final remedy to the problem of hazardous-waste disposal. There will always be a need for treatment and for storage or disposal of some amount of hazardous waste.

8.4.1 Treatment of Hazardous Waste

Hazardous waste can be treated by **chemical, thermal, biological, and physical methods** as discussed below:

(i) **Chemical methods** include ion exchange, precipitation, neutralization, oxidation and reduction.

(ii) **Thermal treatment** includes high-temperature incineration, which not only can detoxify certain organic wastes but also can destroy them. Special types of thermal equipment are used for burning waste in either solid, liquid, or sludge form. These include the fluidized-bed incinerator, multiple-hearth furnace, rotary kiln, and liquid-injection incinerator. One problem posed by hazardous-waste incineration is the potential for air pollution.

(iii) **Biological treatment** of certain organic wastes, such as those from the petroleum industry, is also an option. One method used to treat hazardous waste biologically is called **land farming.** In this technique the waste is carefully mixed with surface soil on a suitable tract of land. Microbes that can metabolize the waste may be added, along with nutrients. In some cases a genetically engineered species of bacteria is used. Food or forage crops are not grown on the same site. Microbes can also be used for stabilizing hazardous wastes on previously contaminated sites; in that case the process is called **bioremediation**.

(iv) The chemical, thermal, and biological treatment methods outlined above change the molecular form of the waste material. **Physical treatment**, on the other hand, concentrates, solidifies, or reduces the volume of the waste. Physical processes include evaporation, sedimentation, flotation, and filtration. Yet another process is **solidification**, which is achieved by encapsulating the waste in concrete, asphalt, or plastic. Encapsulation produces a solid mass of material that is resistant to leaching. Waste can also be mixed with lime, fly ash, and water to form solid, cement like product.

8.4.2 Surface Storage and Land Disposal

Hazardous wastes that are not destroyed by incineration or other chemical processes need to be disposed of properly. For most such wastes, land disposal is the ultimate destination, although it is not an attractive practice, because of the inherent environmental risks involved. Two basic methods of land disposal include **land filling** and **underground injection**. Prior to land disposal, surface storage or containment systems are often employed as a temporary method.

Temporary on-site waste storage facilities include open waste piles and ponds or lagoons. New waste piles must be carefully constructed over an impervious base and must comply with regulatory requirements similar to those for landfills. The piles must be protected from wind dispersion or erosion. If leachate is generated, monitoring and control systems must be provided. Only non-containerized solid, non-flowing waste material can be stored in a new waste pile, and the material must be land filled when the size of the pile becomes unmanageable.

A common type of temporary storage impoundment for hazardous liquid waste is an open pit or holding pond, called a **lagoon**. New lagoons must be lined with impervious clay soils and flexible membrane liners in order to protect groundwater. Leachate collection systems must be installed between the liners and groundwater monitoring wells are required. Except for some sedimentation, evaporation of volatile organics, and possibly some surface aeration, open lagoons provide no treatment of the waste. Accumulated sludge must be removed periodically and subjected to further handling as a hazardous waste.

Many older, unlined waste piles and lagoons are located above aquifers used for public water supply, thus posing significant risks to public health and environmental quality. A large number of these old sites have been identified and scheduled for clean-up, or remediation.

8.4.3 Secure Landfills

Land filling of hazardous solid or containerized waste is regulated more stringently than land filling of municipal solid waste. Hazardous wastes must be deposited in so-called **secure landfills**, which provide at least 3 metres (10 feet) of separation between the bottom of the landfill and the underlying bedrock or groundwater table. A secure hazardous-waste landfill must have two impermeable liners and leachate collection systems. The double leachate collection system consists of a network of perforated pipes placed above each liner. The upper system prevents the accumulation of leachate trapped in the fill, and the lower serves as a backup. Collected leachate is pumped to a treatment plant. In order to reduce the amount of leachate in the fill and minimize the potential for environmental damage, an impermeable cap or cover is placed over a finished landfill.

One option for the disposal of liquid hazardous waste is deep-well injection, a procedure that involves pumping liquid waste through a steel casing into a porous layer of limestone or sandstone. High pressures are applied to force the liquid into the pores and fissures of the rock, where it is to be permanently stored. The injection zone must lie below a layer of impervious rock or clay, and it may extend more than 0.8 km (0.5 mile) below the surface. Deep-well injection is relatively inexpensive and requires little or no pre-treatment of the waste, but it poses a danger of leaking hazardous waste and eventually polluting subsurface water supplies.

A facility for the treatment of hazardous waste will, in addition to an incineration plant has other plants for the treatment of waste oils and inorganic.

The essential plants and the percentage of waste treated by each are:

(i) The incineration ~ 53%

(ii) The inorganic plant ~ 14%

(iii) The waste oil plant ~ 32 % → to incinerator

8.4.4 Incineration

Here the waste is oxidized in an oxygen rich environment at elevated temperatures. Destruction and removal efficiency (DRE) of 99.99% is required. The DRE for an incineration is calculated by

$$DRE = \frac{1 - W_{out}}{W_{in}} \times 100$$

Where

W_{in} = mass feed rate of specific organic component to incinerator

W_{out}= mass emission rate of the same organic component in the exhaust prior to release to the atmosphere

Satisfactory DRE requires

(i) Adequate residence time for gas.

(ii) Adequate solid retention time.

(iii) High temperature.

(iv) Enough oxygen (or hydrogen if incinerating chlorine or bromine).

(v) Sufficient turbulence for mixing waste and oxygen.

8.4.4.1 Chemistry of Incineration

Controlled high temperature oxidation of primarily organic compounds to produce carbon dioxide and water

$$\text{Organic waste} \xrightarrow{\text{Incineration}} CO_2 + H_2O + \text{inoffensive by-products}$$

During incineration, the chemical bonds between the various elements of the reacting mixtures are broken and free radicals are formed, producing the complete combustion products of CO_2 and H_2O.

N_2 will form different forms of oxides of Nitrogen, i.e. NO_2, NO and N_2O depending on the circumstances (temperature, etc.).

NO_2 is reduced in the flue gas by introduction of NH_3 with the following reaction:

$$6NO_2 + 8\ NH_3 \rightarrow 12\ H_2O + 7N_2$$

The reaction will demand a catalyst and temperatures greater than 300°C.

Example 8.1 Find out the combustion and destruction and removal efficiency of an incinerator from the following data. Input phenolic waste 18.2%, outlet phenol 0.04%, outlet gas concentration carbon monoxide 0.1% and CO_2 concentration 25.7%

Solution. $\text{DRE} = \dfrac{W_{in} - W_{out}}{W_{in}} \times 100$ $\quad W_{in} = 18.2\%$

For phenol $\quad W_{out} = 0.04\%$

$$\text{DRE} = \frac{18.2 - 0.04}{18.2} \times 100 = 99.98\%$$

$$\text{CE} = \frac{C_{CO_2} - C_{CO}}{C_{CO_2}} \times 100 = \frac{25.7 - 0.1}{25.7} \times 100 = 99.6\%$$

8.4.5 Inorganic Waste Treatment

The treatment of the waste takes place in stages, starting with detoxification of the cyanide-bearing waste (oxidation with sodium hypochlorite). Followed by a redox process, reducing hexavalent chromium to trivalent chromium, while bivalent iron is oxidized to trivalent iron. The pH level is then adjusted to 9.8 to 10. Here the major part of the heavy metals is precipitated as soluble metal hydroxides. The hydroxides which mainly consist of iron, chromium, nickel, copper, aluminum and zinc are filtered off in a filter press (Fig. 8.2). The filter water is then neutralized and analyzed. If the results are satisfactory (corresponding to the outlet criteria set in the operational permit), the water is drained off into the municipal sewage works. The filter cakes are disposed of at a secured landfill (for hazardous wastes). The cakes are covered with a plastic membrane to prevent precipitation exposure. The filter cakes may potentially in the future form the raw materials for a recovery process extracting the heavy metals.

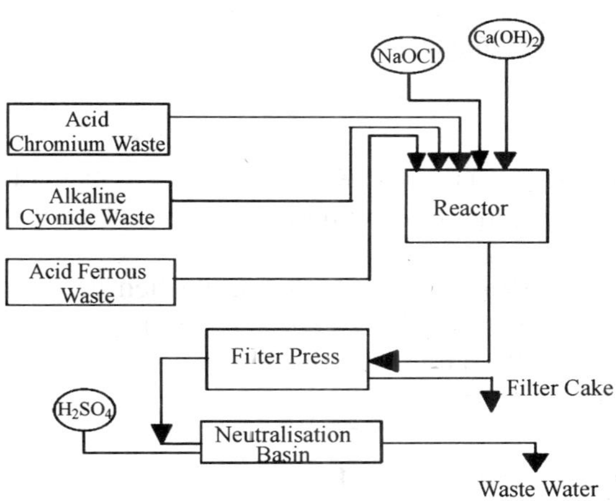

Fig. 8.2 Processes in an Inorganic plant

8.4.6 Waste Oil Plant

Ingoing waste oil mixtures are separated into three fractions by thermal treatment: **oil, oil-contaminated water** and **sludge**. The contaminated water and the sludge are transferred directly to the incineration plant, whereas the oil phase is further treated (filtering, reheating), ending up as a fuel oil suitable for use as supplementary fuel for the incineration plant (Fig. 8.3).

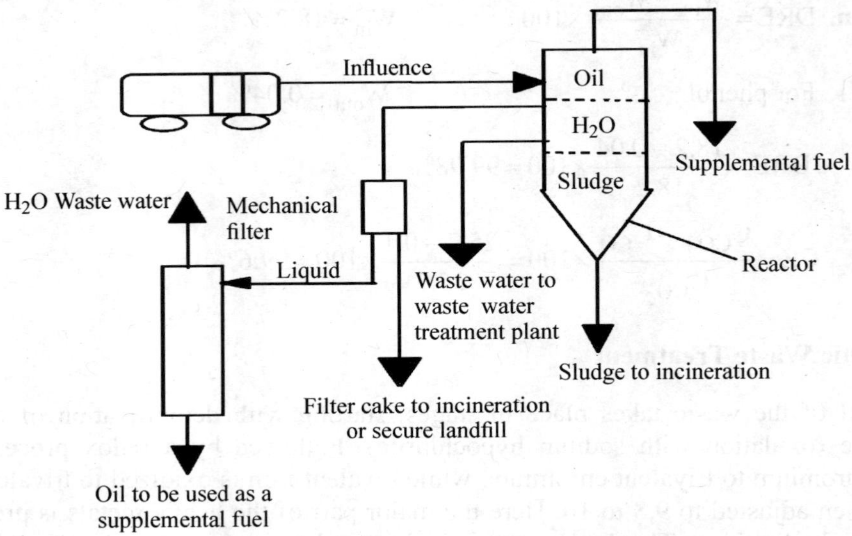

Fig. 8.3 Waste oil plant

8.5 ENVIRONMENTAL IMPACT ASSESSMENT

The uncontrolled human activities have emphasized the developed nations to adopt impact assessment as mandatory steps for environmental protection. Environmental impact assessment (EIA) aims to predict environmental impacts at an early stage in project planning and design, find ways & means to reduce adverse effects, shape projects to suit the local environment & present predictions & options to decision makers.

8.5.1 Definition

Environmental impact assessment is a process that requires consideration of the environment and public participation in the decision – making process of project development.

> Environmental impact assessment rules were notified in 2001 by Government of India under Environment Protection Act, 1986. The GOI established the Ministry of Environment and Forests (MoEF) in the 1980s and made EIA mandatory to all developmental projects prior to site selection and investment decisions.

8.5.2 Environmental Impact Statement (EIS)

EIS is a review document prepared for assessment in the EIA process

8.5.3 Stages of EIA

(i) Screening

Screening is done to determine whether or not a proposal should be subject to EIA and, if so, at what level of detail. Guidelines for whether or not an EIA is required are country specific depending on the laws or norms in operation. Legislation often specifies the criteria for screening and full EIA. Development banks also screen projects presented for financing to decide whether an EIA is required using their set criteria.

The output of the screening process is often a document called an **Initial Environmental Examination or Evaluation** (IEE). The main conclusion will be a classification of the project according to its likely environmental sensitivity. This will determine whether an EIA is needed and if so, to what detail.

(ii) Scoping

The aim of EIA is not to carry out exhaustive studies on all environmental impacts for all projects. Scoping is used to identify the key issues of concern at an early stage in the planning process. The results of scoping will determine the scope, depth and terms of reference to be addressed within the Environmental statement. Scoping is done to:

(a) Identify concerns and issues for consideration in an EIA

(b) Ensure a relevant EIA

(c) Enable those responsible for an EIA study to properly brief the study team on the alternatives and on impacts to be considered at different levels of analysis

(d) Determine the assessment methods to be used

(e) Identify all affected interests

(f) Provide an opportunity for public involvement in determining the factors to be assessed, and facilitate early agreement on contentious issues

(g) Save time and money

(h) Establish terms of reference (TOR) for EIA study

(iii) Baseline data collection

The term baseline refers to the collection of background information on the biophysical, social and economic settings proposed project area. Normally, information is obtained from secondary sources, or the acquisition of new information through field samplings, interviews, surveys and consultations with the public. The task of collecting baseline data starts right from the period of project inception; however, a majority of this task may be undertaken during scoping and actual EIA.

Baseline data is collected for two main purposes:

(a) To provide a description of the current status and trends of environmental factors (e.g., air pollutant concentrations) of the host area against which predicted changes can be compared and evaluated in terms of significance, and

(b) To provide a means of detecting actual change by monitoring once a project has been initiated

Only baseline data needed to assist prediction of the impacts contained in the ToR and scoping report should be collected.

(iv) Impact analysis and prediction

Predicting the magnitude of a development likely impacts and evaluating their significance is core of environmental assessment process. Prediction should be based on the available environmental baseline of the project area. Such predictions are described in quantitative or qualitative terms.

(v) Public hearing and analysis of alternatives

Analysis of alternative is done to establish the preferred or most environmentally sound, financially feasible and benign option for achieving project objectives.

The World Bank directives requires systematic comparison of proposed investment design in terms of site, technology, processes etc in terms of their impacts and feasibility of their mitigation, capital, recurrent costs, suitability under local conditions and institutional, training and monitoring requirements . For each alternative, the environmental cost should be quantified to the extent possible and economic values attached where feasible, and the basic for selected alternative stated. The analysis of alternative should include a NO PROJECT alternative.

(vi) Mitigation and impact management

Mitigation is done to avoid, minimize or offset predicted adverse impacts and, where appropriate, to incorporate these into an environmental management plan or system. For each potential adverse impact the plan for its mitigation at each stage of the project should be documented and costed, as this is very important in the selection of the preferred alternative.

The objectives of mitigation therefore, are to:

find better alternatives and ways of doing things; enhance the environmental and social benefits of a project, minimise or remedy adverse impacts; and ensure that residual adverse impacts are kept within acceptable levels

(vii) Environmental Management Plan (EMP)

An Environmental Management Plan (EMP) is a detailed plan and schedule of measures necessary to minimize, mitigate, etc. any potential environmental impacts identified by the EIA (World Bank 1999). Once the EIA the significant impacts have been identified, it is necessary to prepare an Environmental Management Plan.

An EMP should consist of a set of mitigation, monitoring and institutional measures to be taken during the implementation and operation of the proposed project to eliminate adverse environmental impacts, offset them or reduce them to acceptable levels. The EMP should also include the actions needed to implement these measures, including the following features:

(a) Mitigation based on the environmental impacts reported in the EIA, the EMP should describe with technical details each mitigation measure.

(b) The EMP should then include monitoring objectives that specifies the type of monitoring activities that will be linked to the mitigation measures. Specifically, the monitoring section of the EMP provides:

(c) A specific description, and technical details, of monitoring measures that includes the parameters to be measured, the methods to be used, sampling locations, frequency of measurements, detection limits (where appropriate), and definition of thresholds that will signal the need for corrective actions;

(d) Monitoring and reporting procedures to ensure early detection of conditions that necessitate particular mitigation measures and to furnish information on the progress and results of mitigation.

(e) The EMP should also provide a specific description of institutional arrangements i.e. who is responsible for carrying out the mitigating and monitoring measures (for operation, supervision, enforcement, monitoring of implementation, remedial action, financing, reporting, and staff training).

(f) Additionally, the EMP should include an estimate of the costs of the measures and activities recommended.

(g) It should consider compensatory measures if mitigation measures are not feasible or cost effective.

(h) EMP must be operative throughout the whole Project Cycle.

(viii) Decision making

At each stage of EIA, interim decisions are made. These decisions influence final decisions made about the EIA. The EIS is submitted to designate authority for scrutiny before the final decision. The authority, together with technical review panel determines the quality of EIS and gives the public further opportunity to comment. Based on the outcome of the review, the designated authority or lending institution will accept, reject or make further modifications to avoid future confrontation. If the EIS is accepted, an EIA license is issued and if otherwise, additional studies or recommendations are made before issuance of a license. The decision making process should be autonomous so that the outcome of the review is seen as fair enough. The duration of this process is usually set in the EIA legal framework.

An environmental impact assessment covers two broad parts:

- **Environmental impact statement (EIS):** It includes a brief description of the projects; the existing environment; likely impact of the proposed projects, mitigation, protection and enhancement measures; consideration of alternatives and no change scenario; and summary and conclusions.
- **Environmental management plan (EMP):** It is an implementation plan for mitigation and protection measures that are recommended in the EIS. It includes a work plan, an

implementation schedule, manpower requirement for implementing the plan and monitoring the impacts and protocols for monitoring the impacts.

Convention on biological diversity in its Article 14-16 has also made it clear that the transport and introduction of genetically modified organisms (GMOs) require an EIA.

8.5.4 EIA Procedure

There are three options to consider in establishing EIA procedures as discussed below:

8.5.4.1 Legislative option

(i) The legislative option is the more formal legal approach in which the EIA procedures becomes law and are enforced by the courts.

(ii) National Environmental Policy Act (NEPA) (1969) specifies what is to be included in the EIS, the responsible agency can be challenged in the courts for failure to comply with the legal requirements.

(iii) An agency, Council of Environmental Quality (CEQ) is established in 1970 to oversee the implementation of NEPA.

8.5.4.1.1 Advantages

(i) Mandatory procedures.

(ii) Regulations developed to direct and control activities.

(iii) Enforceable requirements.

8.5.4.1.2 Disadvantages

(i) The cost of bureaucratic machinery required for administration.

(ii) The time cost when the law is challenged in the courts.

(iii) The loss of flexibility in dealing with unique types of projects and or environments.

8.5.4.2 Policy option

(i) The policy option as a basis for EIA procedures means that systems are developed and incorporated within the administrative machinery of government.

(ii) In 1973, the Canadian government implemented the Environmental Assessment Review Process (EARP) which was based on a decision of the federal cabinet.

(iii) Under this policy all agencies of the federal government are to take account the potential environmental and social consequences of development early in project planning.

(iv) The Federal Environmental Assessment Review Office (FEARO) was established to oversee the implementation of EARP.

8.5.4.2.1 Advantages

(i) Greater direct control over the process.

(ii) More opportunity to alter procedures in the light of experience.

(iii) Less administrative costs.

(iv) Avoidance of long delays on projects due to legal arguments.

8.5.4.2.1 Disadvantages

(i) The entire system is more vulnerable to political whims against the process.

(ii) It may be difficult to force agencies to take responsibilities seriously.

8.5.4.3 Middle Ground Option

The middle ground option clearly takes elements of both legislative and policy options.

(i) The EU Environmental Assessment Directive, which was agreed in July 1985, is arguably the most significant of all the existing pieces of community environmental legislation and is an example of middle ground option.

(ii) It simply provides the information on the environmental effects of the project in accordance with the directive must be taken into consideration in the development consent procedure.

(iii) The directive consciously addresses an all–encompassing range of potential environmental assets, from nature protection to cultural heritage.

(iv) Once a project has been determined to fall within the scopes of the directive, the assessment takes the form of five obligatory steps:

  (a) The supply of information on environmental effects by the developer.

  (b) Consultation with other environmental agencies.

  (c) Consultation with the public.

  (d) Consultation with other member states where the project has transnational implications.

  (e) The requirement on the part of the competent authority to take into consideration information.

8.5.4.4 Sensitivity Area Criteria

Two approaches can be considered:

(i) The first refers to the assimilation capacity of the environment and implies a predetermined set of values which describe the resilience of the environment to known levels of disturbance.

(ii) The second reties on the determination of environmentally sensitive areas in terms of features of interest which in combination warrant the designation of an EIA.

8.5.4.5 Matrices

(i) The matrix compares project activities listed under the main headings of site investigation and preparation, construction, operation and maintenance, future and related activities with the environmental consequences which may be physical / chemical, ecological, aesthetic, social.

(ii) Matrix interactions may have no significant effect, be mitigated through an appropriate design measure have unknown effects, have potentially adverse effects and not have significant effects.

Other various methods of EIA

- **Inventorying:** Environmental inventories are prepared that include checklists of various species of plants and animals in a given area. Information containing different habitat types that are found in the site and surrounding. Topography, vegetation, sources of water also included in the inventory.
- **Monitoring:** It refers to the consecutive assessment of the changes that te site and surrounding landscape undergo after the project is launched. It also assess effectiveness of implementation of the EMP.
- **Geographical information system (GIS):** It is a computer-driven package that allows storing and retrieving a range of environmental information. It reveals environmental information about topography, distribution of landscape features including water bodies, natural habitats, agriculture and human habitat.

8.5.5 Initial Environmental Evaluation (IEE)

(i) IEE is 'mini EIA.'

(ii) It requires a description of the environment and the development and the identification of environmental impacts that are anticipated.

8.5.5.1 Project Screening for EIA

The success of EIA is only as effective as the coverage of projects to which it is applied. This has often led to the development of case law which indicates to the agencies developments for which an EIA will be required.

8.5.5.2 Methods of Selecting a Project

(i) The use of positive and negative lists

(ii) Project thresholds

(iii) Sensitive area criteria

(iv) Matrices

(v) Initial environmental evaluation

.5.5.3 Positive and Negative Lists

(i) These are the lists of project types which are considered to be candidates for EIA (positive) or projects for which an EIA will not be required (negative).

(ii) This is further refined by having two (positive) lists:

(a) Projects for which EIA will be required in every circumstance.

(b) Projects for which EIA is required only in certain circumstances.

3.5.5.4 Project thresholds

(i) Thresholds can assist in project screening, where projects exceeding the predetermined thresholds are considered to be candidates for EIA.

(ii) Thresholds can relate to size of project, production or area of land required.

3.5.6 Preparation of EIS

In the preparation of environmental impact statement (EIS) a number of considerations need to be addressed as discussed below:

3.5.6.1 Planning

(i) There are considerable benefits to be gained in pre – planning the preparation of a document before drafting.

(ii) Three issues should be considered: the purpose of the report, its intended audience and the report's structure.

3.5.6.2 Purpose

(i) The purpose of project report is to communicate its findings.

(ii) In the case of environmental study projects, it is to set out in a clear and logical manner the likely environmental consequences of pursuing certain courses of action with the intention that eventual decisions will be better informed and appraised of the possible implications of that action.

3.5.6.3 Audience

(i) Having stated that the purpose is to communicate findings, an understanding and appreciation of the intended audience must be gained.

(ii) The general public represents a second audience that needs to be accommodated within the overall style of the document.

3.5.6.4 Structure

(i) Documents should have a logical structure and avoid repetition.

(ii) Each contributor is required to address the development in terms of:

(a) Baseline conditions (such as air quality, water quality, noise, traffic etc.)
(b) Development features
(c) Anticipated effects
(d) Mitigation
(e) Residual impact

A typical EIS document would comprise the following sections:

(i) Planning framework
(ii) Development proposal
(iii) Archaeology
(iv) Ecology
(v) Landscape
(vi) Water quality
(vii) Air quality (if applicable)
(viii) Waste
(ix) Traffic
(x) Social effects

8.5.7 Non – technical Summary

It is a desirable element of a report, especially in the case of technical assessments which require to be communicated to a wide audience.

(i) They tend to be little more than a précis of the complete document, rather than a summary of the key issues and salient features of a proposal.
(ii) It should restrict its attention to the issues most relevant to the eventual decision.

8.5.8 Review of EIS

There is a need for an impartial, scientific and independent review. The functions of the review authority may include:

(i) The 'scope' of the assessment, i.e. which projects should be subjected to a full or partial EIS.
(ii) General or specific guidelines and advice on methods of EIS.
(iii) Formulate the terms of reference and initiate a detailed EIS.
(iv) Ensure that the EIS has been adequately completed within the terms of reference.

8.6 ENVIRONMENTAL AUDITING (EA)

Environmental audit is listing of all resources of the area. It is defined as a basic management tool which comprises a systematic, documented, periodic and objective evaluation of how well organization, management systems and equipments are performing.

8.6.1 Need for an Environment Audit

(i) EA may be viewed as a management tool internally, and liasion externally with public and the regulatory bodies.

(ii) It helps in pollution control, improves production, safety and health and conservation of natural resources by way of:

  (a) Ensuring waste prevention and reduction.

  (b) Assessing compliance with regulatory requirement.

  (c) Facilitating control of environmental practices by the company's management.

  (d) Placing environmental information to the public.

8.6.2 Components of EA

It has two components such as assessment and verification as discussed below:

8.6.2.1 Assessment

(i) It provides expert judgment/opinion on environmental hazards, associated risks and management and control measures.

(ii) It identifies hazards, estimate the significance of risks, assess current practices and capabilities and provide the basis for recommendations to improve the organizational approach to environmental management.

8.6.2.2 Verification

(i) It determines the documents performance by evaluating the application of and adherence to environmental policies and procedures certifies the validity of environmental data and reports and evaluates the effectiveness of the management systems.

(ii) It can confirm whether the management control systems are in place to manage environmental hazards, verify that regulations and policies are being adhered to assist in identifying gaps in organizational policies. EA objectives include:

Identify documents

  a. Compliance status

  b. Compliance discrepancies with regard to

    - National laws and regulations
    - Company policies and procedures

c. Shortcomings at sites

d. Pattern of deficiencies that may emerge overtime

e. Effectiveness of the compliance management system

(iii) Improve overall Environmental performance.

Review questions

1. Write four major source of hazardous waste generation.

Ans:

1. Electroplating :	Cyanide, heavy metals
2. Pharmaceuticals :	Organic complex
3. Pesticides :	Toxic organic complex
4. Paint industry :	Complex chemical sludge & solvents

2. Enlist the type of incinerators for management of hazardous waste.

Ans:

(i) Rotary kiln incinerator

(ii) Liquid injection incinerator

(iii) Plasma arc incinerator

(iv) Wet air incinerator

(v) Fluidized bed incinerator

3. What are the elements involved in effective hazard control system?

Ans:

1. Estimate the hazard

2. Eliminate the hazard altogether (eg. To get rid of the dangerous machine)

3. Substitute the hazard with a safer alternative (eg. Replace the machine with safer one)

4. Isolate the hazard from anyone who could be harmed

5. Use engineering controls to reduce the risk (eg. To attach guards to machine to protect users)

6. Use administrative controls to reduce risk (eg. Trained workers)

7. Use of personal protective equipment (PPE) eg. To wear gloves and goggles when using such machine.

■■■

Chapter 9

Occupational Health & Safety

9.1 SAFETY

The word **safety** refers to the precautions people take to prevent accidents, harm, danger, damage, loss and air/water/environment pollution. Safety also deals with improvements in working conditions for better health. Safety is the state of being **safe.**

S = Spot the hazard

A = Asses the risk

F = Find a safer way

E = Everyday

Industrial safety is the science and art of identifying, evaluating and controlling workplace hazards. **Hazard** is a state of physical or chemical potential to injure the person or impairment of health.

Risk or danger arises out of hazard. Risk is the product of two factors: that is probability of an event which might occur and severity of the event if it occurs. While analyzing risks, the severity of the events that might occur is more important than the probability of the event to occur. In other words:

Risk = probability of the event that might occur × severity of the event

Probability of each event that might occur may remain constant, whereas the severity of each event that might occur varies and therefore when analyzing risk we are more concerned about the severity of the event that might occur and not the probability of the event that might take place.

9.1.1 General Safety Rules

Some of the general safety rules are listed below which have been formulated not as restrictive measures, but as guides to workers and supervisors, in their cooperative efforts to prevent accidents has being discussed below:

(i) Smoking should be strictly prohibited.

(ii) Act like running, tripping others intentionally, throwing things, directing compressed air at others or such acts which may startle or distract other employees should not be tolerated.

(iii) Personal protective equipment such as safety glasses, respirators etc. must be used.

(iv) Many times red and yellow safety tags are placed on electric switches, valves and other controls to protect personnel and equipment. Unauthorized removal of such tags, misuse of such tags or operation of a control that has a tag attached should be considered as a serious offence.

(v) Maintain good housekeeping, i.e. employees should keep their machines or benches and area surrounding it clean all times. Employees should realize that it is safer to work in an orderly plant.

(vi) Prompt first-aid attention should be obtained for any injury, in order to prevent the infection from developing.

9.1.2 Safety Performance Measurement

Safety performance may be considered with following significant criteria as discussed below:

(i) Capability for limiting any serious consequences leading to accident and mishaps.

(ii) Quantification of measurement.

(iii) Measurement data should consider harmful as well as other consequences.

(iv) Conformance with Bureau of Indian Standard (BIS) so that it can meet the needs for a standard measure of safety performance.

9.1.3 Computation of Injury Rates

(i) Frequency rate of disabling injury (F) relates injuries to working period in hours and expressed in million hour unit.

$$F = \frac{\text{Number of disabling injuries} \times 10{,}00{,}000}{\text{Total employees} - \text{hours worked}}$$

Unit – Disabling injuries per million employee hour exposure.

(ii) Severity rate of disabling injuries (S) relates the days charged to the hours worked during a period and expresses them in terms of million hours.

$$S = \frac{\text{Total days charged} \times 10{,}00{,}000}{\text{Total employees} - \text{hours worked}}$$

(iii) Incident rate (I) relates the number of occupational injuries and illness on lost work days per thousand full time employees.

$$I = \frac{Number\ of\ injuries \times 10{,}00{,}000}{Total\ number\ of\ employees}$$

(iv) Frequency severity index (FSI):

$$FSI = \sqrt{\frac{F \times S}{1000}}$$

FSI has been standardized internationally. Safety performance can be measured on this in India and can be compared with others.

9.2 OCCUPATIONAL HEALTH & SAFETY

Occupational health and safety is getting lot of importance in most of the industries. It is concerned with well-being of all employees and individuals in a plant / factory/ work place and accordingly deals with conditions and factors that have effect on health and safety. In order that all organizations adopt uniform practices, British Standards Institution have brought out a standard called Occupation Health and Safety Assessment Series (OHSAS) 18001: 1999 dealing with this important subject of occupational health and safety. This standard provides requirements for occupational health and safety (OH&S) management system.

9.2.1 OHSAS – 18001: 2007

OHSAS 18001 is the internationally recognized assessment specification for occupational health and safety management systems. The revised OHSAS 18001: 2007 standard is more result oriented than the earlier version. It is better aligned with ISO 9001:2000 and ISO 14001:2004, assisting management systems integration. The highlights of the changes are as follows:

(i) Increased focus on health.

(ii) The accident is now included in incident.

(iii) The term hazard no longer refers to damage to property or damage to work place environment, rather the risk of damage.

(iv) New requirements of hazards identification, risk assessment and selection of controls.

(v) Explicit requirements for management of change.

(vi) New requirements for external consultation and worker participation.

(vii) New definitions such as tolerable risk replaced with acceptable risk.

(viii) New requirements for allocating, documenting and communicating Accountabilities.

(ix) Accountability of management appointee referred.

(x) Addition of new clause/s and review of other clauses.

9.2.2 Objectives and Benefits of OH & S Management System

(i) Reduce workplace injuries and illness

(ii) Reduce cost

(iii) Reduce staff absence

(iv) Increase productivity

(v) Improve insurance liability rating

(vi) Positive response from customers

(vii) Demonstrate due diligence

(viii) Regulatory compliance.

9.2.3 Purpose of OHSAS: 18001 Standards

OHSAS 18001 facilitates the integration of quality, environment and occupational health and safety management systems by any organization. It is compatible with ISO 9001 (quality) and ISO 14001 (environment) management system standards. OHSAS specification is applicable to any organization that wishes to establish an OH & S management system to eliminate or minimize risks to the employees. It specifies requirements for occupational health and safety (OH&S) management systems to enable an organization to control occupational health and safety risks and improve its performance. It enables an organization for continual improvement of OH & S management systems and assures itself of its conformance with its stated OH & S policy. It also deals in detail about the expectations on occupational health standards by any external certifying agency. This serves as broad guide for any medical officer in an industry in formulating various types of medical examinations based on the micro environment and can be used in arriving at a check list before any organization in contemplating of OHSAS certification. It also provides scope for self determination and declaration of conformance with this OHSAS specification.

9.2.4 Occupational Health Hazards

Occupational health hazards mean:

(i) Conditions that cause legally compensable illness.

(ii) Any conditions in the workplace that impair the health of employees enough to make them lose time from work or to work at less than full efficiency.

The various health hazards that may cause sickness, impaired health or significant discomfort or inefficiency in workers are:

(i) Chemical hazards (dust, fumes, fibers, gases, vapors, mists and smoke)

(ii) Physical hazards (noise, vibration, thermal stress, lighting, ionizing radiations and non – ionizing radiations)

(iii) Biological hazards

(iv) Ergonomic hazards

(v) Mechanical hazards

(vi) Psychological hazards.

9.2.5 Hazard Rating

Hazard rating refers to the categorization of the workstations on the basis of different hazard exposures. Hazard rating is a method to classify the percentage of the situations exceeded, ranging or distributing them below and above the threshold limit value, rating the magnitude of the existing hazard and labeling the workstation with the hazard status.

Hazard rating is a tool to industrial hygiene and safety management that throws light upon the criticality of the workstations and the magnitude of the hazards in the environment. It also helps management to take necessary remedial measures on the basis of hazard rating by developing OH & S management programmes and operational control procedures (OCPs). Hazard rating is also a yardstick to rate the improvement activity done to minimize or eliminate or control the hazard exposures / levels.

9.3 ACCIDENTS

Accident is an unexpected, unplanned and uncontrolled event in which the action or reaction of an object, substance, person or radiation results in personal injury or the probability thereof (Fig. 9.1). Injury is any harmful condition sustained by the body as the result of an accident. Hazard is a condition in a work situation with a potential of causing an injury. Danger expresses a relative exposure to the hazard.

The possibility of an accident occurring is present in every sphere of human life: at home, while travelling, at play and at work. Safety performance index (SPL) is a measure of accident occurrence.

$$\text{SPI} = \sqrt{\frac{\text{Accident Frequency} \times \text{Severity Rate}}{1000}}$$

Frequency rate = Number of notified injuries per million employee hour worked.

Severity rate = Total man days lost due to number of notified injuries per million employee hours worked.

9.3.1 Causes of accidents

It is estimated that around 98% of the accidents are caused either by unsafe actions or unsafe mechanical or physical conditions, the break up in general is (Figs 9.1 to 9.3):

Unsafe acts	-	75%
Unsafe conditions	-	23%
Natural calamities	-	2%

Unsafe acts

(i) Working without approval of relevant authority

(ii) Operating at unsafe speed

(iii) Unsafe loading, unloading, removing placing, intermixing etc.

(iv) Taking unsafe position or postures

(v) Working under or near heavy overhead loads without taking due care

(vi) Carelessness during the maintenance and repair work on hazardous machines and gadgets

(vii) Distracting, teasing, abusing etc.

(viii) Failure to use safety devices and personal protective equipments.

9.3.2 Unsafe conditions

Unsafe conditions arise from the following sources:

(i) **Machines:** It includes improperly guarded machines, sudden breakdown or improper upkeep or maintenance of machines, defective or improper construction, design and controlling devices for machines, faulty layout and locations of machines, lack of supervision, hazardous arrangements.

(ii) **Improper working conditions:** It includes defective ventilation, noise pollution coupled with poor housekeeping.

(iii) **Inherent accident proneness in the nature of the job itself:** It includes certain jobs which are inherently hazardous and dangerous, like a blast furnace worker compared to a worker working in the sheet metal section (Fig. 9.1 and 9.2).

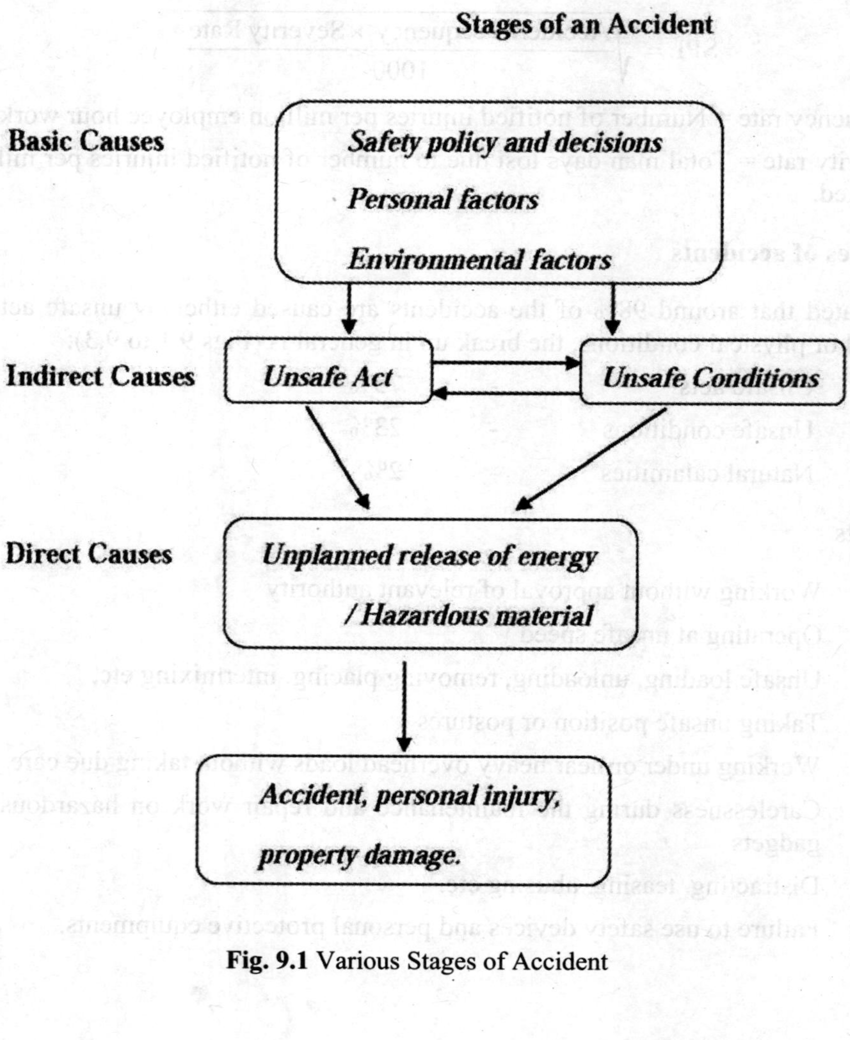

Fig. 9.1 Various Stages of Accident

(iv) **Shift duty :** The pattern of work schedule affects the occurrence of accidents. Fewer accidents occur during the early hours of the day, while they are more frequent during the late hours of the day and especially in the night shifts. Fatigue coupled with lack of rest is the cause of frequent accidents during the night shifts.

(v) **Psychological environment of the working zone:** The alertness and concentration are greatly jeopardized in case of emotionally disturbed and preoccupied workers. The normal behavior of workers is adversely affected as a result of fatigue, monotony, hard labour, frustration, distrust etc. Under such psychological stress, workers are more prone to accidents.

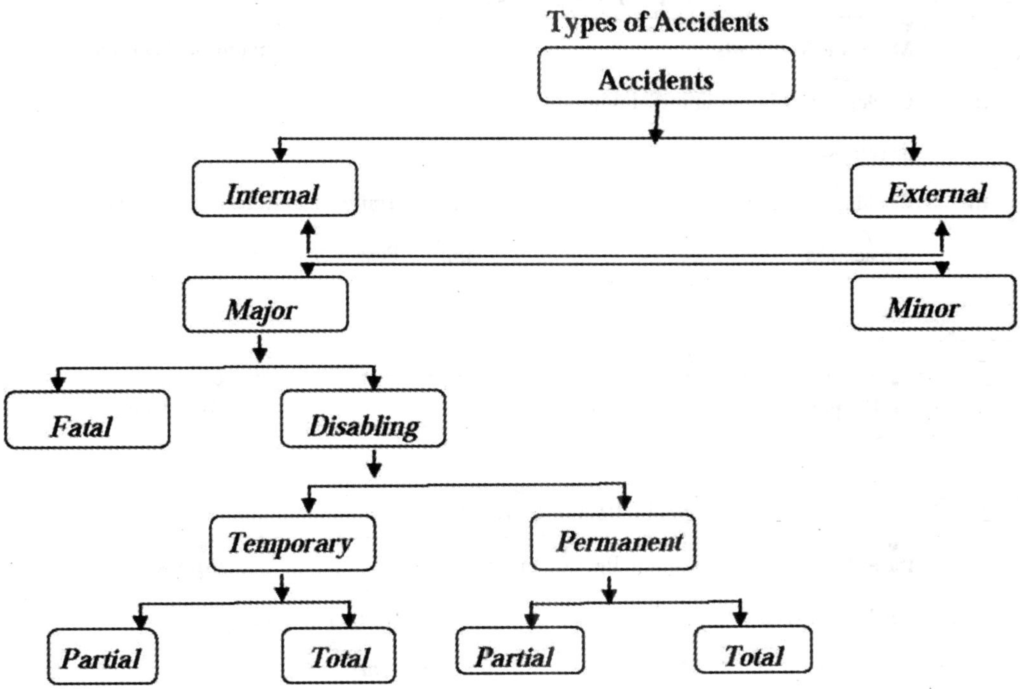

Fig. 9.2 Classification of Accidents

There are several methods of classifying accidents as discussed follows:

(i) Nature of events

(ii) Cause of events

(iii) Damage caused

(iv) Nature of injury

(v) Length of recovery

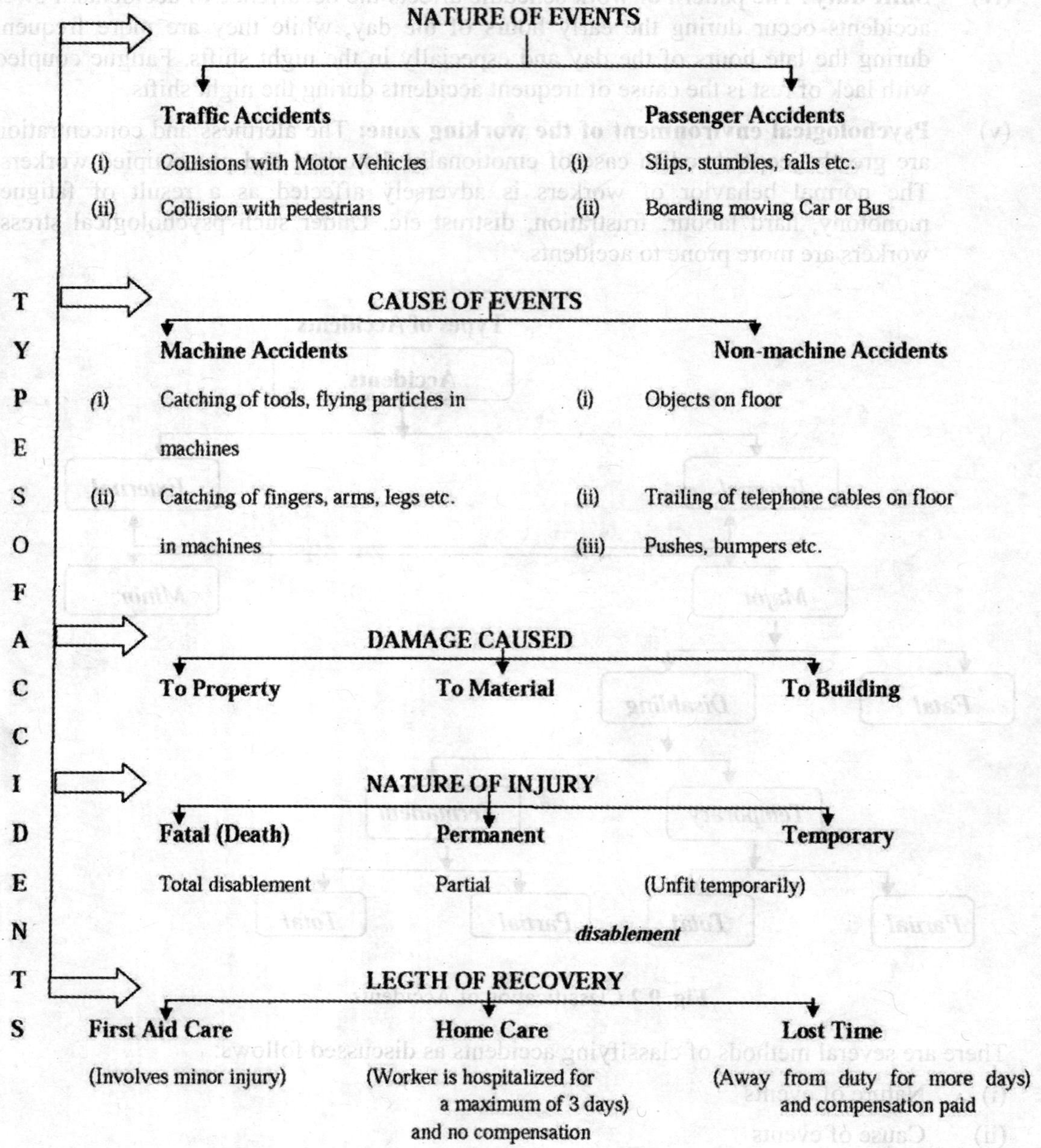

Fig. 9.3 Types of Accident

9.4 CHEMICAL BURNS

Industrial processes involve use of chemicals and hazardous materials, for which safety considerations and controls are highly warranted. The raw materials used in industry could be hazardous due to toxicity inherent in materials. The products, intermediates and finished by products can cause serious problems of chemical safety. The storage, handling, manufacture and use of chemicals call for strict surveillance on the part of managers, supervisors and employees.

A chemical burn occurs when living tissue is exposed to a corrosive substance. It is defined as the transfer of potential energy from chemical reaction leading to tissue damage.

9.4.1 Corrosive Substances

The corrosive substances such as strong acids, alkalis, oxidizing solvents, reducing agents etc.

9.4.2 Symptoms of chemical burn

Itching, bleaching or darkening of skin, burning sensations, trouble in breathing, coughing, tissue necrosis etc. Chemical burns may occur through direct contact on body surfaces including:

(i) Through skin absorption

(ii) By inhalation through respiratory system

(iii) By swallowing

(iv) By eyes

9.4.3 Effects of chemical burns

(i) Chemicals can cause asphyxia or suffocation.

(ii) It causes irritation to respiratory tract or other vital organs like liver, kidney etc.

(iii) May also cause semi consciousness or unconsciousness.

9.4.4 Factors influencing chemical burns

The effects of toxic substances on the workmen depend upon the following considerations:

(i) Absorption: The effects are dependent upon the rate absorption of toxic substances in the human body. Quicker the absorption more is the risk to the workmen.

(ii) Concentration and time of exposure: More the concentration and the time of exposure, more is the amount of toxic substances getting entry into the human body.

(iii) Personal tolerance level: Certain persons can tolerate a particular toxic substance while others cannot.

(iv) Personal hygiene and behavior: Personal hygiene and behaviour including proper maintenance of clothings, cleanliness, tidy habits etc. play an important role.

(v) State of the matter: The state of the matter of toxic substances (whether solid, liquids or gases) plays an important role.

9.4.5 Control measures

(i) Immediate first aid for burns

(a) First, stop the burning to prevent a more severe burn.

(b) Chemical burns: Natural foods such as chili, peppers, which contain a substance irritating to the skin, can cause a burning sensation. When a chemical burn occurs, find out what chemical caused the burn. Call your local **Poison Control Center** or the **National Poison Control Hotline (1-800-222-1222)** for more information about how to treat the burn.

(c) Next, look for other injuries. The burn may not be the only injury.

(d) Remove any jewelry or clothing at the site of the burn. If clothing is stuck to the burn, do not remove it. Carefully cut around the stuck fabric to remove loose fabric. Remove all jewelry, because it may be hard to remove it later if swelling occurs.

(ii) Prepare for an evaluation by a doctor

If you are going to see your doctor soon:

(a) Cover the burn with a clean, dry cloth to reduce the risk of infection.

(b) Do not put any salve or medicine on the burned area, so your doctor can properly assess your burn.

(c) Do not put ice or butter on the burned area, because these measures do not help and can damage the skin tissue.

(iii) Home treatment for minor burns

For home treatment of first-degree burns and sunburns are discussed as follows:

(a) Use cool cloths on burned

(b) Take frequent cool showers or baths.

(c) Apply soothing lotions that contain aloe vera to burned areas to relieve pain and swelling. Applying 0.5% hydrocortisone cream to the burned area also may help.

(d) Do not use the cream on children younger than age 2 unless your doctor tells you to. Do not use in the rectal or vaginal area of children younger than age 12 unless your doctor tells you to.

(e) There isn't much you can do to stop skin from peeling after a sunburn—it is part of the healing process. Lotion may help relieve the itching.

(f) Other home treatment measures, such as chamomile, may help relieve your sunburn symptoms.

9.4.6 Heat (Thermal) Burn

(i) **Heat burns (thermal burns):** Smother any flames by covering them with a blanket or water. If your clothing catches fire, do not run : stop, drop, and roll on the ground to smother the flames.

(ii) **Liquid scald burns (thermal burns):** Run cool tap water over the burn for 10 to 20 minutes. Do not use ice.

(iii) **Electrical burns:** After the person has been separated from the electrical source, check for breathing and a heartbeat. If the person is not breathing or does not have a heartbeat, call 911.

(iv) **Tar or hot plastic burns**: Immediately run cold water over the hot tar or hot plastic to cool the tar or plastic.

CONCLUSION

High-reliability organizations are the prime examples of the system approach. They anticipate the worst and equip themselves to deal with it at all levels of the organization. It is hard, even unnatural, for individuals to remain uneasy over the long term, so their organizational culture takes on a profound importance. Individuals may forget to be afraid, but the culture of a high-reliability organization provides them with both the reminders and the tools to help them remember. For these organizations, the pursuit of safety is not so much about preventing isolated failures, either human or technical, as about making the system as robust as is practicable in the face of its human and operational hazards. High-reliability organizations are not immune to adverse events, but they have learned the knack of converting these occasional setbacks into enhanced resilience of the system.

Review questions

1. **What do you mean by occupational safety?**

   Ans. Occupational safety means safety measures at the working place. It is the science and art of identifying evaluating and controlling hazard at working place.

2. **What is the importance of occupational health?**

   Ans. Improper environmental affect the workers on heath condition. The ill health force the workers to be absent from work.

3. **What is OSHA and what is its aim?**

   Ans. OSHA means occupation safety and health act, established in 1970. The aim of OSHA is to provide the comprehensive and specific heath standard to the workers working in different industries.

4. **What are the factors influencing health of workers?**

   Ans. Apart from hazardous substance, the working environment like lightening, cleanliness, temperature, noise, dust, working space etc. influence the heath of the workers.

5. **What is safety procedure?**

   Ans. The aim of safety procedure is minimization of hazard, safety procedure in working place, for new and hazard equipments and use of PPE.

6. **Define accidents?**

   Ans. Accident may be defined as an unexpected and unplanned occurrence, which may or may not involve injury. The victims of an accident may not be directly involved in the activity that caused accident.

7. What are the causes of accident?

Ans. The causes of the accidents may be Unsafe act, unsafe condition or Multiple causes

8. How the accident rate can be computed?

Ans. The accident rate can be computed by the following

$$\text{Accident rate} = \frac{\text{No. of record accident}}{\text{No. of hours working by all employees}} \times 2000000$$

9. What is chemical burn? What are its treatments?

Ans. Chemical burn is caused by exposure to strong acid strong alkali or chemical irritants. The treatment includes: Remove the victim immediately from the accident site, wash the affected area with sterilized/tepid water and seek medical treatment.

10. What are hazard of electricity?

Ans. The hazard of electricity includes electric shock, electric flashover, fire hazard, explosion in electric equipments etc.

11. What are corrosive substances?

Ans. The corrosive substance is one that which destroy tissue or metals. The common corrosive substances are acid, base, dehydrating agent etc.

12. How to prevent fire hazard?

Ans. Some steps to prevent fire hazard are separate storage of inflammable / explosive materials, following of precaution during storage transport of inflammable material, special precautions during welding etc.

13. What are PPEs? Write the PPEs of head?

Ans. PPEs means personal protective equipments. Various types of PPE and clothing's are extensively used by the personnel's working in industry. The PPEs for head one helmets, hand hats, caps etc.

14. What is a breathing apparatus?

Ans. It is used by the workers working at quite hazard environment. The apparatus contain a compressed oxygen cylinder that protects the users completely from hazardous gases, dust, fames etc.

15. What is a respirator?

Ans. It is used by the workers who are working at the comparative less hazardous conditions, there are various types of respirators are used by the workers are: cartridge type, canister type, mask type etc.

■■■■

Chapter 10

Hazard Control Measures

10.1 HUMAN ERROR AND HAZARD ANALYSIS

It has been estimated that up to 90% of all workplace accidents have human error as a cause. The problem of human error can be analyzed in two ways: individual approach and the system approach. Each has its model of error causation, and each model gives rise to different philosophies of error management. Understanding these differences has important practical implications for coping with the ever-present risk of mishaps in clinical practice.

10.1.1 Individual Approach

The long-standing and widespread tradition of the person approach focuses on the unsafe acts—errors and procedural violations—of people on the front line: nurses, physicians, surgeons, anaesthetists, and pharmacists. These unsafe acts are due to aberrant mental processes such as forgetfulness, inattentiveness, poor motivation, carelessness, negligence, and recklessness and other variation in human behaviour. These methods include poster campaigns that reveal people's fear, writing another procedure (or adding to existing ones), disciplinary measures, threat of litigation, retraining, naming, blaming and shaming. Followers of these approaches tend to treat errors as moral issues, assuming that bad things happen to bad people—what psychologists have called the "just-world hypothesis."

10.1.2 System Approach

The basic principle of this approach is that humans errors can be expected, even in the best organizations. Errors are seen as consequences rather than causes, having their origins not so much in the perversity of human nature as in upstream systemic factors. These include recurrent error traps in the workplace and the organizational processes that give rise to them.

Counter measures are based on the assumption that we cannot change the human condition; however we can change the conditions under which humans work. A central idea is that of system defences. All hazardous technologies possess barriers and safeguards. When an adverse event occurs, the important issue is not who blundered, but how and why the defences failed. Human error includes:

(i) Failing to perform or omitting a task

(ii) Performing the task incorrectly

(iii) Performing an extra or non-required task

(iv) Performing tasks out of sequence

(v) Failing to perform the task within the time limit associated with it

(vi) Failing to respond adequately to a contingency

10.1.3 Hazard Analysis

We can logically classify hazards in the same way we classify disaster events. It is essential for programmatic and operational planning to understand the nature of hazards for defence purpose. It is important to understand that a hazard is a potential cause of a disaster. Not every hazard will result in a disaster, but every disaster will result from a hazardous condition, whether or not that condition was recognized in advance of the event.

10.1.3.1 Analysis process

It is important to have a standard approach to conduct hazard analyses. Although the approach can be improved, if we constantly employ different methods in each analysis it would be very difficult to compare results over time and to identify trends. The model for this presentation is a simple one:

(i) Identify possible hazards.

(ii) Characterize each hazard (in other words define its parameters).

(iii) Apply a rating or assessment scheme to analyse as a basis for comparison with other hazards.

(iv) Communicate the results.

10.1.3.2 Identifying Hazards

Hazard analyses typically develop a predictable list for a locality or community that are reasonably obvious. But if we are going to identify hazards, we need to think globally. First, incidences of disaster events (conducted in the Disaster Database Project) show that these events can occur almost anywhere. It is evident; some types of events are more frequent in some locations than others. Plate boundaries spawn more earthquakes than do mid-plate areas. However, earthquakes do occur mid-plate, tornadoes do occur in mountainous areas and in cities, airplanes drop from the sky almost anywhere, and if there is navigable water, a ship can sink in it.

10.1.3.3 Characterizing Hazards

Few hazards have identical causes, magnitudes and intensities and identical patterns of development. Classes of hazards are not uniform in their range of occurrence, impacts or timelines and in most cases they are neither exclusive nor transposable. The occurrence of a tropical cyclone does not preclude the occurrence of an earthquake (the idea that made changes in atmospheric pressure associated with a hurricane may trigger seismic events). One hazard generally will not substitute for another, at least at the higher level of classification. For example, in hazard analysis an earthquake will not have the same characteristics as a railway accident. It should be noted that a lower level, a railway passenger train accident involving mass casualties can occur in the same fundamental way that freight (or goods) train accident involving hazardous materials. Because hazards are not the same, we must examine each hazard and its characteristics in the context of the time and place in which it is likely to impact our locality or our organization.

In characterizing hazards there are a number of indicators, and these tend to be related:

(i) What is the range of potential magnitudes and intensities that you may experience?

(ii) Is there a time, day, season, or other factor that constrains when this type of event can happen? Earthquakes happen at any time of the day, but mass fatality school shootings happen only when students and staff are in the school buildings. Can we identify the duration of the event, including pre-event, event impact, and diminishing after-impacts? Is there a timeline for how the hazard turns into an actual impact?

(iii) Can we identify the physical extent of the impact zone? Floods inundate their flood plain; therefore, the parts of the community within the flood plain are in the hazard impact zone. Examine hazards in neighboring localities to determine if their impact zones cross into your area.

(iv) How frequently does the hazard develop into an actual event? Frequencies, even if only approximate, become probabilities for decisions under conditions of risk. If none can be established, another key to prediction is the presence of prodromes (signs that a disaster may be imminent). For example, in human systems failures, delayed or nonexistent maintenance is a standard sign of increasing potential for accidents.

(v) Finally are there related hazards (for example, tropical cyclones in North America are known to generate relatively weak tornadoes)? And can the hazard create a cascade of other hazards and impacts as it progresses (serial dam failures are an example)?

10.1.3.4 Hazard control system

A hazard control program consists of all steps necessary to protect workers from exposure to a substance or system, the training is required to monitor exposure and their health to hazards such as chemicals, materials or substance, or other types of hazards such as noise and vibration. The main ways to control a hazard include:

(i) **Elimination (including substitution)**: Remove the hazard from the workplace, or substitute (replace) hazardous materials or machines with less hazardous ones.

(ii) **Engineering controls**: Includes designs or modifications to plants, equipment, ventilation systems and processes that reduce the source of exposure.

(iii) **Administrative controls:** Controls that alter the way the work is done, including timing of work, policies and other rules, and work practices such as standards and operating procedures (including training, housekeeping, equipment maintenance and personal hygiene practices).

(iv) **Personal protective equipment**: equipment worn by individuals to reduce exposure such as contact with chemicals or exposure to noise.

These methods are also known as the **hierarchy of control** because they should be considered in the order presented (it is always best to try to eliminate the hazard first, etc).

10.1.3.5 Outcomes

The outcomes of key indicators allow us to estimate the relative hazard any event poses.

(i) The more severe the magnitude of the event or the intensity of impact, the higher the hazard.

(ii) Events which can occur at any time of day throughout the year pose a different hazard than those that are seasonal or that commonly occur during only part of the day. The individual event in a restricted season may be of catastrophic intensity. For example, the **super typhoon** is a greater national hazard than a crash of a passenger aircraft. However, we must maintain readiness to deal with the worst case aircraft crash impact throughout the year.

(iii) The longer the event duration the greater the hazard. Consider the both the impact and recovery periods, which may be prolonged for weeks, months, or years. A major earthquake happens in seconds, but the combined response and recovery effort may take years.

(iv) The shorter the timeline for development of the event the higher the hazard. In the earthquake example, absent foreshocks to trigger increased readiness, the time does not allow pre-impact response actions to preserve life and property. On the other hand, given good weather forecasting, a developing winter storm may provide hours or days of warning.

(v) The wider the area the disaster impacts may cover, the greater the hazard the event poses. The presence of critical facilities in the probable impact zone increases the hazard. This includes facilities necessary for the coordination of disaster response or the operation of emergency government.

(vi) The more often events happen the greater the hazard because of cumulative effects. This is not universally so; in some cases relatively minor events may release energy that could lead to catastrophic failures.

(vii) Events that are difficult to predict create a higher hazard because of the lack of reliable signals of protective measures. For example, earthquake prediction is currently a very uncertain activity; as a result evacuation and other protective actions are difficult to implement. However, volcanic eruption prediction is becoming reasonably accurate, with evacuation now being a very successful protective tactic.

(viii) Finally events that trigger other events or that cause cascading effects are higher in their hazard potential than events that do not typically do so. The combination of earthquakes and tsunamis is a clear example. In an event such as the Good Friday Alaska Earthquake, causes extensive damage, followed by a tsunami that damages coast lines even at some distance from the quake epicenter.

10.1.3.6 Assessment

Undertaking a hazard analysis is made the more difficult by the wide variability that exists in:

(i) What type of hazards we will actually face versus those that are potential and those that is possible (but unlikely)?

(ii) Where the hazard will strike if it does actually become a threat?

(iii) When the hazard is likely to occur?

(iv) With what intensity it will be experienced?

(v) This is complicated when data is partial, and we may partially understand the problem and the probabilities.

(vi) Finally, the completion of an assessment is often time constricted by an established cycle for the generation of emergency operations plans. This may preclude in depth technical analysis that might improve the overall quality of the analysis. However, this is not an unusual problem in emergency decision making, as many such decisions are based on sufficient, but not perfect, data.

10.1.3.7 Two approaches

(i) **Qualitative assessments** provide practitioner expert judgment based on the best available data. These analyses commonly use terms such as high, medium, or low to communicate an approximate range of the hazard posed by a type of event. This approach is useful when insufficient information is available to develop a detailed quantitative assessment, or when the relationships between the various parameters cannot be reliably established.

(ii) **Quantitative assessments** have the advantage of providing numbers that can be used to express outcomes. Even if the numbers are used to develop broad categories such as high, medium, and low, the presence of numbers adds a comforting sense of exactness to what is essentially a vague process. Quantitative approaches should be regarded with caution; the history of disasters is full of events that were not identified in their actual form by the best analytical tools of the day. One way to perform a quantitative analysis is to:

(a) Define a range for planning and operational needs – more than 5 levels may be confusing, fewer than 3 may not differentiate sufficiently.

(b) Accumulate all the information available on the hazard and its relationship to our locality.

(c) Invite expert participation from various knowledge and skills domain.

(d) Describe the purpose of the process and provide an overview of what is usually known about the hazards.

(e) Develop consensus through discussion and balancing of this hazard against the others we face.

The danger is that a large number of hazards will be seen as being highly hazardous. As a result there may be insufficient differentiation of the hazards to support planning and resource prioritization. This means there must be substantial agreement on the rating range for each event.

10.1.3.7.1 Quantitative analysis

One approach to rating qualitatively is to establish general ranges of impact for key indicators. In this example, factors most commonly reported in media such as deaths, injuries, extent of property damage and the degree to which community services still function. The range of each factor is variable based on the size of the community. For example, a school bus accident that kills 15 and injures 20 in a rural community might well be a high impact event. In a major city, the same casualty count may be a normal weekend evening loading generated by the routine level of urban violence.

We have developed a simple formula to examine the same event. Obviously there are many ways to do this, and this particular way may not be valid, but it will serve as an example for thought:

Event value = {[(magnitude factor x 2 x months) + (area/community area x 10)] x (24 divided by the warning time in hours + prediction constant)} x frequency in 100 years / 100

The quantitative approach may use a formula to develop numerical values for a hazard. In establishing the final format of the analysis, how we structure our list of hazards depends on a number of factors.

If the analysis supports an organization with an all-hazards focus, it makes sense to list the most hazardous to least hazardous in one single list, regardless of the general classifications. However, if our organization has separate budgetary allocations for natural, human systems, and conflict hazards, it may make more sense to list hazards from worst to least within those classifications. And if our doctrine assigns one classification as being the most important, then those hazards sit at the top of the list regardless of their rankings.

10.1.3.7.2 Prioritizing

Whether based on a qualitative or a quantitative approach, the result is a rating for each hazard. This suggests that the hazards should be rank ordered from highest hazard to lowest hazard. The highest hazard is positioned first so that readers will see it first; a reverse order makes it likely the reader will not read the end of the list. If doctrine or political factors require that general classes of hazards are reported separately, the rank order is within the class. However, a more realistic approach is to rank order the entire list of hazards.

In case of emergency this list is considered as a guide. The difference between two potentially catastrophic hazards is not material – we have to prepare for both. We may be constantly reordering the list in our mind based on more current information. The hazard analysis must be observed as a formal written report. It documents the hazards as a written baseline from which our plans are prepared. If a disaster event happens, we have a document to refer for action analysis, to determine if our assessment of the hazard was accurate. The report is equally important as an educational tool for decision makers. The hazard analysis may be the first community leaders to understand the potential threats. Therefore there is value in including explanations of the hazards and examples with which the reader may be familiar.

The report identifies and ranks hazards, and explains the various methods used to determine hazards and rankings. This allows the reader to evaluate the validity and applicability of the assessment, and provides a base for the next hazard assessment. The report should also clearly identify the planning horizon for which it is valid, assumptions that guided preparation, and factors which restricted scope.

10.1.3.7.3 Communication

The hazard analysis should be distributed to everyone involved in emergency planning and governmental decision makers responsible for emergency management and proper execution. It forms the basis for the vulnerability assessment. The extent of further distribution depends on our philosophy regarding citizen access to government information. It is believed that making such documents available is part of government's responsibility for the protection of its citizens.

10.2 HAZARD CONTROL MEASURES IN INTEGRATED STEEL INDUSTRY

The EHS guidelines for integrated steel mills include information relevant to the manufacture of pig iron and raw or low-alloy steel from iron ore and iron-based alloys. It is applicable to the manufacture of metallurgical coke; primary iron and steel production in blast and basic oxygen furnaces (BF and BOF); scrap metal recycling in the electric arc furnace (EAF) process; the production of semi finished products' and hot and cold rolling activities. It does not include extraction of raw materials and further processing of the semi finished products into finished products. Guidance applicable to lime kilns, which may be present in integrated steel mills, is presented in the EHS guidelines for cement and lime manufacturing. This document is organized according to the following sections:

10.2.1 Environmental

Environmental issues associated with steel manufacturing primarily include the following.

10.2 .2 Air Emissions

In addition to the process related to air emissions discussed below, air emissions may be generated from captive power plants fueled with by-product gases (e.g. coke oven gas [COG], blast furnaces [BF] gas, and basic oxygen furnace [BOF] gas). Guidance for the management of emissions from small combustion sources with a capacity of up to 50 mega watt thermal (MWth), including guidelines for exhaust emissions, is provided in the **general EHS guidelines.** Guidance for facilities with power generating capacities in excess of 50 MWth is provided in the **EHS guidelines** for **thermal power.**

10.2.3 Particular Matter

Particulate matter (PM) may be generated in each of the process steps, and may contain varying concentrations of mineral oxides, metals (e.g. arsenic, cadmium, mercury, lead, nickel, chromium, zinc, manganese), and metal oxides. Sources include melting and refining activates (BF, BOF, EAF) and heating furnaces (depending of type of fuels used); mechanical actions (e.g. scarfing and grinding); and handling of materials (e.g. raw materials, additive, recycled and waste materials, and by-products). Additional sources of particulate matter (PM) emissions include coal storage, conveying, charging, coking, pushing, and quenching.

(i) Thermal processes: Particulate matter emissions may arise from thermal processes including coke making, sintering, pelletizing, and direct reduction.

Coke oven plants are another significant source of dust emission. Continuous particulate matter emissions may result from the under-firing process through the combustion stack. Intermittent and fugitive emissions may arise from a large number of sources including oven and leveling doors, valves, and charging holes. Other emissions may arise from pushing, quenching and screening (discontinuous emissions) and from coke oven gas (COG) treatment. Recommended measures to prevent and control particulate matter emissions from coke oven plants include:

(a) Installation of collection hoods for coke oven batteries;

(b) Maintenance and cleaning of all fugitive emissions sources associated with the coke oven (e.g. oven chamber, oven doors, leveling doors, valves and charging holes, and frame seals ascension pipes) are essential for clean and safe operation;

(c) Good operational management to achieve steady state operation to, for example, avoid green push;

(d) Adoption of smokeless charging measures;

(e) Adoption of coke dry quenching (CDQ) system;

(f) Adoption of non recovery- coke battery;

(g) Quick cooling of gas followed by bag filters. The bag filters can be primed with absorbent (e.g. lime or carbon) to further capture volatile impurities;

(h) Use of direct off-gas extraction and canopy hood enclosures and cleaning.

In the casting area (ingots and continuous casting), particulate matter and metals arise from the transfer of molten steel to the mold and from the cutting to length of the product by oxy-fuel torches during continuous casting. Exhausts should be fitted to filters and other relevant abatement equipment, especially in the casting and rolling, and finishing shops, where relevant.

Baghouse filters and ESP have higher particulate collection efficiency, whereas wet scrubbers also allow capturing water-soluble compounds (e.g. sulfur dioxide [SO2] and chlorides). Bag filters are typically installed to control melting shop emissions. They are often preceded by cyclones, which are installed to act as spark separators.

(ii) Mechanical actions: Scarfing and grinding activities may generate particulate matter emissions. Exhausts should be fitted to filters chosen based on the specified activity.

(iii) Raw material handling: To reduce fugitive emissions of particulate matter during handling of materials, the following prevention and control techniques are recommended:

(a) Enclose conveyer transfer points with dust-controls;

(b) Clean return belts in the conveyor belt systems to remove loose dust;

(c) Implement routine plant maintenance and good housekeeping to keep small leaks and spills to a minimum;

(d) Use indoor or covered stockpiles or, when open-air stockpiles are unavoidable, use water spray system, dust suppressants, windbreaks, and other stockpile management techniques;

(e) Design a simple, linear layout for material handling operations to reduce the need for multiple transfer points;

(f) Maximize use of enclosed silos to store bulk powder;

(g) Implement correct loading and unloading practices.

Fugitive emissions of coal dust are a significant concern. Recommendations to prevent and control fugitive coal dust emissions during coal transfer, storage, and preparation include the following:

(a) Minimize the height of coal drop to the stockpile;

(b) Use of water spray systems and polymer coatings to reduce the formation of fugitive dust from coal storage (e.g. on stockpiles);

(c) Use of bag filter or other particulate control equipment for coal dust emissions from crushing / sizing activities;

(d) Installation of centrifugal (cyclone) collectors followed by high efficiency venture aqueous scrubbers on thermal dryers;

(e) Installation of centrifugal (cyclone) collectors followed by fabric filtration for pneumatic coal cleaning equipment and activities;

(f) Use of enclosed conveyors combined with extraction and filtration equipment to prevent the emission of dust at conveyor transfer points;

(g) Rationalizing transport systems to minimize the generation and transport of dust on site.

10.2.4 Nitrogen Oxides

Nitrogen oxides (NO_x) emissions are caused by high furnace temperature and the oxidation of nitrogen. NO_x emissions are associated with sinter operations; pelletization plant operations; Reduction of the coke charge in the blast furnace, including use of pulverized coal injection.

Sinter plant may generate the most significant quantity of particulate matter emissions in integrated steel mills. Emissions in the sinter plant arise primarily from materials-handling operations, which result in airborne particulate matter and from the combustion reaction on the stand. Recommended measures to prevent and control particulate matter emissions from the sinter plant include:

(i) Implement partial or total recirculation of waste gas in the sinter plant, according to sinter quality and productivity;

(ii) Use of electrostatic precipitator (ESP) pulse systems, ESP plus fabric filter, or adoption of pre-dedusting (ESP or cyclones) in addition to high pressure wet scrubbing system for waste gas dedusting. The presence of fine dust, which consists mainly of alkali and lead chlorides, may limit the efficiency of ESPs.

The pelletization of iron ore (an alternative to sintering) may generate dust and particulate matter emissions from grinding of raw material; from the firing zone of the indurations strand; and from screening and handling activities. Recommended measures to prevent and control particulate matter emissions from pelletization are similar to those for sinter operations, as above.

(iii) Melting activities- particulate matter emissions generated by the blast furnace (BF) plant include emissions from the cast house (primarily iron oxide particulates and graphite) and the cleaning of BF gas leaving the top of the furnace. Measures to prevent and control particulate matter emissions from the blast furnace include use of dedusting systems, typically including scrubbers and electrostatic precipitators (ESP), before reuse of the off-gas.

In direct reduction (direct reduction is an alternative route in primary steel production and may reduce overall emissions of dust and other pollutants significantly), dust releases are of similar character, though less than those of blast furnaces.

Particulate matter emissions from the basic oxygen furnace (BOF) arise from hot metal pre-treatment (including hot metal transfer, desulphurization and deslagging processes); charging operations; oxygen blowing to reduce carbon level and oxidation of impurities; and tapping operations. Recommended measures to prevent and control particulate matter emissions from the BOF include:

(a) Use of primary controls for the flue gas of the BOF, including venture scrubbers with or without complete combustion techniques;

(b) Installation of secondary controls to capture off-gas escaping from the BOF process;

(c) Encapsulation of metal pouring lines with fitted extractors.

Electric arc furnaces (EAFs) generate particulate matter during melting; oxygen injection and decarbonizing phases (primary off gas emissions); and charging/tapping (secondary off-gas emissions).

10.3 PETROLEUM REFINERY

The petroleum industry is organized into four broad sectors: exploration and production of crude oil and natural gas; transport; refining; and marketing and distribution. This chapter addresses only petroleum refining.

Crude oil is fractionated into liquefied petroleum gas, naphtha (used to produce gasoline by blending with octane boosters), kerosene/aviation turbine fuel, diesel oil, and residual fuel oil. Catalytic cracking and reforming, thermal cracking, and other secondary processes are used to achieve the desired product specifications. Certain refineries also produce feed stocks for the manufacture of lubricating oils and bitumens. Some refineries also manufacture coke.

10.3.1. Waste Characteristics

Boiler, process heaters, and other process equipment are responsible for the emission of particulates, carbon monoxide, nitrogen oxides (NO_X), sulfur oxides (SO_X), and carbon dioxide. Catalyst changeovers and cookers release particulates. Volatile organic compounds (VOCs) such as benzene, toluene, and xylene are released from storage, product loading and handling facilities and oil water separation systems and as fugitive emissions from flanges, valves, seals, and drains. For each ton of crude processed, emissions from refineries may be approximately as follows:

(i) **Particulate matter:** 0.8 kilograms (kg), ranging from less than 0.1 to 3 kg.

(ii) **Sulfur oxides**: 1.3 kg, ranging 0.2-06 kg; 0.1 kg with the Claus sulfur recovery process.

(iii) **Nitrogen oxides**: 0.3 kg, ranging 0.06-0.5 kg.

(iv) **Benzene, toluene and xylene**: 2.5 grams (g), ranging 0.75 to 6 g; 1 g with the Claus sulphur recovery process. Of this, about 0.14 g benzene, 0.55 g toluene, and 1.8 g xylene may be released per ton of crude processed.

(v) VOC emissions depend on the production techniques, emissions control techniques, equipment maintenance, and climate conditions and may be 1 kg per ton of crude processed (ranging from 0.5 to 6 kg/t of crude).

Petroleum refineries use relatively large volumes of water, especially for cooling system. Surface water runoff and sanitary wastewaters are also generated. The quantity of wastewaters generated and their characteristics depend on the process configuration. As a general guide, approximately 3.5-5 cubic meters (m^3) of wastewater per ton of crude are generated when cooling water is recycled. Refineries generate polluted wastewaters, containing biochemical oxygen demand (BOD) and chemical oxygen demand (COD) levels of approximately 150-250 milli grams per liter (mg/1) and 300-600 mg/l, respectively; phenol levels of 20-200mg/l; oil levels of 100-300 mg/l in desalter water and up to 5,000 mg/l in tank bottoms; benzene levels of 1-100 mg/l; benzopyrene levels of less than 1 to 100 mg/l; heavy metals levels of 0.1 -100 mg/l for chrome and 0.2-10 mg/l for lead; and other pollutants. Refineries also generate solid wastes and sludges (ranging from 3 to 5 kg per ton of crude processed), 80% of which may be considered hazardous because of the presence of toxic organics and heavy metals.

Accidental discharges of large quantities of pollutants can occur as a result of abnormal operation in a refinery and potentially pose a major local environmental hazard.

10.3.2 Pollution Prevention and Control

Petroleum refineries are complex plants, and the combination and sequence of processes is usually very specific to the characteristics of the raw materials (crude oil) and the products. Specific pollution prevention or source reduction measures can often be determined only by the technical staff. However, there are a number of broad areas where improvements are often possible, and site specific waste reduction measures in these areas should be designed into the plant and targeted by management of operating plants. Areas where efforts should be concentrated are discussed here:

Reduction of Air Emissions

(i) Minimize losses from storage tanks and product transfer areas by methods such as vapor recovery systems and double seals.

(ii) Minimize SO_x emissions either through desulfurization of fuels, to the extent feasible, or by directing the use of high-sulfur fuels to units equipped with SO_X emissions controls.

(iii) Recover sulfur from tail gasses in high-efficiency sulfur recovery units.

(iv) Recover non-silica-based (i.e. metallic) catalysts and reduce particulate emissions.

(v) Use low NO_X burners to reduce nitrogen oxide emissions.

(vi) Avoid and limit fugitive emissions by proper process design and maintenance.

(vii) Keep fuel usage to a minimum.

Elimination or Reduction of Pollutants

(i) Consider reformate and other octane boosters instead of tetraethyl lead and other organic lead compounds for octane boosting.

(ii) Use non-chrome-based inhibitors in cooling water, where inhibitors are needed.

(iii) Use long-life catalysts and regenerate to extend the catalysts life cycle.

Recycling and Reuse

(i) Recycling cooling water and, where cost-effective, treated wastewater.

(ii) Maximize recovery of oil from oily wastewaters and sludges. Minimize losses of oil to the effluent system.

(iii) Recover and reuse phenols, caustics, and solvents from their spent solutions.

(iv) Return oily sludges to coking units or crude distillation units.

Operating Procedures

(i) Segregate oily wastewaters from storm water systems.

(ii) Reduce oil losses during tank drainage carried out to remove water before product dispatch.

(iii) Optimize frequency of tank and equipment cleaning to avoid accumulating residue at the bottom of the tanks.

(iv) Prevent solids and oily wastes from entering the drainage system.

(v) Institute dry sweeping instead of wash down to reduce wastewater volumes.

(vi) Establish and maintain an emergency preparedness and response plan and carryout frequent training.

(vii) Practice corrosion monitoring, prevention, and control in underground piping and tank bottoms.

(vii) Establish leak detection and repair programs.

10.3.3 Target Pollution Loads

Implementation of pollution prevention measures can yield both economic and environmental benefits. However, a balance on energy usage and environmental impacts may have to e struck. The production-related targets described below can be achieved by measures such as those detailed in the previous section. The values relate to the production processes before the addition of pollution control measures.

New refineries should be designed to maximize energy conservation and reduce hydrocarbon losses. A good practice target for simple refineries (i.e. refineries with distillation, catalytic reforming, hydrotreating, and offsite facilities) is that the total quantity of oil consumed as fuel and lost in production operations should not exceed 3.5% of the throughput. For refineries with secondary conversion units (i.e. hydrocrackers or lubrication oil units), the target should be 5-6% (and, in some cases, up to 10%) of the throughput. Fugitive VOC emissions from the process units can be reduced to 0.05% of the throughput, with total VOC emissions of less than 1 kg per ton of crude (or 0.1% of throughput). Methods of estimating these figures include emissions monitoring, mass balance, and inventories of emissions sources. Design assumptions should be recorded to allow for subsequent computation and reduction of losses. Vapour recovery systems to control losses of VOCs from storage tanks and loading areas should achieve 90-100% recovery.

Plant operators should aim at using fuel with less than 0.5% sulfur (or an emissions level corresponding to 0.5% sulfur in fuel). High-sulfur fuels should be directed to units equipped with SO_X controls. Fuel blending is another option. A sulfur recovery system that achieves at least 97% (but preferably over 99%) sulfur recovery should be used when the hydrogen sulfide concentration in tail gases exceeds 230 mg/Nm^3. The total release of sulfur dioxide should be below 0.5 kg per ton for a hydro skimming refinery and below 1 kg per ton for a conversion refinery.

A wastewater generation rate of 0.4 m^3/sec of crude processed is achievable with good design and operation, and new refineries should achieve this target as a minimum. The generation rate of solid wastes and sludges should be less than 0.5% of the crude processed, with a target of 0.3%.

10.3.4 Treatment Technologies

(i) **Air emissions:** Control of air emissions normally includes the capture and recycling or combustion of emissions from vents, product transfer points, storage tanks, and other handling equipment. Boilers, heaters, other combustion devices, covers, and catalytic units may require particulate matter controls. Use of a carbon monoxide boiler is normally a standard practice in the fluidized catalytic cracking units. Catalytic cracking units should be provided with particular removal devices. Steam injection in flaring stacks can reduce particulate matter emissions.

(ii) **Liquid effluents**: Refinery wastewaters often require a combination of treatment methods to remove oil and contaminants before discharge. Separation of different streams, such as storm water, cooling water, process water, sanitary, sewage, etc., is essential for minimizing treatment requirements. A typical system may include sour water strip per, gravity separation of oil and water, dissolved air flotation, biological treatment, and clarification. A final polishing step using filtration, activated carbon, or chemical treatment may also be required. Achievable pollutant loads per ton of crude processed include BOD, 6 g; COD, 50 g; suspended solids, 10 g; and oil and grease, 2 g.

(iii) **Solid and hazardous waste:** Sludge treatment is usually performed using land application (bioremediation) or solvent extraction followed by combustion of the residue or by use for asphalt, where feasible. In some cases, the residue may require stabilization prior to disposal to reduce the leachability of toxic metals. Oil is recovered from slops using separation techniques such as gravity separators and centrifuges.

10.3.5 Emissions Guidelines

Emissions levels for the design and operation of each project must be established through the environmental assessment (EA) process on the basis of country legislation and the pollution prevention and abatement handbook, as applied to local conditions (Table 10.1). The emissions levels selected must be justified in the EA and acceptable to the World Bank Group.

Table 10.1 Emissions from Petroleum Industry

Emissions from the Petroleum Industry
(milligrams per normal cubic meter)

Parameter	Maximum value
PM	50
Nitrogen oxides	460
Sulfur oxides	150 for sulfur recovery
Nickel and vanadium	Units; 500 for other units
(combination)	2
Hydrogen sulfide	152

The guidelines given below present emissions levels normally acceptable to the World Bank Group in making decisions regarding provision of World Bank Group assistance. Any deviations from these levels must be described in the World Bank Group project documentation. The emissions levels given here can be consistently achieved by well-designed, well-operated, and well-maintained pollution control systems.

The guidelines are expressed as concentrations to facilitate monitoring. Dilution of air emissions or effluents to achieve these guidelines is unacceptable. All of the maximum levels should be achieved for at least 95% of the time that the plant or unit is operating, to be calculated as a proportion of annual operating hours.

Air emissions

The emissions levels presented in Table 10.1 should be achieved.

Liquid effluents

The emissions levels presented in Table 10.1 should be achieved. Effluent requirements are for direct discharge to surface waters. Discharge to an offsite wastewater treatment plant should meet applicable pretreatment requirements.

Solid wastes and sludges

Wherever possible, generation of sludges should be minimized to 0.3 kg per ton of crude processed, with a maximum of 0.5 kg per ton of crude processed. Sludges must be treated and stabilized to reduce concentrations of toxics (such as benzene and lead) in leachate to acceptable levels, for example, below 0.05 milligram per kg.

Ambient noise

Noise abatement measures should achieve either the levels given below or a maximum increase in background levels of 3 decibels.

Monitoring and Reporting

Frequent sampling may be required during startup and upset conditions. Once a record of consistent performance has been established, sampling for the parameters listed in this document should be as described below. Air emissions from stacks should be monitored once every shift, if not continuously, for opacity (maximum level, 10%). Air emissions of hydrogen sulfide from a sulfur recovery unit should be carried out for sulfur oxides (sulfur content of the fuel monitored on a supply-tank basis) and for nitrogen oxides.

Liquid effluents should be monitored daily for all the parameters listed above, except that metals should be monitored at least monthly. Monitoring data should be analyzed and reviewed at regular intervals and compared with the operating standards so that any necessary corrective actions can be taken. Records of monitoring results should be kept in an acceptable format. The results should be reported to the responsible authorities and relevant parties, as required.

Important key issues

The key production and control practices that will lead to compliance with emissions guidelines can be summarized as follows:

(i) Use vapor recovery systems to reduce VOC emissions.

(ii) Install sulfur recovery systems, where feasible.

(iii) Use low NO_X burners.

(iv) Maintain fuel and losses to 3.5% for simple refineries and below 6% (with 10% as the maximum) for refineries with secondary processing.

(v) Recover and recycle oily wastes.

(vi) Regenerate and reuse spent catalysts and solvents.

(vii) Recycle of cooling water and minimizing waste waters.

(viii) Segregate storm water from process waste water.

(ix) Use non-chrome-based inhibitors (use only to the extent needed tin cooling water).

(x) Minimize the generation of sludges.

(xi) Install spill prevention and control measures.

10.4 LPG BOTTLING

(i) **Bottling plant** is a plant where beverages are put into bottles with a cap. These bottles are widely used for storage purposes.

(ii) **Liquefied petroleum gas (LPG)** consists mainly of propane, propylene, butane, and butylene in various mixtures. However, in the United States, the mixture is mainly propane. It is produced as a by-product of natural gas processing and petroleum refining. The components of LPG are gases at normal temperatures and pressures.

(iii) **LPG bottling plant** is a plant where LPG is put into bottles for storage. The plant has the facility to receive bulk LPG by pipeline from a reliable source or any area. This is the safest and the cheapest way of transporting bulk LPG.

The plant is equipped with hydraulic cylinder testing equipment for periodic testing of the cylinders. To ensure safe and accurate filling of the cylinders well experienced technical people should only be employed in the working of the plant.

10.4.1 Tips While Working with LPG Bottling Plant

(i) The filling machines should be provided with pneumatic cut off system to automatically cut of the LPG supply when the cylinder is filled.

(ii) The filling pressure should not exceed design pressure of the liquid.

(iii) Cylinders have a standard weight and all cylinders should be checked for their weight after they are filled.

(iv) The compact valve tester should be calibrated to detect leakage from the cylinder.

(v) Defective cylinders should be evacuated and evacuation should be done either through one or two vessels. Form these vessels LPG is transferred to the storage vessels through pump.

(vi) Security caps should be provided while performing any test with LPG.

(vii) The sequence of operation should be strictly followed.

(viii) To avoid undue accumulation the cylinders should be immediately shifted to storage after filling.

The bottling plant has its own system of stringent safety and quality control checks built in at different stages. The quality control checks at the bottling plants are carried out at 3 stages:

(i) **Pre-filling quality control checks**: The pre-filling checks include segregation of damaged cylinders, spurious cylinders, cylinders with missing O-Rings and cylinders due for pressure testing.

(ii) **Post-filling quality control checks**: In the post-filling checks, it is ensured that all the cylinders filled on the automatic cut-off weighing scale/filling machines are subjected to 100% cross – checks for weight.

(iii) **Statistical quality control checks**: These checks include all pre-filling and post-filling checks in addition to checking the quality of the seal.

10.4.2 The most important LPG Bottling Plants are located in

(i) Cudappah in Andhra Pradesh
(ii) Cherlapalli in Andhra Pradesh
(iii) Madurai in Tamil Nadu
(iv) Mayiladuturai in Tamil Nadu
(v) Ennore in Tamil Nadu
(vi) Thimapur in Andhra Pradesh
(vii) Bangalore in Karnataka
(viii) Vijayawada in Andhra Pradesh
(ix) Calicut in Kerala

10.4.3 Location and Spacing

(i) Inspection – criteria

(a) Site: above ground and area/space well ventilated. Separation distance, required between vessels. Vessels and compound boundary. Boundary and fixed source of ignition.

(b) Ground below the bullet is concreted and given slope towards the catchment pit.

(c) LPG yard is free from grass, shrubs, etc.

(d) Catchment pit, sufficient capacity to entail larger credible leak and is ventilated.

(e) Gas detector are given in the pits.

(f) Thermal radiation of burning LPG in the pit should not cause danger to the bullet.

(g) Storage vessels not placed in the space susceptible to flooding.

(h) Minimum separation distance between vessels is maintained.

(i) LPG vessels are not stored near any other flammable.

(j) Any vessels in one group have a minimum separation distance of 7.5 m from the nearest vessel in another group unless a fire wall is oriented between two group.

(k) Long axis of the horizontal storage vessel does not point towards nearby occupied buildings/hazardous materials.

(ii) Mechanical integrity

(a) Vessels are designed, constructed, tested & certified to appropriate standards.

(b) Maximum safe operating pressure is 14.5 kg's (210 psi.) per design standard.

(c) Minimum safe operating pressure is 480 absolute (7 psi) per design standard.

(d) Vessels are marked with minimum safe operating pressure.

(e) Safe operating limit of pressure, temperature at unloading.

(f) There are certified by competent person after installation of the vessel and before filling it.

(g) Periodic examinations are done by competent person in accordance with the prescribed codes.

(h) Maximum permissible fill (contents) and maximum permissible load on supports is included in the documentation.

10.5 PHARMACEUTICAL INDUSTRY

The **pharmaceutical industry** develops, produces, and markets drugs licensed for use as medications. Pharmaceutical companies can deal in generic and/or brand medications. They are subject to a variety of laws and regulations regarding the patenting, testing and marketing of drugs.

Nowadays it is clear that an accident damages the factory and their reputation besides having to dole out huge compensation and even face the prospect of closure. In the pharmaceuticals industry, various chemicals are being, handled through several equipment, machinery and processes using electricity and mechanical devices. Various microorganisms are also being handled both in the laboratory and in production process. So it is required to study the whole setup, machinery, process, raw materials, chemicals, microorganisms, etc. thoroughly. It is also required to identify all possible hazards and then take preventive and corrective measures to prevent the accident. In view of these, it is necessary to consider all these aspects for safety purpose during planning and implementation of project. This helps in prevention of accidents during running of plant.

10.5.1 Evaluation of Pharmaceutical Safety Aspects

The following steps should be discussed for safety aspects as described below:

(i) Source of hazard.

(ii) Type of hazard

(iii) Control procedure

(iv) Contingency plan

10.5.2 Possible Sources of Hazard in Pharmaceutical Industry

(i) Handling and processing of inflammable and combustible raw materials

(ii) Handling of compressed gases

(iii) Handling and processing of toxic substance

(iv) Handling of corrosive substances Handling of oxidizing substances

(v) Operating electrical & mechanical – operated machineries for manufacture and testing of products

(vi) Handling of dust-producing equipment and also dust

(vii) Handling of boiler

(viii) Handling of various microorganisms in manufacture and also in testing of drugs

(ix) And several other operations in Pharmaceutical Industry for production and testing of drugs.

10.5.3 Types of Hazards

(i) Electrical shock

(ii) Injury and death

(iii) Fire

(iv) Chemical burn

(v) Infection

(vi) Intoxication.

10.5.4 Control Procedure

The framing of the control-procedure is an essential criterion in manufacturing houses so as to provide safe work environment in working place as well as to prevent accident.

In pharmaceutical industry the manufacturing control procedures are prescribed in elaborate way as per the provision of Schedule M of the Drugs & Cosmetics Acts & Rules. Besides this the provisions of GMP (Good Manufacturing practices) and GLP (Good Laboratory Practices) are also adopted in to which covers total aspects of the factory i.e. building, plant setup, process, personnel, quality assurance, manufacturing control procedures, documentation etc. which certainly provide safe work environment. However the lapses or deviation in procedures lead to accidents hence the concerned personnel should always be alert to avoid lapse / deviation of even slight nature.

Hence for ensuring total safe work environment and to prevent accident, the under mentioned control procedures may be adopted and discussed below:

(i) Standard Operating Procedure (SOP): In the pharma industry SOP is prepared in more elaborate manner for manufacture of each formulation. All the processes are carried out within the prescribed operational parameters. It is of utmost importance to adhere to the SOP and all parameters, which automatically provide safety. Hence for total safety it is required to include critical points and relevant corrective measures for all processes and also include the facilities provided in SOP with operating instructions. The SOP complete in all respects, ensures total safety in Pharmaceutical industry.

(ii) Safety policy : Besides processing, there are several sources (stated above) which may create accident. Hence it is imperative to define a safety policy and assigned the task to a safety officer. The safety policy is to be made as accident prevention and control programme to provide safety in factory and in office, too to each employee. This is drawn by considering all phases of company's activities with the followings:

(a) Acceptable level of operational risks

(b) Standard of industrial hygiene

(c) Delegation of responsibility to employees as staff function

(d) Process of alternation, modification or change in procedure in case of deviation or apprehension of accident.

(e) Steps to be taken in case of accident

(f) The Company's safety policy should be hanged – up in notice board for information to all employees.

(iii) Monitoring: It is essential to monitor each & every operation for getting success. Hence all the processes & safety aspects must be monitored. It is also required to closely monitor the effectiveness of all safety tools on regular basis. On getting deviation or lapses, the cause is detected and the remedial measures are recommended for immediate implementation.

(iv) Safety audit: Audit is an essential step to check the application of systems in true manner. Hence safety audit is to be done at regular interval so as to ensure the following-up of SOP with parameters, critical points, safety aspects and all steps for prevention of accidents. In this respect, it is better to draw a safety audit format which will be helpful during auditing by safety officer or their authorized representative.

(v) Risk analysis: There are chances of occurring of several types-of unwanted accidents in factory premises including office which may be fatal. Analysis in systemic manner by considering all aspects covering men, machine, processing, activities, building and all events including hazard due to human error or failure of any of the operation or tool. After evaluation of risk analysis and the safety measures are developed. Accordingly the safety aspects are incorporated in SOP and safety policy.

(vi) Preventive maintenance: Several accidents are avoidable by taking precautions and timely maintenance of machineries, building, pipelines etc. So, effective maintenance programme will certainly eliminate accident. Accordingly preventive maintenance must be planned to overtone any dangerous situation. These aspects should be incorporated in SOP, Safety Audit format, safety Monitoring format and safety policy.

(vii) Involvement of personnel: Most of the accidents are caused due to lapses on the part of person concerned. So it is imperative that the person concerned must be alert during work, which will present accident certainly.

Hence it is most essential to involve all personnel in safety programme. All employees must be informed of safety policy and the concerned employees should be made fully aware about concerned safety aspects involved.

Review questions

1. **What are the different types of industrial hazard?**

   Ans. The industrial hazard may be grouped into chemical hazard, mechanical hazard, biological hazard, environmental hazard, electrical hazard etc.

2. **What is human error?**

   Ans. Human error is one of the major factors for accident in working places. Prevention, recognition and mitigation of human error will reduce the accident rate. The human error can be minimized by,

   (i) Proper design of work system.

   (ii) Creation of safety culture.

3. **What do you mean by hazardous analysis?**

   Ans. Hazardous analysis is important for mitigate accident. It can be done by based on past data, identification and recognition of hazard in the system, evaluation and assessment of probability and severity of accident etc.

4. **What are hazards in pharmaceutical industry?**

   Ans. The types of hazard in pharmaceutical industry are infection, intoxication, chemical burn, fire hazard, injury & illness & electrical shocks.

■■■■

Chapter 11

Product Safety

11.1 INTRODUCTION

Product safety (PS) refers to physical health and safety of citizens with respect to non – food products such as household appliances, toys, cars, cosmetics etc. It is designed to reduce the number of injuries and facilities caused by non – food products. The dangerous products need to be rapidly identified in order to be taken off from the market before they reach the people. The most common types of danger are the risk of the electric shock, the risk of toxic substances and the risk of injury. The general product safety directive (GPSD) lays down the product safety requirements for all other non-consumer products. It provides a general definition of a safe product and establishes an alert system on dangerous products.

11.1.1 Basic Elements of Product Safety & Management

Basic elements of product safety programming are designed to identify and evaluate potential product hazards for systematic control using the techniques of safety management and safety engineering. Generally, basic elements of product safety programming include (but are not limited to):

(i) A clear, explicit, and documented statement of product safety policy.

(ii) A clear, explicit, and documented assignment of individual responsibility for the conduct of product safety activity.

(iii) A clear, explicit, and documented product safety program plan outlining the specific steps, procedures, and techniques to be followed on conducting product safety activity during the product design and marketing processes to achieve product safety goals.

(iv) As a starting point, a documented search for authoritative literature and relevant standards relating to a potential safety concerns associated with the product to be designed or marketed.

(v) The conduct of explicit and documented activity giving attention to the systematic discovery or identification of reasonably anticipated potential product or system hazards, followed by an evaluation of those hazards in terms of associated risk factors (likely loss event probability and severity).

(vi) The documented use of the core concepts and principles of safety management and safety engineering, and the cardinal rules of hazard control, to reasonably eliminate or minimize unacceptable product hazards (though, in order of preference and effectiveness, use of design, safeguarding, or warning means).

(vii) To summarize, an effective product safety program must:

  (a) Formally declare to all personnel that product safety is important.

  (b) Assign responsibility to specific individuals (or heads of departments) to assure product safety during the product design, manufacturing, and marketing process.

  (c) Establish specific activity to identify and evaluate potential product hazards based on reasonably foreseeable conditions of product use.

  (d) Utilize reasonably well established and available safety standards and guidelines to design hazards out of products on the drawing board.

  (e) Add components or devices to products to safeguard the remaining hazard.

  (f) Provide adequate product warnings and instructions that address hazards that must remain part of the product design (cannot be reasonably eliminated or controlled through design means) or are inherent to product use.

(viii) Two standards regarding the basic elements of product safety programs published in the 1970's are **Guidelines for Organizing a Product Safety Program**, published by the American National. Standards Institute (ANSI) in 1972 (revised in 1978) and *Handbook & Standards for Manufacturing Safer Consumer Products* (1977), a publication of the Consumer Product Safety Commission (CPSC).

11.1.2 Product Safety Program

Product safety programs should appropriately include the following:

(i) Policy statement: A widely publicized explicit formal statement, as a matter of record, regarding top management's commitment to state-of-the-art product safety and the pre-eminent importance of product safety during product (system) design, production, and distribution.

(ii) Organization, staffing, and program planning: Specific assignment of product safety responsibility and authority should be clearly established beginning at the executive level and reflected in various position descriptions and performance appraisals. Specific procedures should be developed for each element of the product safety program. Department plans should encompass product safety as it would relate to design, purchasing of parts, manufacture, marketing, use, service, and disposal. Means should be established to provide administrative surveillance and technical product safety assistance and guidance to various operating departments.

(iii) Communications and training: Appropriate training should be conducted to ensure that those participating in the product safety program know what their responsibility and role is and have the appropriate skills to properly carry out their contribution to product safety activity. An adequate and current reference library of applicable texts, periodicals, and related safety standards should be maintained.

(iv) Design review: Design review involves the formal examination of product materials, components, configurations, packaging, and labelling (instructions and warnings) to identify, evaluate, and control potential product hazards. Hazard identification and evaluation criteria should include objective estimations of the conditions under which the product will be used, including such things as the age levels and physical limitations of users, and potentialities that might occur as a result of reasonably foreseeable product misuse. Comparisons should be made to applicable authoritative guidelines and standards. Adequate laboratory and field-tests, addressing normal use and reasonably foreseeable misuse, should be conducted to verify product safety design factors throughout the product's useful life and disposal. Hazard analysis should be conducted using currently accepted techniques. Appropriate corrective action must be taken when product safety hazards are identified. Findings (details regarding potential hazards) and decisions (regarding corrective actions) should be documented. Group design reviews should be chaired by a designated senior official having overall product safety authority over all participating departments.

(v) Documentation and change control: Documentation and change control involves the documentation of design decisions and the technical basis or reasons for those decisions. After all reasonable design changes have been made, remaining hazards and the required actions that users must follow to protect themselves from danger (for later production of adequate product warnings and instructions) should be clearly identified in these documents.

(vi) Purchase product control: Purchase product control involves control over purchase procedures and decisions to include steps to ensure that safe materials, components, and finished products (free of hazards that will ultimately affect the overall safety of products sold in the marketplace) are obtained from suppliers, including the following: (a) preparation or purchase documents with clear and precise statements regarding safety requirements (adherence to state-of-the-art product safety techniques), (b) selection of suppliers with proven ability to provide safe products, and (c) the unequivocal understanding regarding the responsibilities of suppliers to declare and report substantial product hazards.

(vii) Production control: Production control involves the control of production processes to assure product safety. Special concerns should include assurance that (a) raw materials, semi-finished, or finished materials conform to required specifications, (b) work instructions affecting product safety are described in writing, including inspection and test procedures, (c) the precision and accuracy of equipment and tooling meet established production tolerances and quality (including appropriate equipment tests and inspections), (d) out-of-spec parts or finished products are properly handled, (e) work environments enhance products safety, and (f) finished work is safely handled and stored to preclude damage and the introduction of safety hazards.

(viii) Quality control: This step includes proper and adequate product sampling, inspection, testing, and repair or segregation of non-conforming (unsafe) products to remove unsafe products from the production process.

(ix) Distribution: This step includes the design and selection of proper packaging and product distribution methods to prevent hazards from being introduced during the shipping and handling process. This also includes the provision, where necessary, of current and adequate product assembly and test instructions.

(x) Customer service: Customer service includes the development of various manuals and warnings notices (including product labels) that provide- (a) adequate product assembly and test instructions, (b) safe operating and maintenance instructions, and (c) adequate warnings concerning specific product hazards. Service after-the-sale includes the provision of up-dated product hazard information to previous product buyers (current users).

(xi) Records: An effective product safety program requires the (pro-active) compiling and keeping of records in sufficient detail and format to permit timely detection of safety hazards and trends. This includes the maintenance of records related to (a) safe product design, (b) production quality control (results of tests and inspections), (c) consumer complaints, comments, or accidents, (d) actions taken to correct product deficiencies, and (e) the tracking of products within the production and distribution (and Ownership) chain to facilitate potential product recall (or issuance of updated product warnings).

(xii) Audits: This involves the planned formal examination of product safety procedures and operations to ensure that the overall product safety program (established policies and procedures) is being properly implemented.

11.2 PRODUCT SAFETY MANAGEMENT SYSTEM (PSMS)

The PSMS serves multiple purposes, and must address the several phases of the product life cycle:

(i) Ensure the product development process includes both implicit and derived requirements for the Product (reflecting how it is used, maintained, and disposed of).

(ii) Ensure the product development process can be applied safely (as regards physical activities of manufacture, integration, test, disposal).

(iii) Ensure that what is built is what was intended.

(iv) Contribute to maintaining the intended level of safety for the delivered product.

(v) The PSMS must provide the means to address and resolve.

(vi) Cultural and organizational issues.

(vii) Hierarchies of governance & influence both internal and external to the organization, and.

(vii) The tendency of all systems to move towards a less safe state.

11.2.1 Principles of Product Safety

The Product Safety Principles are discussed below:

11.2.1.1 Accountability

We shall work with our customers and others to ensure that there is, at all times through the life of every Product, accountability for its unintended effects on the safety of people:

(i) We are and remain accountable for those aspects of our Products that are under our control or for which we are legally responsible;

(ii) We shall make reasonable efforts to maintain accountability when we no longer have control of, or responsibility for, our Products; and

(iii) The Company's accountability will be delegated to individuals.

(iv) Accountability – establishing accountability, including delegations of authority = an Actuator

11.2.1.2 Level of Safety

(i) We shall work with each customer to agree the level of safety that is to be achieved by each Product through its life.

(ii) We shall seek the highest level of safety of those who might be unintentionally harmed by the Product that is compatible with the Product's required performance, cost and schedule and the way that it will be used.

(iii) Level of safety – represents the identification of customer needs and required functionality (in terms of safety provisions and features) including derived requirements and requirements originating in other controllers –an actuator

11.2.1.3 Conforming product

We shall ensure that our products conform to their definition:

(i) With internal and, where necessary external, approvals for the organization and product.

(ii) By deploying suitably qualified and experienced people; and

(iii) By applying independent assurance.

(iv) Conforming product – represents the assessment of the design/Product to "validate" that the required level of safety has been met via conformance to the established design requirements – a Sensor

11.2.1.4 Learning and sharing information

We shall work with our customers and suppliers through the life of each product to:

(i) Provide topical information on safety so that each customer may determine how the product is used, and obtain information on the use and performance of the product to assess the consequences for safety;

(ii) Understand the cause of significant accidents and incidents involving our products, where appropriate with independent accident investigators, to reduce the probability of recurrence.

(iii) We shall seek to learn from other parts of the company, organizations and domains.

(iv) Learning and sharing information – since it is used for influencing product development and deriving knowledge during the development/deployment & use of existing products = both actuator & sensor

11.3 FIRE

Fire is an integral part of human life system. Under controlled use, it is useful but very destructive when breaks down. It can create great loss to human life and property. Whatever may be the reason, incidents of fire hazards are very common in day to day life. Every year fire takes thousands of lives and destroys property worth crores of rupees all over the country. Fire is an exothermic chemical reaction from which huge amount of energy is liberated in the form of light, heat, sound and energized smoke (Fig. 11.1). Active rapid burning and oxidation process are accompanied by heat, light and poisonous gases/ smoke/ carbon dioxide gas due to combustion. Fire ignites at a hot spot and spreads along the combustible materials to neighboring areas subject to availability of: combustible materials, air and ignition temperature.

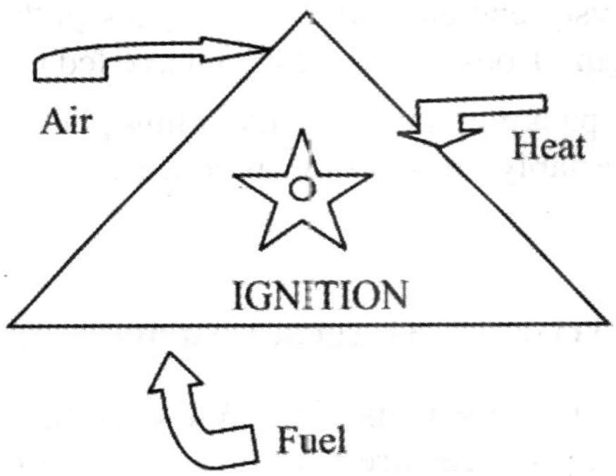

Fig. 11.1 Fire Triangle

11.3.1 Classification of Fire

According to the flash point and auto ignition temperature of combustible materials, fires are classified into four major groups:

(i) **Class A**: Fire involving materials normally of an organic nature in which combustion occurs with the formation of glowing embers where the cooling effect of water is essential for extinction of fires e.g. wood, paper, cloth, rubber, plastics etc.

(ii) **Class B**: Fire involving flammable liquids e.g. oils, solvents, petroleum products, varnishes, paints etc.

(iii) **Class C**: Fires involving gaseous substances under pressure where it is necessary to dilute the burning gas at a very fast rate with an inert gas or powder.

(iv) **Class D:** Fires involving metals, where the burning metal is reactive to water and which requires special extinguishing media or technique, e.g. magnesium, aluminum, zinc, potassium etc.

11.3.2 Causes of Fire

The principal causes of outbreak of fire are:

(i) Careless smoking / bad housekeeping

(ii) Electrical spark / electrical faults

(iii) Naked light / flame heaters / sigris inside the building

(iv) Failure of equipments / wrong operation of equipment

(v) Ignored safety devices

(vi) Over heating / friction / lightening etc.

(vii) Spontaneous ignition (of oil/paint smeared cotton waste thrown underneath the stacks after cleaning/ maintenance work)

(viii) External sparks

(ix) Welding/ cutting, spray painting etc. within the building

(x) Use of highly combustible materials like cork, expanded foam insulations (of polyurethane, polystyrene etc.) In cold storages (although they operate in very low temperatures ranging from 50°C to 20°C), have led to major fires

(xi) Large amounts of packing materials (including plastic expanded foam), paper, straw etc. which are invariably present in warehouses.

11.3.3 Fire Prevention

11.3.3.1 The Fire Prevention Triangle – Oxygen, Heat and Fuel

A fire needs three elements - heat, oxygen and fuel. Without heat, oxygen and fuel a fire will not start or spread. A key strategy to prevent fire is to remove one or more of heat, oxygen or fuel. The risk assessment should include detail on all three elements to minimise the risk of a fire starting/ spreading. A fire prevention strategy and a fire risk assessment should include detail and a full consideration of all of the issues - including issues arising from heat, oxygen and fuel.

11.3.3.2 Oxygen

Nearly all combustion processes require the presence of oxygen. Futhermore, the higher the oxygen concentration, the more rapidly will the burning process be. Inerting is often used to reduce the concentration of oxygen to a safe concentration. This process involves the addition of an inert gas, usually nitrogen or carbon dioxide. Sometimes, steam is also used.

11.3.3.3 Heat

Burning is an exothermic process. The very small fire started by a tiny heat source supplies to its surroundings more heat than it absorbs, thus enabling it to ignite more fuel and oxygen mixture. The combustion started will then propagate to initiate more fires. The various sources of ignition are open flames, electric sources, hot surfaces, spontaneous ignition, sparks, static electricity and friction. Ensuring that there is no contact of the heat source to possible flammable fuel-oxygen mixture can prevent the occurrence of a fire.

11.3.3.4 Fuel

Combustion takes place most readily between oxygen and a fuel in its vapour or other finely divided state. Solids are most easily ignited when reduced to powder or vaporized by the application of heat. For liquids, some will give off dangerous quantities of flammable vapors at below room temperature. Preventing the vaporization of a flammable chemical and its accumulation to form dangerous concentration are the two principles of fire prevention.

11.3.4 Preventive Measures Against Fire

(i) Flammable liquids should be handled in well- ventilated areas.

(ii) Effective ventilation should be applied to prevent accumulation of flammable aerosols or vapors.

(iii) Flammable liquids should be stored in safety containers. The safety containers should be properly labeled and kept in a properly constructed fireproof store when not in use.

(iv) Grounding wires should be used when transferring a flammable solvent from one container to another. This would prevent generation of static electricity, which could ignite the vapors.

(v) Safety devices such as automatic temperature control sensors could be used to warn operators of overheating in process vessels so that appropriate measures can be promptly applied.

(vi) Spark proof or non-sparking tools and materials should be used in areas where flammable materials are likely to be present.

(vii) Regular and thorough housekeeping should be practiced to minimize flammable dust accumulation.

(viii) Smoking, welding, flame cutting and other hot work should be prohibited where flammable materials are stored or handled.

(ix) All used flammable liquid soiled materials including oily rags should be disposed of in a container placed at a designated area. Smoking should also not be allowed in this area.

11.3.5 Methods of Fire Fighting

(i) **Cooling method**: Cooling of burning material by application of water.

(ii) **Smothering method**: Covering the burning material to cut-off the supply of air.

(iii) **Starvation method**: Removal of material from zone of fire.

11.3.6 Fire Extinguishers

Fire fighting is the process of extinguishing the fires. It aims at:

(i) Extinguishing fire

(ii) Stop spreading of fire and smoke/poisonous gases

(iii) Saving life

(iv) Saving property

(v) Minimizing the risks to adjacent buildings and premises.

Fire fighting has the following actions of defences:

(i) Immediate use of portable fire extinguishers

(ii) Use of fire hydrant and its accessories

(iii) Use of automatic fire extinguishing systems installed in the plant

(iv) Call fire brigade simultaneously

(v) Leave the location.

11.3.6.1 Types of fire extinguisher

11.3.6.1.1 Water fire extinguishers

The cheapest and most widely used is **fire extinguishers**. Used for Class A fires. Not suitable for Class B (Liquid) fires, or where electricity is involved.

11.3.6.1.2 Foam fire extinguishers

It is more expensive than water, but more versatile. It is used for Classes A & B fires. Foam spray extinguishers are not recommended for fires involving electricity, but are safer than water if inadvertently sprayed onto live electrical apparatus.

11.3.6.1.3 Dry powder

It is often termed the multi-purpose extinguisher, as it can be used on classes A, B & C fires. It is best for running liquid fires (class b). It will efficiently extinguish class C gas fires, but beware; it can be dangerous to extinguish a gas fire without first isolating the gas supply. Special powders are available for class D metal fires.

For metal fires- A specialist fire extinguisher for use on Class D fires - metal fires such as sodium, lithium, manganese and aluminium when in the form of swarf or turnings.

Warning: When used indoors, powder can obscure vision or damage goods and machinery. It is also very messy.

Carbon dioxide is ideal for fires involving electrical apparatus, and will also extinguish class B liquid fires, but has no post fire security and the fire could re-ignite

11.3.6.1.4 Colour coding

Prior to 1st Jan 1997, the code of practice for fire extinguishers in the UK was BS 5423, which advised the colour coding of fire extinguishers as follows:

(i) **Water – Red, Foam – Cream, Dry Powder – Blue, Carbon Dioxide** (CO_2) - **Black**

(ii) **Halon-Green** (now 'illegal' except for a few exceptions such as the Police, Armed Services and Aircraft).

Note: New extinguishers should conform to BS EN 3, which requires that the entire body of the extinguisher be coloured red. A zone of colour of up to 5% of the external area can be used to identify the contents using the old colour coding shown above.

11.4 ELECTRICAL SAFETY

The use of electricity is essential in various industrial and other activities. The incidents of electric shocks and fires are on rising path. Awareness, education and training, the effective safety management has become essential during every stage of project installation, testing, commissioning and operation / maintenance of electrical plants and appliances. By adequate precautions, the risks of electrical accidents can be minimized and total safety can be ensured.

11.4.1 Flow of Electricity

Electricity flows more easily through some materials than others. Some substances such as metals generally offer very little resistance to the flow of electric current and are called **conductors**. A common but perhaps overlooked conductor is the surface of the earth. Glass, plastic, porcelain, clay, pottery, dry wood and similar substances generally slow or stop the flow of electricity. They are called **insulators**. Even air, normally an insulator, can become a conductor, as occurs during an arc or lightning stroke.

11.4.2 Electrical Hazards

The primary hazards of electricity are discussed below:

(i) Electric shock

(ii) Electric fire and explosions

(iii) Electric flash

(iv) Electric burns

The secondary hazards are:

(i) Persons falling from height

(ii) Dropping of tools and objects.

11.4.3 Causes of Electric Shocks

Electricity travels in closed circuits, normally through a conductor. But sometimes a person's body – an efficient conductor of electricity – mistakenly becomes part of the electric circuit. This can cause an electrical shock. Shocks occur when a person's body completes the current path with:

(i) Both wires of an electric circuit

(ii) One wire of an energized circuit and the ground

(iii) A metal part that accidentally becomes energized, for example, due to a break in its insulation, or

(iv) Another 'conductor' that is carrying a current

When a person receives a shock, electricity flows between parts of the body or through the body to the ground or the earth.

11.4.4 Effects of Electric Current on Human Body

An electric shock can result in anything from a slight tingling sensation to immediate cardiac arrest. The severity depends on the following (Table 11.1):

(i) The amount of current flowing through the body.

(ii) The current's path through the body.

(iii) The length of time the body remains in the circuit.

(iv) The current's frequency.

Table 11.1 Effects of Electric Current on Human Body

Current	Reaction
Below 1 milliampere	Generally not perceptible
1 milliamperes	Faint tingle
5 milliamperes	Slight shock felt; not painful but disturbing
6-25 milliamperes (Women)	Painful shock, loss of muscular control. If the extensor muscles are excited by the shock, the person may be thrown away from the power source
9 – 30 milliamperes (Men)	The freezing current or 'let – go' range. If the extensor muscles are excited by the shock, the person may be thrown away from the power source
50 – 150 milliamperes	Extreme pain, respiratory arrest, severe muscular contractions, Death is possible
1,000 – 4,300 milliamperes	Rhythmic pumping action of the heart ceases, muscular contraction and nerve damage occur; death likely
10, 000 milliamperes	Cardiac arrest, severe burns, death probable

11.4.5 Kinds of Burns Caused by a Shock

Burns are the most common shock-related injury. An electrical accident can result in an electrical burn, arc burn, thermal burn or a combination of burns.

(i) **Electrical burns** are among the most serious burns and require immediate medical attention. They occur when electric current flows through tissues or bone, generating heat that causes tissue damage.

(ii) **Arc or flash burns** result from high temperature caused by an electric arc or explosion near the body. These burns should be treated promptly.

(iii) **Thermal burns** are caused when the skin touches hot surfaces of overheated electric conductors or other energized equipment. Thermal burns can also be caused when clothing catches fire, as may occur when an electric are is produced.

In addition to shock and burn hazards, electricity poses other dangers. For example, arcs that result from short circuits can cause injury or start a fire. Extremely high – energy arcs can damage equipment, causing fragmented metal to fly in all directions. Even low-energy arcs can cause violent explosions in atmosphere that contain flammable gases, vapours or combustible dusts.

11.4.6 Protection against Electrical Hazards

Most electrical accidents result from one of the following three factors:

(i) Unsafe equipment or installation

(ii) Unsafe environment

(iii) Unsafe work practices

Some ways to prevent these accidents are through the use of insulation, guarding, grounding, electrical protective devices and safe work practices.

11.4.7 Unsafe Equipment & Environment

Insulators such as glass, mica, rubber or plastic used to coat metals and other conductors help stop or reduce the flow of electrical current. This helps prevent shock, fires and short circuits. To be effective, the insulation must be suitable for the voltage used and conditions such as temperature and other environmental factors like moisture, oil, gasoline, corrosive fumes or other substances that could cause the insulator to fail.

11.4.8 Unsafe Work Practices

Electrical accidents are largely preventable through safe work practices. Examples of these practices include the following:

(i) Denergizing electric equipment before inspection or repair.

(ii) Keeping electric tools properly maintained.

(iii) Exercising caution when working near energized lines.

(iv) Using appropriate protective equipments.

11.4.9 Control Measures for Electric Safety

(i) All electrical equipment and installations designed, constructed, installed, maintained, protected, and used to prevent danger

(ii) Get a qualified electrical contractor to carry out installation and repairs to electrical equipment and fittings

(iii) Maintain proper pest control to avoid rodent damage to electric wiring and equipment

(iv) Check electrical equipment and remove defective equipment.

(v) Ensure electrical cords are in good condition.

(vi) Plug appliances and lights into separate electrical outlets.

(vii) Avoid using extension cords. If you require an outlet in an area where there is none, have one installed by a qualified electrician.

(viii) Use extension cords safety - not under carpets or across walking areas.

(ix) Use only one device per outlet.

Review Questions

1. How to prevent fire hazard?

Ans. Some steps to prevent fire hazard are separate storage of inflammable / explosive materials, following of precaution during storage transport of inflammable material, special precautions during welding etc.

2. What are hazard of electricity?

Ans. The hazard of electricity includes electric shock, electric flashover, fire hazard, explosion in electric equipments etc.

■■■■

Chapter 12

Safety Handling and Storage of Hazardous Materials

12.1 INTRODUCTION

Material handling is the preparation, placing and poisoning of materials to facilitate their movement or storage. It is interesting to note that:

(i) About 40% of industrial accidents are in material handling.

(ii) Nearly 50-100 tons of materials are handled and re-handled for every ton of finished product.

(iii) Material handling accounts for 36% of production cost and about 20 to 80% of the total labour cost goes to labour used in material handling.

(iv) About 2/3 of the manufacturing cycle time is spent in handling.

(v) Material handling increases cost of the product without adding any value.

Every type of material handled in a factory must be studied in detail and consideration given to factors such as:

(i) The weight handled

(ii) Character of material handled

(iii) Size

(iv) Rate of handling

(v) Distance moved

(vi) Purpose of moving or handling, etc and suitable methods of handling decided upon.

12.1.1 Material handling can be broadly classified as

(i) Manual handling

(ii) Mechanical handling

12.1.1.2 Manual handling

(i) Lifting, pulling or pushing.

(ii) Dragging or sliding on skids or rollers.

(iii) Use of wheel barrows, trolley, hand truck.

12.1.1.3 Mechanical handling

The mechanical equipments are classified as:

(i) **Hosting equipment:** e.g. jack, pulley, cranes, elevators, winches, hand or power operated hoists.

(ii) **Conveying equipment**: e.g. belt and metallic conveyors. Pipelines, pneumatic/hydraulic devices.

(iii) **Surface and overhead equipment**: e.g. bulldozers, scrapes, narrows gauge cars, power trucks.

The basic fact that the unsafe acts of a person are causative factors in the majority of accidents, justifies the substitution of mechanical for manual operations wherever practicable. This statement implies of course, that the mechanical means used must be of safe design and construction, suited to their purpose, thoroughly guarded, properly maintained and that the personnel selected to operate them are properly qualified and adequately supervised.

12.1.1.2.1 Manual handling – accident prevention

The reasons for accidents in manual handling are:

(i) Lifting, pulling or pushing improperly.

(ii) Carrying too heavy a load.

(iii) Unsafe gripping.

(iv) Lack of co-ordination in team work.

(v) Failure to wear personal protective equipments.

Accidents can be prevented by good training, effective supervision, and periodic re-examination to check physical weakness.

12.1.1.2.2 Manual proper method of lifting

(i) The feet should be placed close to the load and properly spaced for.

(ii) Body balance.

(iii) Keep back straight and as nearly vertical as possible.

(iv) Keep elbows as straight as possible.

(v) Knees bent until the hands reach the proper place for gripping the load.

(vi) Grasp the load firmly, and then lift should be completed by straightening the knees, keeping the load close to the body.

(vii) While putting down the load, the above procedure should be followed in reverse order.

12.1.1.2.3 Manual proper method of pulling

(i) Grip the rope, palm upwards; relax knees, lower hands, elbows into body

(ii) Placing rear foot well back makes effective use of body weight and pull.

12.1.1.2.4 Manual proper method of pushing

(i) Grip the handle, turn palms inwards

(ii) Relax knees, draw elbows into body and push

This gives better thrust and less strain.

Practicing correct way of handling, lifting, pulling and pushing leads to

(i) Right use of the muscles for the job

(ii) Avoiding strain and injury

(iii) Less fatigue

(iv) The correct way or the easiest and safest way

12.1.1.2.5 Safe Handling of specific type of Articles

(i) **Boxes and cartons:** Grasp them at opposite top and bottom corners, draw a corner between the legs.

(ii) **Barrels and drums:** Two men stand on opposite side grasp both chins near the high point up end, press down the bottom end and straighten up with the drum.

(iii) To handle on incline, make use of ropes and tackles. To control motion, pass a rope around a drum, one end of rope fastened to the platform at higher level. One worker keeps a firm grip on the free end and then gradually lowers or raises the drum along the rickers.

(iv) **Sheet metal:** Handle with leather gloves or hand leathers, cover wrists and forearms with long leather sleeves. Leather or canvas apron and guards for feet and ankles should also be worn.

(v) **Long objects:** Long pipes, bars, lumber should be carried over the shoulders, with the front end held as high as possible to prevent striking other persons especially at corners.

(vi) **Metal scrap:** Wear goggles, gloves or hand leathers, safety shoes and shin guards. Person should be cautioned against tripping or slipping on objects which may roll or slide under the feet.

(vii) **Gas cylinders:** Cylinders should be carried in a cradle or in a suitable type of carrying device. Cylinders may be rolled on the bottom edge but never be dragged or rolled on the ground.

12.1.1.2.6 Accessories for Manual Handling

(i) **Hooks:** Danger of glancing off hard objects. When carried in a belt, the point should be covered to prevent injury.

(ii) **Crow bar:** Main hazard is that of it slipping. Never work astride a crow bar. Care should be taken to properly position hand and body to minimize chances of hand pinching or falling if the crow bar slips.

(iii) **Rollers:** Main hazard is the possibility of the fingers or toes getting pinched or crushed between the roller and the floor or the roller and the object, when the direction of the roller is changed. To move a roller under load, use a sledge or bar, never the hand or foot. Use safety shoes with a steel toe cap.

12.1.1.2.7 Safety in the Use of Wheel Barrows, Hand Trucks and Trolleys

(i) Inspect daily and keep in good repair.

(ii) On a two wheeler, center of gravity of load should be placed well forward.

(iii) Provide guards to protect hands from jamming against door frames or other obstructions.

(iv) Load should be so placed that it will not tilt, shift, fall off or block clear view ahead.

(v) Wheel should be under the truck if possible to save injury to toes and feet.

(vi) Provide brakes to avoid holding a truck with a foot on wheel or on axle.

(vii) Avoid collisions especially at corners.

(viii) Do not park trucks in aisle ways so as to obstruct traffic or cause stumbling hazards.

(ix) Special care should be taken to see that the equipment is not striking against a ladder.

12.1.1.3 Mechanical handling

12.1.1.3.1 Safety in hosting equipment

The Factories Act 1948, Section 29 deals with the safety requirements in respect of lifts, hoists, lifting tackle, lifting machines, chains and ropes. It should of good construction, sound material and of adequate strength. These are to be properly maintained and thoroughly examined by a competent person at least once in six months in the case of hoists and lifts and 12 months in the case of other and a register shall be kept containing the prescribed particulars of every such examination.

Lifting machine means a crane, crab, winch, pulley block, gin wheel, transporter or runway. Lifting tackle means chain slings, rope slings, hooks, shackles and swivels.

12.1.1.3.2 Hooks

The hooks should be withdrawn from service if –

(i) The wear at the load point is more than 10%.

(ii) If the throat opening has increased by 15% in comparison to the standard throat opening.

A special load locking attachment may also be fitted as a safety measure to prevent slings coming out accidentally from the hook and load falling down.

12.1.1.3.3 Chains, Steel Wire Rope, Fiber Rope

(i) The examination of these is identical with that for slings.

(ii) Care should be taken that while passing rope over a pulley, the diameter of the pulley is not less than the following dimension.

(iii) Diameter of pulley – not less than 20 times diameter of wire rope; 10 times diameter of fiber rope; 30 times diameter of chain in case of power drive.

12.1.1.3.4 Conveyors

Various types of conveyors and monorail systems are used in many industries to eliminate manual labour, to expedite the movement of materials and also to facilitate the processing or assembling. But conveyors are widely used for handling a large variety of materials. Some important safety measures to be taken are as follows:

(i) Conveyors should be so constructed and installed as to avoid hazardous point between moving and stationary parts or objects.

(ii) Where workers have to cross over conveyors, regular crossing facilities, affording safe passage and adequate lighting should be provided.

(iii) Conveyors should be provided at loading and unloading stations and other convenient places with devices for stopping the conveyor in an emergency.

(iv) When two or more conveyors are operated together, controlling device should be so designed that no conveyor can feed on to a stopped conveyor.

(v) Where conveyors extend to points not visible from the control station, they should be equipped with signaling devices to be used by the operators before starting the machinery so as to warn workers, who might be in a dangerous position.

(vi) Conveyors should be provided with automatic and continuous lubrication system or with lubricating facilities so arranged that oiling and greasing can be performed without the oiler coming into dangerous proximity of the moving parts.

(vii) Persons should not be allowed to ride on conveyors.

(viii) While carrying out maintenance work, electrical permit should be taken.

12.1.1.3.5 Power trucks

Power truckers are extensively used in store houses of factories for handling of materials. The trucks are of many types eg. fixed platform, elevating platform, fork lift, etc. the fork lift truck makes the lift by means of two strong forks and lifts the load up from the floor permitting high piling to conserve space. They can also be fitted with special attachments for handling barrels, paper reels, hot billets and a variety of materials.

Following safety precautions should be observed in operation of fork lift trucks:

(i) The capacity of every lift truck should be marked on the truck and they should not be overloaded.

(ii) Loaded or empty forks should be carried as low as possible but high enough not to strike a ramp.

(iii) Care must be taken to avoid jerking while tilting a load forward or backward, especially when the load is at a height.

(iv) Persons should not be allowed to ride on the forks.

(v) Forks should be driven well under the load, preferably full length or at least 2/3 of the length.

(vi) Canopy guard should be provided if there is a danger of objects falling over the operator.

(vii) Aisles, floors should be maintained in good condition including proper lighting.

12.1.1.3.6 Safety in material storage

The reasons for accidents in materials storage are mainly due to improper stacking and destacking of materials. All stacking should be properly planned to eliminate the hazards and congestion, which accompany haphazard or chance methods. The main factors to be considered are:

(i) The most suitable layout of sites and gangways and defining of these sites.

(ii) The routing of materials in and out.

(iii) The method of stacking to be used.

(iv) Handling facilities required or available.

(v) Provision of equipment, ladders, pallet boards, frames, etc.

(vi) The lighting of the area. This should be such that dark shadows are avoided in the working area of each stack.

(vii) The organization of effective means to ensure that the scheme approved is used.

12.1.1.3.7 Forms of stack

The different types of stacks are classified as follows –

(i) **Column** – A stack comprised of single articles placed one above the other.

(ii) **Square** – Any stack other than a column, which has all its sides vertical.

(iii) **Pyramid** – A stack in which the plan area is reduced in every succeeding tier, regardless of whether it finally reaches an apex or not.

(iv) **Stepped** – A stack in which two, three or more adjacent tiers have the same area and each succeeding group of tiers has a smaller area than the group on which it is **lean – to – stacks** which are in pyramid or stepped form on one side and vertical on the other side.

(v) **Spiral** – A method of close stowing articles of similar form but of shapes not suited to any of the previously mentioned methods.

12.1.1.3.8 Precautions to be taken while stacking materials

(i) Floors used for stacking should be of sound construction and free from structural weakness and leveled.

(ii) Stacks should not be built up within 3 feet of the edge of concrete rafts.

(iii) Stacks should not be built within 1.5 feet of a wall. The leaning of materials against walls should be prohibited.

(iv) Materials must not be stacked in any position where they will provide a footing for anyone within 8 feet of unfenced machinery.

(v) No stack should be erected where any part of it will be within 4.5 feet of a rail track.

(vi) Careful consideration should be given to the stacking of materials in the vicinity of machinery or operation which cause vibrations.

(vii) The aggregate weight to be borne by the articles in the lowest tier should be taken into account, to decide height of stacking.

(viii) If the shape and strength of the goods are suited to self supporting stacking, stability will be ensured, if the height does not exceed three times the narrower width of the base, when all other considerations are favorable.

(ix) The stacks should be tidy. Untidy stacks offend the eye and are rarely safe by construction.

12.1.1.3.9 Stacking specific types of goods

(i) **Bagged goods** – These should be stacked in cross tiers with mouths inside. When the pile is 5 feet high, step back by one row for each additional 3 feet.

(ii) **Pipe and bar stock** – Small size bars and tubes should be stored in racks inclining towards the back. Large size pipes can be stacked in layers with strips of wood in between layers. Very big size pipes can be stacked in pyramid shape with secured wedge at the end.

(iii) **Barrels and kegs** – These should be stacked in pyramid shape with secured wedging. Pyramid shape is safer. When piling on ends plank should be laid between rows.

(iv) **Sheet materials** – Flat stacking should be employed. Sheets having a width of 18 inches or less should be stacked to heights not more than 3 times their width. Others may be stacked safely up to 4 times their width. Removable or fixed retaining structures may be used as an additional safety measure.

(v) **Rolled steel joists** – These should be stacked on their flanges in pyramid style if their dimensions are such that the ratio of flange width to depth does not exceed 1:2. When the flange width / depth of web ratio exceed 1:2, such joists should be laid with the web horizontal and with bearing spacers between the tiers.

12.2 CORROSIVE SUBSTANCES

The word corrosion is derived from the Latin verb corrodere which means to gnaw. A corrosive substance is one that will destroy or irreversibly damage another surface or substance with which it comes into contact (Table 12.1). These are substances which by chemical action, will cause severe damage when in contact with living tissue or in case of leakage, will materially damage or even destroy other goods or the means of transport; they may also cause other hazards.

Corrosive chemicals can be solids e.g. sodium hydroxide, liquids e.g. hypochlorite solution or gases e.g. chlorine and ammonia. Some chemicals become corrosive when they come into contact with water or moisture e.g. benzyl chloride and chlorosilanes on contact with sweat on the skin.

12.2.1 Classification

Table 12.1 Corrosive chemicals commonly used in industry may be grouped into the following classes:

Group of Corrosive Chemicals	Examples
Acids and anhydrides	Sulphuric acid, hydrochloric acid, nitric acid, acetic acid, acetic anhydride, phosphoric acid, phosphorous trioxide
Alkalis or bases	Potassium hydroxide, sodium hydroxide, organic amines such as ethanolamine
Halogens, halogen salts, organic halides	Chlorine gas, ferric chloride, chlorite solutions, acetyl iodide
Other corrosive substances	Ammonium polysulphide, peroxides, hydrazine

Acids can be classified as inorganic or mineral acids and organic acids. Examples of mineral acids are sulphuric acid, hydrochloric acid and nitric and examples of organic acids are acetic acid, ascorbic acid and salicylic acid.

Anhydrides are substances which have properties similar to those of acids. They react with water forming the corresponding acids e.g. acetic anhydride reacts with water to form acetic acid.

12.2.1.1 Properties of Acids and Alkalis

(i) Acids and alkalis (bases) have a common property; they are corrosive. Besides attacking living tissue, they also attack many other materials. They react with metals to produce hydrogen which is highly flammable.

(ii) Acids and bases may have toxic properties and some are also flammable. Some acids like nitric acid and hydrochloric acid, release highly corrosive vapours at room temperature when in concentrated form.

(iii) When an acid and a base are mixed, they neutralise each other producing a salt and water which also produces heat.

(iv) Mixing an acid or a base with water also produces heat. This may create hazards, such as splashes and formation of dangerous mists when water is added to a concentrated acid.

12.2.1.2 Acidity and alkalinity

The pH is the scale used to compare the strength or level of acidity and alkalinity of acids and bases and their diluted solutions.

The pH is related to the amount of hydrogen ions present in the solution. It is expressed on a scale from 0 to 14. The pH of a neutral solution is 7. Acids have a pH of <7 and bases of >7. The limits of acidity and alkalinity may be classified as follows:

pH	0 to 2	strongly acidic pH
	3 to 5	weakly acidic pH
	6 to 8	neutral
pH	9 to 11	weakly basic pH
	12 to 14	strongly basic

Acids with pH values 0 to 2 and bases with pH values 11.5 to 14 may be classified as corrosive.

12.2.2 Health Effects

The health effects of direct contact with acids or alkalis vary from irritation causing inflammation to a corrosive effect causing ulceration and in severe cases, chemical burns. The effects depend on the properties of the acids and alkalis, the concentration and time of contact with these substances.

Some acids and alkalis produce heat when they come into contact with water or moisture. These chemicals may cause both corrosive injuries and burns due to the heat produced. Occupational hazards due to contact with acids or alkalis mostly affect the skin, eyes and respiratory tract. The eyes are most susceptible to rapid, severe and often irreversible damage.

Acid fumes may also corrode the teeth e.g. long-term exposure to low levels of hydrochloric acid fumes can result in erosion of the incisor teeth. The effect of strong acids and alkalis is experienced within moments of exposure. The effect may also be delayed, depending on the substance and the concentration. For example, the effect of diluted hydrofluoric acid may vary from irritation to severe burns of the skin depending on the concentration and duration of exposure. Direct contact of organic anhydrides with the eyes, skin, mucous membranes, or the respiratory system causes irritation and sensitization.

12.2.3 Storage Containers

(i) The construction materials for tanks and containers for storage of acids and alkalis must be able to resist corrosion from inside and outside. These materials must not react with the contents and impurities.

(ii) Steel is the most common construction material for corrosion-resistant tanks and containers. Carbon steel is not resistant to strong acids and elevated temperatures. Aluminum tank containers should not be used for acids or alkalis.

(iii) Reinforced plastic is light and chemically resistant to mineral acids. This material may be used at temperatures up to 80°C. The corrosive effect of an organic acid should be checked before putting it into a reinforced plastic tank.

(iv) Thermoplastic materials such as polyvinyl chloride (PVC), polyethylene (PE), polypropylene (PP) and polytetrafluorine ethylene (PTFE) are usually chemically resistant to acids and alkalis.

12.3 GAS CYLINDERS

Gas cylinders can cause serious injury or even death. Gas cylinders used in adverse or extreme conditions, such as under water, may require special precautions.

12.3.1 Uses of Gas Cylinders

Gas cylinders are a convenient way to transport and store gases under pressure. These gases are used for many different purposes including: chemical processes, soldering, welding and flame cutting; breathing (eg diving, emergency rescue), medical and laboratory uses, dispensing beverages, fuel for vehicles (e.g. fork-lift trucks), extinguishing fires, heating and cooking, water treatment. The main hazards are:

(i) Impact from the blast of a gas cylinder explosion or rapid release of compressed gas.

(ii) Impact from parts of gas cylinders that fail, or any flying debris.

(iii) Contact with the released gas or fluid (such as chlorine).

(iv) Fire resulting from the escape of flammable gases or fluids (such as liquefied petroleum gas).

(v) Impact from falling cylinders.

(vi) Manual handling injuries.

12.3.2 The Main Causes of Accidents

Inadequate training and supervision, poor installation, poor examination and maintenance, faulty equipment and/or design (e.g. badly fitted valves and regulators), poor handling, poor storage, inadequately ventilated working conditions, incorrect filling procedures, hidden damage.

All gas cylinders must be designed and manufactured to an approved specification to withstand everyday use and to prevent danger. They must be periodically examined at appropriate intervals to ensure that they remain safe in service. To reduce the risks of failure we need to know, and act on, the following precautions.

12.3.2.1 Training

Anyone who examines, refurbishes, fills or uses a gas cylinder should be suitably trained and have the necessary skills to carry out their job safely. They should understand the risks associated with the gas cylinder and its contents.

12.3.2.2 Filling

Anyone carrying out the filling of gas cylinders should wear suitable personal protective equipment as appropriate. This may include safety shoes, protective overalls, gloves, and ear and eye protection. Before filling a gas cylinder it should be checked that:

(i) From the markings on the cylinder, it has been properly examined by a relevant inspection body and is still within its due test date.

(ii) It shows no sign of damage, external corrosion, falsification (defacing) of markings or illicit repairs that may affect integrity.

(iii) Valves, fittings and regulators (where fitted) are:

 (a) correctly fitted and not leaking

 (b) not damaged and in good working condition

 (c) suitable for their intended purpose

 (d) not contaminated, e.g. with incompatible lubricants.

If any of the above conditions are not met, then the cylinder must not be filled. After filling a gas cylinder it should be checked that:

(i) It is within its safe operating limits

(ii) It is not overfilled or over pressurised. In the event of inadvertent overfilling, any excess gas must be removed in a safe manner and the cylinder checked for further fitness for service

(iii) The cylinder's valves, fittings and regulators are not leaking, for example by using special equipment, such as 'sniffers' or manometers.

(iv) If appropriate, the simple method of using a soapy water solution can be adopted.

12.3.2.3 Handling and use

(i) Use gas cylinders in a vertical position, unless specifically designed to be used otherwise

(ii) Securely restrain cylinders to prevent them falling over

(iii) Always double check that the cylinder/gas is the right one for the intended use

(iv) Before connecting a gas cylinder to equipment or pipe work make sure that the regulator and pipe work is suitable for the type of gas and pressure being used

(v) When required, wear suitable safety shoes and other personal protective equipment when handling gas cylinders

(vi) Do not use gas cylinders for any other purpose than the transport and storage of gas

(vii) Do not drop, roll or drag gas cylinders

(viii) Close the cylinder valve and replace dust caps, where provided, when a gas cylinder is not in use

(ix) Where appropriate, fit cylinders with residual pressure valves (non-return valves) to reduce the risk of back flow of water or other materials into the cylinder during use that might corrode it (e.g. beer forced into an empty gas cylinder during cylinder change-over)

12.3.2.4 Lifting

(i) Use suitable cradles, slings, clamps or other effective means when lifting cylinders with a hoist or crane

(ii) Do not use valves, shrouds and caps for lifting cylinders unless they have been designed and manufactured for this purpose

(iii) Gas cylinders should not be raised or lowered on the forks of lift trucks unless adequate precautions are taken to prevent them from falling

12.3.2.5 Transport

(i) Fit suitable protective valve caps and covers to cylinders, when necessary, before transporting. Caps and covers help prevent moisture and dirt from gathering in the valve of the cylinder, in addition to providing protection during transport

(ii) Securely stack gas cylinders to prevent them from moving or falling. This is normally in the vertical position, unless instructions for transport state otherwise

(iii) Disconnect regulators and hoses from cylinders whenever practicable

(iv) do not let gas cylinders project beyond the sides or end of vehicles (e.g. fork-lift trucks)

(v) Ensure gas cylinders are clearly marked to show their contents (including their UN number) and the danger signs associated with their contents

(vi) It may be necessary to take special measures with certain types and quantities of compressed gases and fluids in order to ensure their safe carriage

(vii) the transport of gas cylinders is subject to carriage requirements. For example, that: - the vehicle is suitable for the purpose;

(a) The vehicle is suitably marked to show that it is carrying dangerous goods

(b) The driver is suitably trained

(c) The driver carries the appropriate documentation about the nature of the gases being carried

12.3.2.6 Storage

(i) Store gas cylinders in a dry, safe place on a flat surface in the open air. If this is not reasonably practicable, store in an adequately ventilated building or part of a building specifically reserved for this purpose.

(ii) Gas cylinders containing flammable gas should not be stored in part of a building used for other purposes.

(iii) Protect gas cylinders from external heat sources that may adversely affect their mechanical integrity .

(iv) Gas cylinders should be stored away from sources of ignition and other flammable materials.

(v) Avoid storing gas cylinders so that they stand or lie in water.

(vi) Ensure the valve is kept shut on empty cylinders to prevent contaminants getting in.

(vii) Store gas cylinders securely when they are not in use. They should be properly restrained, unless designed to be freestanding.

(viii) Gas cylinders must be clearly marked to show what they contain and the hazards associated with their contents.

(ix) Store cylinders where they are not vulnerable to hazards caused by impact, e.g. from vehicles such as fork-lift trucks.

The following is based on guidance issued by the health and safety executive (June 2004).

Accidents involving gas cylinders can cause serious injury or even death. This guidance provides simple practical advice on eliminating or reducing the risks associated with using gas cylinders. The guidance is aimed at anyone who manufactures, owns, fills, repairs or uses gas cylinders at work, and especially at those who own or manage small businesses.

12.4 HYDROCARBONS

Organic chemicals that contain only carbon and hydrogen atoms are called hydrocarbons. Organic compounds can be grouped according to the functional groups attached to the carbon skeleton. Some of the major classes of organic chemicals and their effects are as follows:

12.4.1 Alkanes

Alkanes are paraffinic (saturated) hydrocarbons having a general structural formula C_nH_{2n+2} where n is the number of carbon atoms.

Examples: methane (CH_4), butane (C_4H_{10}), hexane (C_6H_{14})

Alkanes are flammable but relatively non-toxic, except n-hexane which is known to be neurotoxic.

12.4.2 Alkenes

Alkenes are olefinic (unsaturated) hydrocarbons having a general formula C_nH_{2n}.

Examples: ethylene (C_2H_4), propylene (C_3H_6). Alkenes are flammable but relatively non-toxic.

12.4.3 Cyclic Hydrocarbons

These are hydrocarbons having a ring structure, saturated or unsaturated with hydrogen.

Example: cyclohexane (C_6H_{12}). Cyclic hydrocarbons are of low toxicity. The un-saturated cyclic hydrocarbons generally are more irritating than the saturated forms.

12.4.4 Aromatics

Aromatic hydrocarbons contain a 6-carbon ring structure. Aromatic compounds are unsaturated.

Examples: benzene [C6H6], toluene [C6H5CH3]. The aromatic hydrocarbons, in general, are irritants and potent narcotics. The main health effects of the common aromatic solvents other than benzene are dermatitis and effects on the central nervous system. Benzene is notorious for its effects on the blood forming tissues of the bone marrow. It is a leukemogenic agent.

12.4.5 Alcohols

Alcohols are characterized by the presence of a hydroxyl group (-OH).

Examples: methanol (CH3OH), ethanol (C2H5OH), propanol (C3H7OH). Alcohols can depress the central nervous system. Exposure to high air levels may lead to unconsciousness and even death. Methanol is especially noted for being able to cause impairment of vision due to injury to the optic nerve.

12.4.6 Aldehydes

Aldehydes contain the double bonded carbonyl group, C=O with only one hydrocarbon group on the carbon.

Examples: formaldehyde (HCHO), acetaldehyde (CH3CHO). The aldehydes are well known for skin and mucosal irritation and their action on the central nervous system. Dermatitis from aldehydes is common. Aldehydes are also characterized by their sensitizing properties. Allergic responses are common.

12.4.7 Ketones

Ketones contain the double bonded carbonyl group, C=O with two hydrocarbon groups on the carbon.

Examples: acetone (CH3COCH3), methyl ethyl ketone (CH3COC2H5). The common ketones generally exert a narcotic effect. All are irritating to the eyes, nose and throat. Methyl n-butyl ketone is especially noted for its effects on peripheral nerves.

12.4.8 Ethers

Ethers contain the C-O-C linkage.

Example: ethyl ether (C2H5OC2H5).

Ethers are volatile hydrocarbons. Their primary effect is anesthetic and irritating.

12.4. 9 Esters

Esters are formed by the reaction of an organic acid with an alcohol.

Example: ethyl acetate (CH3COOC2H5). Esters are noted for their irritating effects on the exposed skin and to the respiratory tract. They are also potent anesthetics.

12.4.10 Glycols

Glycols contain double hydroxyl (-OH) groups.

Example: ethylene glycol [$C_2H_4(OH)_2$]. Glycols are non-volatile due to their low vapor pressures. Inhalation exposures are not likely unless they are heated or sprayed. The mists and vapors are irritating.

12.4.11 Halogenated Hydrocarbons

Halogenated hydrocarbons are compounds with one or more hydrogen atoms replaced by halogen atom(s) – fluorine (F), chlorine (Cl), bromine (Br) or iodine (I).

Examples: carbon tetrachloride (CCl_4), trichloroethylene ($CHCl{=}CCl_2$), methylene chloride (CH_2Cl_2), bromochloromethane ($BrCH_2Cl$). Halogenated hydrocarbons are stable and non-flammable. The chlorinated hydrocarbons, in general, are more toxic than the common fluorinated hydrocarbons. Specific effects and toxicities vary widely, but the most common effects from the chlorinated hydrocarbons are depression of the central nervous system, dermatitis and injury to the liver.

12.5 PERSONAL PROTECTIVE EQUIPMENT (PPE)

It is clothing and equipment worn by employees, students, contractors or visitors to protect or shield their bodies from workplace hazards.

Section 19 of the occupational safety and health act 1984 sets out the obligation of an employer where it is not practicable to avoid the presence of hazards at the workplace, to provide the employees with, or otherwise provide for the employees to have, such adequate personal protective clothing and equipment as is practicable to protect them against those hazards, without any cost to the employees'.

In the hierarchy of controls (elimination, substitution, engineering, administration and PPE), personal protective equipment is considered the least satisfactory method in the prevention of work-related injury or illness and is only to be used when other measures are not feasible or cannot be implemented immediately. PPE should be used, however, to supplement or augment other means of hazard control, to further minimise the risk of injury. When selecting and using PPE:

(i) Choose equipment that suits the user – consider the size, fit and weight of the PPE. If the users help choose it, they will be more likely to use it.

(ii) If more than one item of PPE is worn at the same time, make sure they can be used together, e.g. wearing safety glasses may disturb the seal of a respirator, causing air leaks.

(iii) Instruct and train people how to use it, e.g. train people to remove gloves without contaminating their skin. Tell them why it is needed, when to use it and what its limitations are.

(iv) Never allow exemptions from wearing PPE for those jobs that 'only take a few minutes.

(v) Check with your supplier on what PPE is appropriate – explain the job to them.

(vi) If in doubt, seek further advice from a specialist adviser.

12.5.1 Maintenance

PPE must be properly looked after and stored when not in use, e.g. in a dry, clean cupboard. If it is reusable it must be cleaned and kept in good condition. Think about:

(i) Using the right replacement parts which match the original, e.g. respirator filters

(ii) Keeping replacement PPE available

(iii) Who is responsible for maintenance and how it is to be done

(iv) Having a supply of appropriate disposable suits which are useful for dirty jobs where laundry costs are high, e.g. for visitors who need protective clothing. Employees must make proper use of PPE and report its loss or destruction or any fault in it.

12.5.2 Eyes

(i) Hazards: Chemical or metal splash, dust, projectiles, gas and vapour, radiation

(ii) Options: Safety spectacles, goggles, face screens, face shields

(iii) Note: Make sure the eye protection chosen has the right combination of impact/ dust/splash/molten metal eye protection for the task and fits the user properly.

12.5.3 Head and Neck

(i) Hazards: Impact from falling or flying objects, risk of head bumping, hair getting tangled in machinery, chemical drips or splash, climate or temperature

(ii) Options: Industrial safety helmets, bump caps, hairnets and fire fighters' helmets

(iii) Note:

(a) Some safety helmets incorporate or can be fitted with specially-designed eye or hearing protection.

(b) Don't forget neck protection, e.g. scarves for use during welding.

(c) Replace head protection if it is damaged.

12.5.4 Ears

(i) Hazards: Noise – a combination of sound level and duration of exposure, very high-level sounds are a hazard even with short duration

(ii) Options: Earplugs, earmuffs, semi-insert/canal caps

(iii) Note:

(a) Provide the right hearing protectors for the type of work, and make sure workers know how to fit them.

(b) Choose protectors that reduce noise to an acceptable level, while allowing for safety and communication.

12.5.5 Hands and Arms

(i) **Hazards:** Abrasion, temperature extremes, cuts and punctures, impact, chemicals, electric shock, radiation, vibration, biological agents and prolonged immersion in water.

(ii) **Options:** Gloves, gloves with a cuff and sleeving that covers part or all of the arm.

(iii) **Note:**

(a) Avoid gloves when operating machines such as bench drills where the gloves might get caught.

(b) Some materials are quickly penetrated by chemicals – take care in selection

(c) Barrier creams are unreliable and are no substitute for proper PPE.

(d) Wearing gloves for long periods can make the skin hot and sweaty, leading to skin problems. Using separate cotton inner gloves can help prevent this.

12.5.6 Feet and Legs

(i) **Hazards:** Wet, hot and cold conditions, electrostatic build-up, slipping, cuts and punctures, falling objects, heavy loads, metal and chemical splash, vehicles.

(ii) **Options:** Safety boots and shoes with protective toe caps and penetration-resistant, mid-sole wellington boots and specific footwear, e.g. foundry boots and chainsaw boots.

(iii) **Note**

(a) Footwear can have a variety of sole patterns and materials to help prevent slips in different conditions, including oil- or chemical-resistant soles. It can also be anti-static, electrically conductive or thermally insulating.

(b) Appropriate footwear should be selected for the risks identified.

12.5.7 Lungs

(i) **Hazards:** Oxygen-deficient atmospheres, dusts, gases and vapours

(ii) **Options – respiratory protective equipment (RPE)**

(a) Some respirators rely on filtering contaminants from workplace air. These include simple filtering face pieces and respirators and power-assisted respirators.

(b) Make sure it fits properly, e.g. for tight-fitting respirators (filtering face pieces, half and full masks).

(c) There are also types of breathing apparatus which give an independent supply of breathable air, e.g. fresh-air hose, compressed airline and self-contained breathing apparatus.

(iii) Note

(a) The right type of respirator filter must be used as each is effective for only a limited range of substances.

(b) Filters have only a limited life. Where there is a shortage of oxygen or any danger of losing consciousness due to exposure to high levels of harmful fumes, only use breathing apparatus – never use a filtering cartridge.

(c) You will need to use breathing apparatus in a confined space or if there is a chance of an oxygen deficiency in the work area.

12.5.8 Whole Body

(i) **Hazards:** Heat, chemical or metal splash, spray from pressure leaks or spray guns, contaminated dust, impact or penetration, excessive wear or entanglement of own clothing.

(ii) **Options:** Conventional or disposable overalls, boiler suits, aprons, chemical suits.

(iii) **Note:**

a. The choice of materials includes flame-retardant, anti-static, chain mail, chemically impermeable, and high-visibility.

b. Don't forget other protection, like safety harnesses or life jackets.

12.6 RESPONSIBILITIES FOR EQUIPMENTS

Management must ensure that:

(i) The need for PPE are assessed by a person who is competent to judge whether other methods of risk control can offer better protection of safety and health than the provision of PPE.

(ii) Professional advice is obtained, where necessary, to identify the most suitable types of PPE for the tasks to be carried out.

(iii) Training is provided to supervisors and employees to enable them to ensure the proper selection, fit, use, cleaning and maintenance of PPE.

(iv) Supervision and enforcement of the PPE policy is undertaken.

(v) Evaluation of the effectiveness of the PPE program is carried out on a regular basis.

(vi) Suitable PPE is provided for visitors who may be exposed to hazards in the workplace

Review Questions

1. **What are hazards of electricity?**

   Ans. The hazard of electricity includes electric shock, electric flashover, fire hazard, explosion in electric equipments etc.

2. **What are corrosive substances?**

   Ans. The corrosive substance is one that which destroy tissue or metals. The common corrosive substances are acid, base, dehydrating agent etc.

3. **What are PPEs? Write the PPEs of head?**

   Ans. PPEs means personal protective equipments. Various types of PPE and clothing's are extensively used by the personnel's working in industry. The PPEs for head one helmets, Hand hats, Caps etc.

4. **What is a breathing apparatus?**

   Ans. It is used by the workers working at quite hazard environment. The apparatus contain a compressed oxygen cylinder that protects the users completely from hazardous gases, dust, fames etc.

5. **What is a respirator?**

   Ans. It is used by the workers who are working at the comparative less hazardous conditions, there are various types of respirators are used by the workers are: cartridge type, canister type, mask type etc.

6. **What are the elements involved in effective hazard control system?**

   Ans.

   1. Estimate the hazard
   2. Eliminate the hazard altogether (eg. To get rid of the dangerous machine)
   3. Substitute the hazard with a safer alternative (eg. Replace the machine with safer one)
   4. Isolate the hazard from anyone who could be harmed
   5. Use engineering controls to reduce the risk (eg. To attach guards to machine to protect users)
   6. Use administrative controls to reduce risk (eg. Trained workers)
   7. Use of personal protective equipment (PPE) eg. To wear gloves and goggles when using such machine.

■■■■

Glossary

A

Abiotic environment: The part of an ecosystem that includes the nonliving surroundings.

Activated sludge: An active population of micro-organisms used to treat wastewater, or the process in which the organisms are employed.

Adsorption: A surface phenomena in which a solute (soluble material) concentrates or collects at a surface (the adsorbent).

Advanced wastewater treatment: The removal of any dissolved or suspended contaminants beyond secondary treatment, often this is the removal of the nutrients nitrogen and/or phosphorus.

Air pollution: concentration of foreign matter in air in excessive quantity which is harmful to the health of man.

Aeration: Intimate contact of the atmosphere and water to add air (oxygen) to the water. The term is also applied to gas stripping where an undesirable gas is removed from the water. See

Aerobes: Organisms which require molecular oxygen as an electron acceptor for energy production.

Aerobic process: A process which requires molecular oxygen.

Alcohol: An organic compound with one or more hydroxyl "-OH" groups.

Aldehyde: An organic compound with a carbonyl at one end of a hydrocarbon group.

Alkalinity: The capacity of a water to neutralize acids.

Amine: A functional group consisting of "-NH2."

Amino acid: A functional group which consists of a carbon with a carboxylic acid, "- COOH" and an amine, "-NH2." These compounds are the building blocks for proteins. Anabolism Biosynthesis, the production of new cellular materials from other organic or inorganic chemicals.

Anaerobes: A group of organisms that do not require molecular oxygen. These organisms, as well as all known life forms, require oxygen. These organisms obtain their oxygen from inorganic ions such as nitrate or sulfate or from protein.

Anaerobic process: A process which only occurs in the absence of molecular oxygen.

Anoxic process: A process which occurs only at very low levels of molecular oxygen or in the absence of molecular oxygen.

Anthropogenic: Of, made, or caused by human activity or actions.

Aromatic: A form of bonding in which ring compounds share electrons over more than two atoms. The electrons are delocalized. This leads to unusual ring stability.

Attached growth reactor: A reactor in which the microorganisms are attached to engineered surfaces within the reactor. Examples of attached growth reactors are the trickling filter and the rotating biological contactor.

Autotrophic: Organisms which utilize inorganic carbon for synthesis of protoplasm. Ecologists narrow the definition further by requiring that autotrophs obtain their energy from the sun. In microbiologist parlance, this would be a photoautotroph.

Autotrophs: A group of organisms capable of obtaining carbon for synthesis from inorganic carbon sources such as carbon dioxide and its dissolved species (the carbonates). This group includes plants and algae.

Acute effects: with in twenty four hours of sudden exposure to polluted air illness would occur.

Aerosols: Small solid or liquid particles (fine drops or droplets) that are suspended in air.

B

Bacteria: One celled microorganisms which do not have a nuclear membrane.

Baghouse filter: A fabric filter device used to remove particulate air pollutants.

Biochemical oxygen demand (BOD): The amount of oxygen required to oxidize any organic matter present in a water during a specified period of time, usually 5 days. It is an indirect measure of the amount of organic matter present in a water.

Biofilm: A film of microorganisms attached to a surface, such as that on a trickling filter, rotating biological contactor, or rocks in natural streams.

Biogeochemical cycle: The cycle of elements through the biotic and abiotic environment.

Biosynthesis: Catabolism, the production of new cellular materials from other organic or inorganic chemicals.

BOD: See biochemical oxygen demand.

C

Carbonaceous biochemical oxygen demand (CBOD): The amount of oxygen required to oxidize any carbon containing matter present in a water.

Carbonyl: A functional group with an oxygen atom double bonded to a carbon atom.

Catabolism: The production of energy by the degradation of organic compounds.

Cell: A unit of varying dimensions in a landfill which is isolated from the environment by 6 to 12 inches of soil cover. A cell is one day's waste or less. A cell is covered with soil at the end of each day.

CFCs: Chlorofluorocarbons, chemicals which result in a depletion of the ozone layer in the upper atmosphere.

Chemical fixation (or stabilization/ solidification): A term for several different methods of chemically immobilizing hazardous materials into a cement, plastic, or other matrix.

Chemical oxygen demand (COD): The amount of oxygen required to oxidize any organic matter in the water using harsh chemical conditions.

Chemoautotrophic: Organisms which utilize inorganic carbon (carbon dioxide or carb-onates) for synthesis and inorganic chemicals for energy.

Chemotroph: Organisms which obtain energy from the metabolism of chemicals, either organic or inorganic.

Chlorofluorocarbons: Synthetic organic compounds used for refrigerants, aerosol propellants (prohibited in the U.S.), and blowing agents in plastic foams. CFCs migrate to the upper atmosphere destroying ozone and increasing global warming. Typica atmospheric residence times are 50 to 200 years.

Clarifier: (sedimentation basin) A tank in which quiescent settling occurs, allowing solid particles suspended in the water to agglomerate and settle to the bottom of the tank. The solids resulting from the settling being removed as a sludge.

Climatology: The study of the climate, how the earth's atmosphere performs over long periods of time.

Closure: The act of preparing a landfill for long term inactivity, including placement of a cover over the landfill to prevent infiltration of surface water.

Coagulation: Particle destabilization to enhance agglomeration.

Colloids: Small particles which have a negligible settling velocity. These particles have a very small mass so gravitational force is low compared to surface frictional forces. Typical colloidal sizes range from 10-3 mm to 1 mm.

Complexation: The ionic bonding of one or more central ions or molecules by one or more surrounding ions or molecules.

Component: A part of a mixture or solution.

Composting: The controlled aerobic degradation of organic wastes into a material which can be used for landscaping, landfill cover, or soil conditioning.

Compound: A substance composed of two or more elements.

Compression settling: Settling which occurs in the lower reaches of clarifiers where particle concentrations are highest. Particles can settle only by compressing the mass of particles below.

Consumers: Organisms which consume protopl asm produced from photosynthesis or consume organisms from higher levels which indirectly consume protoplasm from photosynthesis.

Conversion: The fraction of a species entering a system which is converted to product.

Corrosive waste: A waste that is outside the pH range of 2 to 12.5 or a waste that corrodes steel at a rate greater than 6.35 mm (0.25 in) per year. One of EPA's four hazardous waste properties.

Covalent bond: A bond in which electrons are shared approximately equally by two atoms.

Cybernetic: Systems which change in response to feedback.

D

Decomposers: Organisms which utilize energy from wastes or dead organisms. Decomposers complete the cycle by returning nutrients to the soil or water and carbon dioxide to the air or water.

Denitrification: The anoxic biological conversion of nitrate to nitrogen gas. It occurs naturally in surface waters low in oxygen, and it can be engineered in wastewater treatment systems.

Deoxygenation: The consumption of oxygen by the different aquatic organisms as they oxidized materials in the aquatic environment.

Discrete settling: Settling in which individual particles settle independently, neither agglomerating or interfering with the settling of the other particles present. This occurs in waters with a low concentration of particles.

Disease: Any impairment of the normal function of an organism.

Disinfection: The destruction or inactivation of pathogenic microorganisms.

Dispersion: A stable mixture of particles suspended in a fluid medium.

Dissolved oxygen (DO): The amount of molecular oxygen dissolved in water.

Dump: An illegal and uncontrolled area where wastes have been placed on or in the ground.

E

Ecology: The study of living organisms and their environment or habitat.

Ecosystem: An organism or group of organisms and their surroundings. The boundary of an ecosystem may be arbitrarily chosen to suit the area of interest or study.

Effluent: The fluid exiting a system, process, tank, etc. An effluent from one process can be an influent to another process.

Effluent based standards: Standards which set concentration or mass per time limits on the effluent being discharged to a receiving water.

Electronegativity: The potential of an atom to attract electrons when the atom is bonded in a compound. The scale is 0 to 4 with 0 being the most electropositive (low attraction) and 4 being the most electronegative (high attraction).

Electrostatic precipitator: A device which uses an electric field to trap particulate pollutants.

Elementary reaction: A reaction in which the rate expression corresponds to the stoichiometric equation.

Epilimnion: The top layer of a lake.

Equivalent: The mass of the compound which will produce one mole of available reacting substance. Thus, for an acid, this would be the mass of acid which will produce one mole of H^+, for a base, one mole of OH^-.

Ethers: An organic compound which has two hydrocarbon groups bound by an interior oxygen atom. The general formula is R'-O-R".

Eucaryotic organisms: Organisms which possess a nuclear membrane. This includes all known organisms except viruses and bacteria.

Evaporation: The conversion of liquid water to water vapor. It occurs on the surface of water bodies such as lakes and rivers and immediately after precipitation events in small depressions and other storage areas.

Evapotranspiration: The sum of evaporation and transpiration. Since it is difficult to measure the two terms independently, they are often grouped as one value.

F

Facultative: A group of microorganisms which prefer or preferentially use molecular oxygen when available, but are capable of suing other pathways for energy and synthesis if molecular oxygen is not available.

Fermentation: Energy production without the benefit of oxygen as a terminal electron acceptor, i.e. oxidation in which the net effect is one organic compound oxidizing another.

Fixed solids (FS): are the solids that do not volatilize at 550°C.

Fixed suspended solids (FSS):is the matter remaining from the suspended solids analysis which will not burn at 550°C. It represents the non-filterable inorganic residue in a sample.

Flash point: The lowest temperature at which sufficient vapor is produced to cause combustion if an ignition source is present.

Flocculant settling: Settling in which particle concentrations are sufficiently high that particle agglomeration occurs. This results in a reduction in the number of particles and an increase in average particle mass. As agglomeration occurs higher settling velocities result.

Fumes: are polydispersed fine aerosols consisting of solid particles that often aggregate together, so that many little particulates may form one big particle.

Fluidization: The suspension of particles by sufficient upward velocity of the fluid. During fluidization the gravity force is overcome by a combination of buoyancy and fluid friction.

Flux: The movement of a mass past a surface, plane, or boundary. The units are mass per unit area per unit time or [Kg/m2-hour].

G

Gas stripping: Gas transfer of an undesirable gas from a water stream to the atmosphere.

Global warming: The long-term warming of the plant due to increases in greenhouse gases which trap reflected light preventing it from exiting to space.

Greenhouse gases: Gases which trap solar radiation. Of the solar energy entering the earth's atmosphere a portion is reflected back and a portion penetrates onto the earth's surface. The portion reflected back from the earth's surface is at a different wavelength that when it entered. Carbon dioxide and other gases, which pass solar radiation, absorb this reflected radiation, increasing the earth's temperature. This is much like a greenhouse, hence the name.

Groundwater: Water which is contained in geologic strata. Also properly written as two words, ground water.

H

HAPs: Hazardous air pollutants.

Hardness: The sum of the divalent cation concentrations expressed as meq/L or mg calcium carbonate per liter [mg CaCO3/L]. It is important because hard waters require increased amounts of soap for bathing or washing clothes and because of scale formation on piping, cooking vessels, water heaters, boilers, heat exchangers, etc.

Heterotrophic: A group of organisms which obtain carbon for synthesis from other organic matter or proteins.

Hindered (Zone) settling: Settling in which particle concentrations are sufficient that particles interfere with the settling of other particles. Particles settle together as a body or structure with the water required to traverse the particle interstices.

Hydrocarbon: Any organic compound composed entirely of carbon and hydrogen. Two examples are methane gas and octane.

Hypolimnion: The lower layer of a lake.

I

In situ treatment: Treatment of a waste in place, as opposed to pumping or digging the waste up and then treating it.

Ionosphere: Above the stratosphere is the "ionosphere" the top of which is the border line space.

Infectious disease: A disease caused by pathogenic organisms.

Infiltration: The movement of water from the surface of the land through the unsaturated zone and into the groundwater. This occurs during and immediately after precipitation events. It can also occur at the bottom of lakes and rivers.

Influent: The fluid entering a system, process, tank, etc. An effluent from one process can be an influent to another process.

Ion exchange: An adsorption process in which one ion is exchanged for another ion of like charge. There is an equivalence of exchanged charge.

Irreversible reaction: A reaction in which the reactant(s) proceed to product(s), but the products react at an appreciable rate to reform reactant(s).

Isomers: Two or more different compounds with the same chemical formula but different structure and characteristics.

Inhalable fraction: Particles less than 100 μm that can be inhaled into the respiratory throat (trachea).

K

Kerogen: A fossilized organic material present in oil shale and some other sedimentary rocks.

Ketones: Organic compounds with two hydrocarbon groups bonded to a carbonyl group.

L

Landfill: A legal and controlled area for the placement of wastes into the ground.

Landfilling: The placement of wastes into the land under controlled conditions to minimize their migration or effect on the surrounding environment.

Leachate: A liquid generated in landfills. It is the result of water seeping into and through the wastes. As the water contacts the waste materials it dissolves part of the organic and inorganic matter contained in the landfill. If this leachate is allowed to exit the bottom of the landfill, it will carry contaminants to the groundwater and/or adjoining surface water.

Leaching: The act of dissolving the soluble portion of a solid mixture by some solvent. An example is the dissolving of inorganic or organic contaminants from refuse in a landfill by infiltrating rain water.

Ligand: The ion or molecule which surrounds or complexes with the central atom or ion.

Limnology: The study of freshwater ecosystems.

M

Mass balance: An organized accounting of all inputs and outputs to an arbitrary but defined system. Stated in other terms, the rate of mass accumulation within a system is equal to the rate of mass input less the rate of mass output plus the rate of mass generation within the system.

Maximum contaminant level (MCL): The maximum allowable concentration of a given constituent in potable water.

MCL: See maximum contaminant level.

Mercaptans: See thiols.

Metabolism: The processes which sustain an organism, including energy production, synthesis of proteins for repair and replication.

Metalimnion: The middle layer of a lake.

Meteorology; The study of the atmosphere and weather of the lower atmosphere, below 100 km.

Mixed liquor suspended solids (MLSS): The total suspended solids concentration in the activated sludge tank.

Mixed liquor volatile suspended solids (MLVSS): The volatile suspended solids concentration in the activated sludge tank.

Mesosphere: Above the stratosphere, or the middle layer.

Mist: A cloud or dense collection of droplets suspended in air.

N

Nitrification: The biological oxidation of ammonia and ammonium sequentially to nitrite and then nitrate. It occurs naturally in surface waters, and can be engineered in wastewater treatment systems. The purpose of nitrification in wastewater treatment Systems are a reduction in the oxygen demand resulting from the ammonia.

Nitrogen fixation: The conversion of atmospheric (or dissolved) nitrogen gas into nitrate by microorganisms.

Nitrogenous oxygen demand (NOD): The amount of oxygen required to oxidize any ammonia present in a water.

Nonpoint source pollution (NPSP): Any pollution from a source which cannot be attributed to a particular discharge point, e.g. from agricultural crops, city streets, construction sites, etc.

NPDES: The National Pollutant Discharge Elimination System. The discharge criteria and permitting system established by the U.S. EPA as a result of the Clean Water Act and its subsequent amendments or the permit required by each discharger as a result of the Clean Water Act.

O

Organic compound: Any compound containing carbon except for the carbonates (carbon dioxide, the carbonates and bicarbonates), the cyanides, and cyanates.

Organic nitrogen: Nitrogen contained as amines in organic compounds such as amino acids and proteins.

Oxidative phosphorylation: The synthesis of the energy storage compound adenosine triphosphate (ATP) from adenosine diphosphate (ADP) using a chemical substrate and molecular oxygen.

P

Packed tower: See trickling filter.

Pathogenic organism: An organism capable of causing infection.

Phenol: An aromatic benzene ring with a hydroxyl substituted for one hydrogen.

Phenyl: A benzene ring named as a constituent group, C6H5-.

Phosphorylation: The synthesis of the energy storage compound adenosine triphosphate (ATP) from adenosine diphosphate (ADP).

Photoautotrophic: Organisms which utilize inorganic carbon dioxide for protoplasm synthesis and light for an energy source.

Photochemical pollutants: Chemicals which react photochemically (in the presence of sunlight) to destroy ozone in the stratosphere.

Photophosphorylation: The synthesis of the energy storage compound adenosine triphosphate (ATP) from adenosine diphosphate (ADP) using solar energy.

Phototroph: Organisms which obtain energy from light using photooxidation.

Pollution: Any man made condition which adversely affects the quality of the environment.

Potable water: Water that has does not contain harmful or objectionable impurities and is aesthetically pleasing to drink.

POTW or Publicly Owned Treatment Works: Any municipally owned wastewater treatment facility.

Precipitation: The falling to earth of condensed water vapor in the form of rain, snow, sleet or hail.

Primary standards: Required drinking water quality standards related directly to human health. These standards are required and enforceable by the U.S. EPA.

Primary treatment: Treatment which includes all operation prior to and including primary treatment, e.g., bar screening, grit removal, comminution, and primary sedimentation.

Procaryotic organisms: Organisms which do not have a cellular membrane.

Producers: Autotrophic organisms which produce protoplasm using inorganic carbon and energy from the sun.

Publicly owned treatment works (POTW): A municipal or domestic wastewater treatment facility.

R

Reactive waste: A waste which; 1) reacts violently with water, 2) forms potentially explosive mixtures with water, 3) is normally unstable, 4) contains cyanide or sulfide in sufficient quantity to evolve toxic fumes at high or low pH, 5) is capable of exploding if heated under pressure, or 6) is an explosive compound listed in Department of Transportation (DoT) regulations. One of EPA's four hazardous waste properties.

Reaeration: The dissolving of molecular oxygen from the atmosphere into the water.

Receiving water: A water which receives wastewater (treated or otherwise) discharges.

Receiving water quality standards Standards which require a discharger to maintain a certain quality level in the receiving water.

Recycling: The recovery and reuse of a product which would otherwise be thrown away.

Refuse: All forms of solid waste.

Refuse derived fuel (RDF): A fuel derived from the combustible portion of municipal solid waste. The fuel is often processed into small briquettes, similar in size to charcoal.

Respiration: Energy production in which oxygen is the terminal electron acceptor, i.e. oxidation to produce energy where oxygen is the oxidizing agent.

Reversible reaction: A reaction in which the reactant(s) proceed to product(s), but the products react at an appreciable rate to reform reactant(s).

Runoff: The water that flows overland to lakes or streams during and shortly after a precipitation event.

Respirable range: the greatest penetration and retention of particles is in the range 10.0 to 0.1 µm

S

Saltwater intrusion: The gradual replacement of freshwater by saltwater in coastal areas where excessive pumping of groundwater occurs.

Secondary standards: Recommended drinking water quality standards which relate to aesthetics and/or health. These standards are recommended, not required.

Secondary treatment: In wastewater treatment, the conversion of the suspended, colloidal and dissolved organics remaining after primary treatment into a microbial mass with is then removed in a second sedimentation process. Secondary treatment included both the biological process and the associated sedimentation process.

Secured landfill: A landfill which has containment measures such as liners and a leachate collection system so that materials placed in the landfill will not migrate into the surrounding soil, air and water.

Sedimentation: The gravity settling, and thus removal, of materials more dense than the suspending fluid.

Sedimentation basin: See clarifier.

Shock load Influent wastewater entering the plant which has an unusually high organic content and/or high flow rate.

Site remediation: The process of cleaning up a hazardous waste disposal site that has either been abandoned or that those responsible either refuse to cleanup or are financially unable to cleanup.

Siting: Obtaining government (federal, state, and local) permission to construct an environmental processing, treatment, or disposal facility at a given site.

Sludge: A mixture of solid waste material and water. Sludges result from the concentration of contaminants in water and wastewater treatment processes. Typical wastewater sludges contain from 0.5 to 10 percent solid matter. Typical water treatment sludges contain 8 to 10 percent solids.

Softening: The removal of divalent cations by precipitation or ion exchange.

Solidification: See chemical fixation.

Solids flux: See flux.

Source reduction: The elimination or reduction of the waste at the source by modification of the actual process which produces the waste.

Species: In chemistry, an ion or molecule in solution.

Stabilization: See chemical fixation.

Sterilization: The destruction or inactivation of all microorganisms.

Storage: The short term retention of water after a precipitation event.

Stratosphere: The atmosphere from approximately 12 km to 70 km. The temperature of the atmosphere increases in this region.

Stream based standards: See receiving water quality standards.

Strong acid: An acid that, for practical purposes, ionizes completely under the conditions of interest. Common strong acids are hydrochloric, nitric, and sulfuric. See weak acid.

Substrate level phosphorylation: The synthesis of the energy storage compound adenosine triphosphate (ATP) from adenosine diphosphate (ADP) using organic substrates without molecular oxygen.

Surface water: Water which is contained in lakes, rivers, and oceans.

Suspended growth reactor: A reactor in which the microorganisms are suspended in the wastewater. Examples of suspended growth reactors are activated sludge reactors and anaerobic digesters. See attached growth reactor.

Synergism: is the act of working together. Two chemicals which are synergistic have a greater effect together than the sum of their individual effects. The effect can be either positive or negative.

System: An arbitrarily defined area or volume surrounded by a boundary and possessing specific inputs, outputs, and reactions.

T

Thermocline: The depth at which an inflection point occurs in a lake temperature profile.

Thiols: Organic compounds which contain the "-SH" functional group. Also called mercaptans.

Total dissolved solids (TDS); is the amount of dissolved matter in the water.

Total solids (TS): is the amount of organic and inorganic matter which is contained in a water.

Total suspended solids: (TSS) is the amount of suspended (filterable) matter in a water.

Toxicity: A U.S. EPA hazardous waste characteristic defined with a rigorous test procedure, the TCLP (for Toxicity Characteristic Leaching Procedure). In the procedure, a waste is extracted for 24 hours with an acetic acid solution. The acid extract is then analyzed for the presence of any of the contaminants listed in the procedure.

Trace contaminants: Contamination found in trace (very low) levels.

Transpiration: The loss of water from plants through leaves and other parts. This loss can be a significant amount of water during very dry periods.

Trickling filter: An attached growth biological process in which the microbial film is attached to non-moving rock or plastic media.

Trophic level: A level in the food chain. The first trophic level consists of the primary producers, autotrophs. The second trophic level is vegetarians which consume autotrophic organisms.

Troposphere: The lower atmosphere, from the earth's surface to approximately 12 km. This portion of the earth's atmosphere contains about 95 percent of the atmospheric gases. The temperature gradually declines through this region.

Thoracic fraction: Those particles below 20 μm, that can penetrate into the lungs.

Thermosphere: This is a region of highly ionized gases, extending to about 1600 km.

U

Ultimate biochemical oxygen demand: (BODu) The total amount of oxygen required to oxidize any organic matter present in a water, i.e. after an extended period, such as 20 or 30 days.

Ultimate disposal: The process of returning residuals back to the environment in a form which will have the minimal or reduced negative environmental impacts.

V

Virion: A virus particle. Viral DNA or RNA enclosed in an organic capsule. See virus.

Virus: A submicroscopic genetic constituent which can alternate between two distinct phases. As a virus particle, or virion, it is DNA or RNA enveloped in an organic capsule. As an intracellular virus, it is viral DNA or RNA inserted into the host organisms DNA or RNA.

Volatile: A material which will vaporize easily.

Volatile solids: (VS) is the amount of matter which volatilizes (or burns) when a water sample is heated to 550EC.

Vapour: The evaporated compound in the gas phase.

Volatile suspended solids: (VSS) is the non-filterable residue remaining after firing the total suspended solids at 550EC. See total suspended solids and fixed suspended solids.

W

Waste minimization: The elimination or reduction of a waste prior to its generation. This is accomplished by process changes rather than waste treatment methods.

Wastewater: Consumed or used water from a municipality or industry that contains dissolved and/or suspended matter.

Weak acid: An acid that does not ionize completely under the conditions of interest. Examples include acetic acid, carbonic acid, and hypochlorous acid. See strong acid.

Wetland: Semi-aquatic land, which is land that is either inundated or saturated by water for varying periods of time during each year, and that supports aquatic vegetation which is specifically adapted for saturated soil conditions.

Wind: Is simply air in motion

Z

Zone settling: See hindered settling